Evidence and Procedures for Boundary Location

Evidence and Procedures for Boundary Location

Second Edition

CURTIS M. BROWN
La Mesa, California

WALTER G. ROBILLARD
Atlanta, Georgia

DONALD A. WILSON
East Kingston, New Hampshire

A Wiley-Interscience Publication
JOHN WILEY & SONS
New York • *Chichester* • *Brisbane* • *Toronto* • *Singapore*

Copyright © 1981 by John Wiley & Sons, Inc.

All rights reserved. Published simultaneously in Canada.

Reproduction or translation of any part of this work beyond that permitted by Sections 107 or 108 of the 1976 United States Copyright Act without the permission of the copyright owner is unlawful. Requests for permission or further information should be addressed to the Permissions Department, John Wiley & Sons, Inc.

Library of Congress Cataloging in Publication Data:

Brown, Curtis Maitland.
   Evidence and procedures for boundary location.

   "A Wiley-Interscience publication."
   Includes bibliographical references and index.
   1. Boundaries (Estates)—United States.
I. Robillard, Walter G. (Walter George), 1930-
II. Wilson, Donald A., 1941-   . III. Title.
KF639.B73   1981   346.7304'32   81-11440
ISBN 0-471-08382-8   347.306432   AACR2

Printed in the United States of America

15 14 13

Preface to the Second Edition

Three major additions are found in this edition: (1) an explanation of the liability of the surveyor, (2) a discussion of unwritten rights and what to do about them, and (3) inclusion of unique boundary problems found in the eastern states. In addition, other topics are brought up to date.

The dramatic increase in the liability of the land surveyor has been the result of many states adopting the "discovery rule" and the erosion of "privity of contract." Although many states or their courts have not adopted the discovery rule (a rule whereby the statute of limitations commences to run after the discovery of an error), it is believed that most will. Chapter 4 discusses discovery principles in detail.

Considerable consternation exists concerning what the surveyor should do about unwritten rights. Should he monument lines that he believes to be ownership lines because of possible rights acquired by occupation, or should he only monument lines in accordance with a written conveyance and inform the client of the possibility of an occupancy right? Chapter 3 considers unwritten rights and what to do about them, and Chapter 4 considers the liability that may result.

In the eastern states unique, original survey procedures have created special resurvey problems that vary from state to state and sometimes from locality to locality. To aid in covering this important aspect of boundary location, Walter Robillard of Georgia and Donald Wilson of New Hampshire have agreed to be co-authors. The special eastern procedures are not found in one place but are discussed under appropriate headings.

Winfield H. Eldridge, co-author (with Curtis M. Brown) of the first edition, met with an untimely death in a bicycle accident soon after the publication of the first edition. His original endeavor is still relevant, and it is regretted that Professor Eldridge cannot assist in this revision.

Since the close of World War II a very steady improvement in the professional standing and attitude of land surveyors has occurred, largely because of the efforts of land surveyors themselves. Major indicators of this trend are increases in membership of professional societies and changes for the better in registration laws. Whereas the many changes have been small and gradual, the sum is significant. This revision is directed more toward bringing out professional responsibilities of surveyors than toward the mechanics of instrument operations or how to use tools.

CURTIS M. BROWN
WALTER G. ROBILLARD
DONALD A. WILSON

La Mesa, California
August 1981

Preface to the First Edition

In his famous essay, "Judicial Functions of Surveyors,"[1] Chief Justice Thomas M. Cooley had this to say about the duties of the surveyor:

> If the original monuments are no longer discoverable, the question of location becomes one of evidence merely. It is merely idle for any State statute to direct a surveyor to locate or "establish" a corner, as the place of the original monument, according to some inflexible rule. The surveyor, on the other hand, must inquire into all the facts, giving due prominence to the acts of parties concerned, and always keeping in mind, first, that neither his opinion nor his survey can be conclusive upon parties concerned, and, second, that courts and juries may be required to follow after the surveyor over the same ground, and that it is exceedingly desirable that he govern his action by the same lights and the same rules that will govern theirs.

At the conclusion of the essay, Mr. Justice Cooley speaks of the "quasi-judicial capacity" of surveyors:

> Surveyors are not and cannot be judicial officers, but in a great many cases, they act in a quasi-judicial capacity with the acquiescence of parties concerned; and it is important for them to know by what rules they are to be guided in the discharge of their judicial functions.

Though these words were written in the last century, they are most timely today, and express quite eloquently the responsibilities of the surveyor engaged in property location. It is the opinion of the authors of this book that the proper location of property lines depends upon *evidence* and *procedures*.

At the eighth annual meeting of the American Congress on Surveying and Mapping, held in Washington, D.C., in 1948, the comment was made:

> It is strange and sad that ours is probably the only profession in the country that has no book devoted specifically to the profession. There are a number of textbooks on Surveying, but there is no textbook that is keyed directly to the practicing property

---

[1] Article appeared originally in the *Michigan Engineer*, but has been repeated in many subsequent publications. One of the more recent may be found in *Surveying and Mapping*, Vol. XIV, No. 2, pp. 161–168.

surveyor. Such a book has not as yet been published though recent works have partially closed the gap. Much still needs to be done to provide the profession with comprehensive practical coverage of the field of property surveying.[2]

In 1958 at the annual conference of the California Council of Civil Engineers and Land Surveyors, a request was made of Curtis M. Brown to prepare text material directed toward professional surveyors. A committee, headed by William Pafford and assisted by George Psomas and Howard Young, were to serve as a review team. Part of the material was prepared and constructively criticized by the committee.

Later it was deemed advisable to increase the scope so as to make the text applicable to all areas of the United States. In 1960, Winfield H. Eldridge of the University of Illinois, because of his recognition and active experience, was invited to contribute as co-author. Since then, the material has been revised and adapted to the varied conditions throughout the United States and many new concepts have been added.

The professional property surveyor, in addition to other duties, is often given an exclusive franchise or a shared privilege (1) to locate property lines in accordance with a written description, (2) to locate encroachments on written title lines, (3) to create new land divisions, and (4) to describe by map or writings the divisions created. In performing these duties he utilizes knowledge of measurement techniques, of mathematics, of legal principles, including the law of evidence, and of past history and customs affecting surveys. Such is the scope of this book.

The precise line of demarcation between the practices of law and of property surveying is difficult to distinguish; one seems to merge into the other. Perhaps the best concept is one that states:

> The surveyor must obey the law in locating former divisions of land, in locating new divisions of land and in describing the land divided; hence he must know what the law is pertaining to these functions.

This is the premise on which this book has been developed.

Property surveyors do not determine land ownership. Ownership of land is made up of and dependent on elements such as taxes, liens, mortgages, community rights, heirs, senior rights, restrictions, judgments, and many others. Evaluation of these elements is a question of law beyond the responsibility of the property surveyor and will not be treated in this work.

The court often calls upon the surveyor for expert testimony, and such testimony is a means to introduce evidence and never to expound on the laws of boundaries; the surveyor is an expert on *measurements*. The length of a block may be testified to by him, but whether the record measurement or a proportion of the whole is to be used is a question of law to be decided by the court. The

---

[2]*Surveying and Mapping*, Vol. IX, No. 4, p. 291, Washington: American Congress on Surveying and Mapping, 1949.

surveyor can testify about what he did and the legal reasons why he did it, but he can never tell the court what the judgment should be.

In order to properly locate a boundary, the surveyor must make decisions in accordance with law; if proration is proper, he must use it. The surveyor must apply the law, as it exists, to boundary location as he is practicing it in accordance with the law. Errors in judgment or even the lack of knowledge of some hidden evidence might be excused, but the surveyor is presumed to be competent in measuring. He is an expert because he knows how to measure, how to analyze errors in measurements, and how to control the quality of measurements for a given situation. The subject of measurement procedures is treated more extensively in other works, but its importance should never be minimized.

Measurements can never be made without an error of some magnitude, though it may be insignificant. For every survey, the surveyor should know how much uncertainty can be tolerated. The authors believe that a satisfactory discussion of permissible uncertainty and of how errors should be evaluated by property surveyors is not currently available. A modern treatment and concept, varying from present practice is discussed in Chapter 10.

This is a book of what, why, and how; it is arranged essentially from the surveyor's viewpoint, but other persons, especially title insurance men and attorneys, will find it of beneficial use. In communicating information to the reader, every attempt has been made to reduce the words to a minimum, to include essential useful information related to the location of property lines, and to state precisely the principles involved. This is an exposition designed to make efficient use of the reader's attentions and save him confusion in the areas of science, law, mathematics, and mensuration as they are applicable to the professional property surveyor when he is locating and describing land boundaries.

It is intended that there be omissions from this book. The subject matter of *Boundary Control and Legal Principles*[3] is not duplicated, but this work will augment that book, and together they present a reasonably thorough coverage of the field of property surveying.

Selected references at the close of some of the chapters are provided to enable the reader to delve more deeply into many of the more specific subjects.

CURTIS M. BROWN
WINFIELD H. ELDRIDGE

July 1962

---

[3]By Curtis M. Brown, John Wiley & Sons, New York, 1957.

Contents

1. Introduction, 1
2. Evidence, 9
3. Unwritten Transfers of Land Ownership, 79
4. Professional Liability, 119
5. Professional Stature, 132
6. Historical Development of Property Surveying, 143
7. Location Establishment by Statutory Proceedings, 209
8. Procedures for Locating Written Title Boundaries, 217
9. Apportionment Procedures for Land and Water Boundaries, 242
10. Measurements, Errors, and Computations, 265
11. Preservation of Evidence, 305
12. Guarantees of Title and Location, 318
13. Platting Laws and Original Surveys, 333
14. Survey Plats, 348
15. Writing Descriptions, 364
16. The Surveyor in Court, 404
17. Eminent Domain, 426

Index, 439

Evidence and Procedures
for Boundary Location

CHAPTER 1

Introduction

1-1 Scope of Book

The functions of land surveyors can be divided into (1) locating or relocating described parcels of land and (2) creating new parcels. The first portion of the book covers resurveys or surveys based on the record; the latter part covers new divisions of land.

How original surveys are made is subject to control by the legislative branch of the government, and since there are 50 states and the Federal Government, subdivision laws and regulations of original surveys are extremely variable. After land has been divided and described, the courts interpret where and what the boundaries are. Today, practically all original surveys must start from a survey of an existing parcel. For this reason, the first portion of this book pertains to the location of previously described parcels.

In a survey of a described parcel of land, the only correct location of its boundaries is where a court of competent jurisdiction would locate them. To know where a court would locate property boundaries, the surveyor must have expert knowledge of the laws of boundaries.

A jury trial is divided into two parts: the jury decides what are the facts; the judge applies the law. Note the old saying "Where boundaries are is a matter of fact; what are boundaries is a matter of law." Thus in a trial the jury decides where an original monument position was located; the judge decides whether the monument or measurement controls as a matter of law. In a survey based on the record, the surveyor may be charged with both determinations. Chapter 2 pertains to the laws of evidence needed to prove facts and the order of importance of discovered evidence.

According to the Statutes of Frauds, as first adopted in England and later approved in the United States, land ownership must be accompanied by written evidence. In early cases the court found that the requirement for written deeds sometimes caused a fraud; thus the concept of title passing without writings came into being (called *unwritten title*). Since an unwritten title extinguishes the conflicting written title, this important subject is discussed early in Chapter 3.

Within recent years a marked change has occurred in the court's thinking on the subject of professional liability. The concept of privity of contract and the time of commencement of the running of the statute of limitations are vastly different from what they were 50 years ago. Chapter 4 and some of Chapter 5 ("Professional Stature") treat these subjects.

According to decisions in previous court cases, surveyors, in retracing old property lines, are obligated to follow the "footsteps of the original surveyor"; therefore, it is essential that in their areas of practice they have knowledge of the historical background of land surveys and existing laws under which they were performed. The purpose of surveying history, as given in Chapter 6, is to aid present-day surveyors in understanding why they must follow certain procedures when locating property boundaries. Such history is to be used as a tool—not as a device to present the romance of the past. Exploration of historical background and the study of the development of various survey systems provide the needed background of laws and customs governing property owners' rights and privileges.

Certain phases in the history of surveying have widespread application within the United States, such as (1) the English system that gave rise to English Common Law as used in the colonies and which now forms fundamental rules for most of the remainder of the United States; (2) the Mexican and Spanish Land Grant Systems; (3) the French system used in the Louisiana Purchase area and elsewhere; (4) the sectionalized land system of the public domain; (5) land divisions under state laws, especially in Texas and eastern seaboard states; (6) the various other systems brought about by events in the history of our nation.

The intent and meaning of a deed is always interpreted in the light of laws and conditions existing as of the date of the document. In New York, under Dutch rule, land dedicated for road purposes passed in fee title to the Crown. New York, on acquiring the streets, retained the fee title; hence vacated, former Dutch streets revert to the state. The same can be said of Texas roads dedicated during Spanish or Mexican times (*Mitchell* v. *Bass*, 33 Tex. 260). The ownership of stream beds or bodies of water is often dependent on which nation had jurisdiction at the time of the land's original alienation and on the effect of the laws in force at the time of the grant. An indispensable part of all boundary location is knowledge of past history.

Following the historical development of surveying the remainder of the book explains the procedures used in locating already-described parcels and procedures used to create new parcels, including how to describe parcels by writings. If some of the aspects of property surveying appear to be treated too briefly, or are not mentioned at all, it is because these topics are adequately discussed in other works previously published and available to the practitioner and student.

1–2 Definitions of Surveys and Surveyors

The term "survey" has a broad connotation; it applies to many exacting procedures as well as vague studies. When used with other terms, such as "land," "property," "boundary," or "cadastral," it may be pertinent to the subject of this book. The word *cadastre* is defined as an official register of the quantity, value, and ownership of real estate used in apportioning taxes. *Cadastration* is the act or process of making a cadastre or cadastral survey.¹ In popular use, *land surveying* is defined as the determination of boundaries and areas of a tract of land. A *boundary survey* is understood by some as that which is conducted for the location and establishment of lines between political units, and a *cadastral survey* is confined to the location and subdivision of the public domain.² To minimize confusion, *property surveying* will be employed throughout this book to denote the activity of locating, establishing, and delimiting boundaries of real property. The practice of property surveying is defined in many of the state registration laws. Such definitions usually include the measurements of area, lengths, directions, and the correct determinations of descriptions, especially when such property is to be conveyed or when the instrument of conveyance is to become of public record.

The National Council of State Boards of Engineering Examiners (NCSBEE) defined *land surveying* to mean "the performance or practice of any professional service requiring education, training and experience in the application of special knowledge in the mathematical, physical and technical arts and sciences to such professional services as the establishment or relocation of land boundaries, the subdivision of land, the determination of land areas, the accurate and legal description of land areas and the platting of land subdivisions for record."³

In a model registration law, approved by the same council, the following statement appears:

> The term Land Surveying used in this act shall mean and shall include assuming responsible charge for and/or executing: the surveying of areas for their correct determination and description and for conveyancing; the establishment of corners, lines, boundaries and monuments; the platting of land and subdivisions thereof including as required, the functions of topography, grading, street design, drainage and minor structures, and extensions of sewer and water lines; the defining and location of corners, lines, boundaries and monuments of land after they have been established; and preparing the maps and accurate records and descriptions thereof.⁴

The American Congress on Surveying and Mapping (ACSM) and the American Society of Civil Engineers (ASCE) jointly published *Definitions of Surveying and Associated Terms* in 1978, and in it the following definition of land surveying appears: "Land surveying is the art and science of: (1) Reestablishing cadastral surveys and land boundaries based on documents of record and historical evidence; (2) planning, designing and establishing property boundaries; and (3) certifying surveys as required by statute or local ordinance

---

[1] *Webster's New International Dictionary*, 3d ed.
[2] *Definition of Surveying and Associated Terms*, ACSM and ASCE, 1978.
[3] Proceedings, 34th Annual Meeting of the NCSBEE, 1955.
[4] Model law as approved by NCSBEE, 1960.

such as subdivision plats, registered land surveys, judicial surveys, and space delineation. Land surveying can include associated services such as mapping and related data accumulation; construction layout surveys; precision measurements of length, angle, elevation, area and volume; horizontal and vertical control systems; and the analysis and utilization of survey data."

1-3 Activities of the Property Surveyor

Property surveyors are found in private practice; are employed by federal, state, county, and local government; and are associated with related business. In the past those engaged in private practice and especially those in rural areas often had small organizations composed of the surveyor with one or more helpers. In larger cities and in densely populated areas it is not uncommon to find large surveyor firms preparing subdivisions and locating described parcels. Many land surveyors have found that a small organization in a country setting is a most enjoyable way to live out a lifetime and offers many advantages.

The Federal Government, through the Bureau of Land Management (BLM), is still engaged in the subdivision and resurvey of the public domain, and the U.S. Forest Service employs surveyors who are well versed in land surveying. One of the more important functions of State Highway Divisions is the location of rights-of-way with respect to adjacent property. The counties and cities often have similar problems, although confined to a more local area. In a few states, the county surveyor makes private property locations as a part of his official duties; in other states the county surveyor's responsibility is confined to county government problems.

The surveyor may not necessarily confine his work to property surveys; he often maps topography and stakes the outline of engineering projects such as buildings, sewer lines, water lines, curbs, sidewalks, and paving. Although these are important functions, they will not be treated in this book, which is limited to the location of property boundaries.

1-4 The Surveyor in Society

There is good reason to believe that the practice of property surveying is as ancient as property ownership itself. In Babylon over 3500 years ago inscribed on a boundary stone was the name of the surveyor ("Babylonian Boundary Stones and Memorial Tablets in the British Museum, L.W. King"). In early times, surveyors possessed special skills and talents that were regarded with almost reverent respect; they filled a necessary need in civilization, and they utilized the most advanced sciences known to humankind.

There is little doubt but that the practice of property surveying today should be performed by a profession upholding the standards of the higher meaning of the word. The rapid development of this country's lands has created a need for professional surveyors in greater needs than they are being supplied by the schools. At the present time, many surveyors have reached their professional status by an apprenticeship and self-study program, without the aid of much formal education. Such a regeneration probably will not suffice in the future, and the property surveyors will have to come from schools and colleges.

The practice of property surveying is one that must consider inexact laws in addition to exact engineering sciences. Few problems confronting the property surveyor can be solved completely by applying only exact sciences; they also depend on law, an inexact science. The development of mature judgment is demanded.

Many famous men in history, including several presidents, have been engaged in property surveying during their lifetime. The property surveyor, perhaps more than his associates in engineering, is in constant contact with people. This gives the property surveyor an opportunity to present himself as a member of a profession with superior standing and integrity. Awareness of the responsibilities and a conscientious fulfillment of professional obligations will enable the property surveyor to maintain a status in society.

1-5 The Present Need for Property Surveys

In the United States there are millions of people having many basic needs that involve land. All food and minerals are derived from the soil. Fibers and substances that go into essential goods are a product of the land. Rain must be conserved and controlled as it falls on the earth. Lakes and rivers provide transportation, irrigation, and energy for those who live along their shores.

This century has been highlighted by many spectacular engineering structures, buildings, bridges, dams, and highways, all of which are located on the land—the same land that has been here for so many centuries. When these structures have disappeared the land on which they stood will still be there. When a hundred miles of new highways are constructed, location will depend on more than 200 miles of new property lines!

An exploding population is creating demands for more and more residential lots. New land developments containing 50,000 homesites and small subdivisions with only a few lots are being created. For each parcel of land, there are owners' rights and privileges extending to the limits of their property, both skyward and to the depths of the earth. With expansion, a community requires additional lands for public and municipal purposes, such as waste disposal, service utilities, parks, schools, and rights-of-way.

In the early years of the country, real property sold for $1.25 per acre and less; today, portions of that same land have value in excess of $2,000,000 per acre, and some land is valued in terms of thousands of dollars per front-foot. Even "poor" land in rural areas may bring $1000 per acre or more.

Of the thousands of property surveyors in the country today, many are not prepared to cope with these critical problems, and, even were they able, the numbers of surveyors are not sufficient to satisfy the needs of the growing population. It is the twofold purpose of this book to help those in practice to better their understanding of professional problems and then to aid the beginner in preparing for the profession.

6 / INTRODUCTION

Figure 1-1 Modern development in flatlands.

1-6 The Future Needs for Property Surveyors

No one can accurately predict the problems of the distant future, but some of the needs can be anticipated and prepared for accordingly. Certain needs are obvious; the population will expand, and thus more efficient use of land must be made. As the costs of land spiral upward, the delineation of property lines will become more critical. New divisions of land will continue to be made each day, and the ancient surveys will have to be identified and retraced.

Even in the future, the property surveyor must not ignore the past, for this professional's problems go back as far as land ownership itself. The mistakes of yesterday will provoke headaches tomorrow. Rare are the areas where the value of real property has not risen considerably over the years; new surveys must be performed more accurately.

The property surveyor of tomorrow will need to exercise far more technical skill and judgment than his fraternity is displaying today. Just as the compass and chain are gone, the transit and tape are not sufficient to cope with future problems. Modern technology has developed many marvelous devices and techniques with which to make more accurate surveys and aid in studying and

Figure 1-2 Hillside development.

solving property surveying problems. Such tools as photogrammetry, electronic computers, and microdistance devices are now commonplace.

It is possible that the property surveyor of tomorrow will locate title boundaries many fathoms under the sea or may need to subdivide the antarctic continent. Whatever are his problems, the property surveyor will need to be armed with *all* the tools, knowledge, and education of his profession and have the ability to exercise sound and mature reasoning.

1-7 Land Data System

As of this time the need for revamping and reorganizing land data systems existing in almost every state is urgent. The move toward modernization of land data systems includes pilot projects in various parts of the country, some of which have been in existence for a few years. Although needed land data will vary from place to place, the system in general will be similar in many respects to the Land Registration and Information Services system in use in the Maritime Provinces of Canada.

Systems index parcels of land according to a unique number taken from a cadastral map that is based on a rigid framework of horizontal and vertical control. Data are computerized and can include not only title information and deed descriptions but also information concerning natural resources, zoning, and the like. Most recording offices now use microfilm as a necessity for saving space.

Surveyors should be a integral part of the development and maintenance of land data records; they should be the ones to see that evidence of monuments and their position are included with all other information.

REFERENCE

McEntyre, John J., *Land Survey Systems*, John Wiley & Sons, New York, 1978.

CHAPTER 2

Evidence

2-1 Contents

Evidence can consist of almost any object, thing, action, or verbal statement; the law of evidence considers the admissibility, effect, and relative importance of the evidence produced. A found monument is evidence; the control afforded the found monument is determined by the law of evidence. Evidence and the law of evidence are closely interrelated, and the discussion of one invariably includes a discussion of the other. Both subjects are discussed herein.

In a broad sense, a discussion of evidence and the law of evidence could include most phases of the property surveyor's practice. All initial land transfers must be in writings, and writings are evidence. Measurements are evidence to prove where property lines or corners are located. Evidence of possession may be proof of an unwritten conveyance. Scientific principles are evidence called *judicial notice* by the court.

Although an entire book could be written from the standpoint of evidence, it would not be a logical arrangement for surveyors. This chapter is devoted to the fundamental concepts of evidence and the law of evidence. In later chapters discussions of the amount, kind, and sufficiency of evidence necessary to prove a fact are included where appropriate. The evidence necessary to prove unwritten conveyances is better understood if included along with the subject matter of unwritten conveyances rather than discussing it here.

2-2 Importance of Evidence

An indispensable part of every perimeter survey of an existing conveyance is the discovery and evaluation of evidence. In the orderly process of performing a property survey, the surveyor follows nine steps, usually in this order to perform a complete job. (1) He obtains (often, from the client) written evidence of title in the form of a deed, abstract, or title policy. (2) He seeks evidence of maps, field notes, county and city records of surveys, state and other public agency records of surveys, and all written records that disclose evidence of monument positions pertaining to the survey. (3) He reads adjoiner deeds for evidence of seniority or conflicts. (4) He goes on the land and seeks evidence of existing monuments and evidence of possession and usage. (5) He may seek testimony (evidence) of the existence and location of old monuments. (6) He makes measurements from found monuments to determine search areas or locations to dig for missing monuments, and he makes measurements (also evidence) to tie found monuments together. (7) He makes calculations (also a form of evidence). (8) From the evidence of monuments, measurements, testimony, and computations, he comes to conclusions in accordance with the law of evidence. (9) He uses measurements to set new monuments in accordance with his conclusions.

Before a surveyor obtains knowledge of all the available evidence, it is almost impossible to make a correct boundary location. There is an old saying, change the evidence and you change the law. The important aspects of property surveying are the ability to search, find, and discover *all* available evidence and the ability to arrive at conclusions about where boundaries belong in accordance with the laws of evidence and the laws of boundaries. A surveyor may be able to compute, make drawings, use instruments, and stake engineering projects, but he is not qualified to make property locations until he understands the law of property lines and the law of evidence.

2-3 Arrangement of the Subject Matter

The subject of evidence is comprehensively treated in legal literature but is seldom found in surveying books. Because of this, early space is devoted to defining terms and elementary rules of evidence.

Evidence in itself is not proof of facts; conclusions or inferences that can be drawn from evidence are proof. In defining what inferences or conclusions can be drawn from evidence, the law has evolved rules to aid in evaluating evidence. And the law uses such words or phrases as "presumption," "burden of proof," "extrinsic evidence," and "preponderance of evidence" that must be defined to clearly understand the law of evidence.

Evidence varies in significance, importance, and application. Thus witnesses may give testimony about the location of a property corner. But such evidence is incompetent and irrelevant to overcome the location of an original, undisturbed monument called for. The evidence of a rock mound is of little importance where the original notes called for a oak tree. The evidence of a measurement is incompetent to prove an original monument to be in error.

In coming to conclusions from evidence, the most important need of the surveyor is the ability to recognize and know what is the best evidence of that available. After defining terms and after presenting legal concepts that determine the effect of evidence, the remainder of this chapter is arranged, as best may be, in the descending order of importance of evidence, that is, in the order of the best-available evidence. An outline is as follows:

1. Definitions
2. Effect of evidence
3. Best available evidence (excluding unwritten rights)
    a. Senior rights
    b. Writings
    c. Intent of conveyances

      d. Calls for surveys
      e. Monuments: (1) natural; (2) artificial; (3) record
      f. Measurements
   4. Surveyor's obligations with respect to evidence
   5. Examples

Many misconcepts about where to locate property lines arise from misapplications of rules of evidence concerning what is the best-available evidence. Thus most surveyors understand that a called-for, found, and undisturbed monument is given preference over calls for measurements; but it must be distinctly understood that the monument *must be called for* by the written evidence, either directly or implied, to have primary significance.

The objectives of the arrangement of this subject matter are (1) defining what is acceptable evidence, (2) stressing the order of importance of evidence, that is, what is the best-available evidence, and (3) stating the property surveyor's obligations with respect to evidence.

2-4 Evidence

DEFINITION. *At law evidence is that which is legally submitted to a competent tribunal as a means of ascertaining the truth or untruth of any alleged matter of fact under investigation before it.*

At law there are at *least four kinds of evidence:* (1) *Oral evidence* or *testimony* is evidence given by witnesses. (2) *Written evidence* is evidence in the form of documents. (3) *Real evidence* consists of material objects addressed directly to the senses such as physical monuments. (4) *Judicial notice* is evidence in the form of knowledge. The courts may take judicial notice of certain facts such as *(a)* the true significance and meaning of all English words and phrases, *(b)* whatever is established by law, *(c)* the laws of nature, and *(d)* other well known and commonly accepted facts.

2-5 Evidence, Conclusions, and Proof

PRINCIPLE. *Evidence is not proof. A consideration of all evidence and conclusions to be drawn from evidence, in accordance with the law of evidence, may produce proof.*

Evidence is not proof of a fact; a conclusion or inference that may be drawn from evidence is the proof. A written deed is evidence of ownership; it is not proof of ownership. Land can be gained by unwritten means; hence a paper title does not always prove ownership. A written deed may be void because of state statutes defining limitations or prior court decisions. If a person can prove by evidence that he has a written deed vesting title in him, that the person conveying the land to him was competent to do so, that no one adversely occupies the land described by his deed, that title has not passed by escheating, and that other items making up ownership do not operate against him, he ma have proof of ownership.

Surveyors take a legal description of property and mark it on the ground; the do not and should not, unless specifically requested by legal counsel, conside validity of signatures, possible insanity, whether escheated, competency of th seller, or like considerations. They do consider senior rights and note possessio1 not in agreement with the written deed.

2-6 Classifications of Evidence

Evidence varies in value and dignity. In general evidence may be divided intc the following classifications:

1. *Indispensable evidence* is evidence that is necessary to prove a fact. Conveyance of property must be in writing; hence a conveyance cannot be proved without proof that there was a written document.
2. *Conclusive evidence* is that which the law does not permit to be contradicted. As an example, the contents of conveyance writings (recital of a consideration excluded) are conclusive as between the parties, excepting for pleadings of illegality, fraud, mistake, or reformation. Also, the written document cannot be altered by oral testimony, and, as is commonly stated, everyone is presumed to know the law.
3. *Prima facie evidence* is that which suffices for. proof of a fact until rebutted by other evidence. In the event that an original deed cannot be produced, a recorded deed is *prima facie* evidence of the deed's contents. In many areas the results of the survey of certain official surveyors, such as the county surveyor, are *prima facie* evidence of the location of lines. *Prima facie* evidence may be disproved, but, until it has been proved incorrect, it is assumed to be correct. The law specifies what is or is not *prima facie* evidence.
4. *Primary evidence* is that which is most certain. The contents of a written document are more certain than the oral testimony of what the document contained.
5. *Secondary evidence* is inferior to primary evidence. A copy of the original document is inferior to the original. Secondary evidence is used to prove the contents of lost or unavailable primary evidence.
6. *Direct evidence* proves a fact directly without resorting to presumptions or inference; for example, Jones testified, "I saw the original surveyor drive the particular stake in the ground."
7. *Indirect* or *circumstantial evidence* depends on inferences or presumptions that tend to prove a fact by proving another; for example, Jones testified, "I saw the original surveyor drive similar stakes at other corners."

8. *Partial evidence* is to establish some detached fact. It is often used to corroborate other evidence.
9. *Extrinsic evidence* is derived from sources outside the writings (see Section 2-14).

2-7 Types of Evidence Gathered by Surveyors

Evidence used by surveyors to prove property-line locations can be classified as follows:

1. Written documents, maps, and historical facts.
2. Facts that the court takes judicial notice of (knowledge of the court).
3. Physical objects (real evidence) observed by the surveyor: surveyors' stakes, trees, fences, rivers, street improvements, and the like.
4. Parol evidence. This can be divided into (1) witnesses who observed the former location of physical objects (a monument now destroyed); (2) witnesses who can explain a latent ambiguity; (3) witnesses who can testify about commonly reported facts; (4) witnesses who can describe the customs or conditions existing as of the date of the deed.
5. Measurements of distances, bearings, and angles.
6. Mathematical calculations.

EFFECT OF EVIDENCE

2-8 Scope

The law of evidence includes a definition of evidence, the effect of evidence, and the competency of evidence. A summary of the terms used by the courts to determine the effect of evidence and a summary of what rules determine the amount of evidence necessary to produce proof provide the introduction to the subsequent section on "best available evidence."

2-9 The Law of Evidence

DEFINITION. *The law of evidence is a collection of general rules established by law (1) for declaring what is to be taken as true without proof; (2) for declaring the presumptions of law, both those that are disputable and those which are conclusive; (3) for the production of legal evidence; (4) for the exclusion of whatever is not legal; and (5) for determining, in certain cases, the value and effect of evidence.*

The surveyor uses evidence to assist him in locating and proving property locations, and, unless correctly evaluated, evidence is worthless. The law of evidence—that is, the law that declares what evidence is admissible and what right is to be accorded to admissible evidence—is determined both by statutes and common law. It is the law of evidence that the surveyor relies on to guide him in assigning proper values and weight to discovered and known evidence. Not all jurisdictions have identically the same rules of evidence; hence variations of the law of evidence can be expected in different states.

It has been said that the only correct location for a written deed is in the position that a court of competent jurisdiction would decree to be the correct location. The court bases its decision on admissible evidence and the application of law. Understanding the significance and value of a particular piece of evidence is just as important as understanding the statutory and common laws that pertain to boundary location. The courts sometimes disagree on the value and conclusiveness of evidence. A higher court has been known to reverse the lower court, and, on occasion, two judges with identical circumstances have rendered opposite opinions. In the Supreme Court of the United States there have been numerous five to four decisions. If those who are experts in interpreting the value of evidence are not in complete harmony, it can be expected that surveyors on occasion will not agree. Disagreement in itself is sometimes a desirable thing. Disagreement with existing theories of science has, without doubt, been the cause of new discoveries. Disagreements may bring advancements; but disagreement based on stupidity, lack of knowledge, or plain contrariness is undesirable.

Although it is admitted that sometimes the value of given evidence is debatable, usually evidence is not subject to alternate interpretations. Since the practicing surveyor can expect on occasion that his property-location findings will be tested in court, he should assign values to evidence in the same proportion that it will be accepted in court. For example, it is indeed a difficult task to get an unrecorded map accepted for court evidence, unless it is accompanied by the author's testimony, or unless it is declared ancient by law and is commonly reputed as being correct. Surveys based on unrecorded maps of unprovable origin may be impossible to substantiate in court.

The value of evidence is relative and is subject to general, not definite, rules. In a border-line decision, what the law of evidence is, will not be known until the conclusion of the case. The court then declares what the law is and further takes the attitude that "this is the law as it always was, and this is the law as you should have known it to exist all the time." Fortunately, the majority of boundary surveys are not dependent on doubtful legal considerations.

2-10 Burden of Proof

PRINCIPLE. *The affirmative party of plaintiff to a civil case has the duty of presenting sufficient evidence to convince either the judge or jury of his allegations. A defendant has no obligation to present evidence.*

If the plaintiff fails to present sufficient evidence to prove his contentions, the court will leave the parties in the position it found them and hold for the defendant. In effect the defendant wins, and the status quo is maintained.

If his measurements are incorrect, or if he started from the wrong beginning point, of course his survey and his testimony fall, and respondent's case falls with it.
—*Clark* v. *McAtee*, 253 Mo. 196.

The boundary in dispute is located on the line between Range 71 and Range 72 in Township 50. The land owned by the state was settled upon and homesteaded by John F. Gates in about 1907. He constructed a fence on the east of his land, which appears to be now claimed by the state as the true line. The boundaries of the land were unknown, Gates employed no surveyor, and he testified, in substance, that the fence was built as a fence of convenience, not intended to represent the line to which he made absolute claim, and that he would have changed the fence to the true line, if and when it should become known. The point in dispute in fact is whether the Roberts survey and the United States Government survey approved in 1927 on the one hand or the Harvey survey on the other is correct. Harvey, who surveyed the same township did not run a straight line clean through, but only from the southeast corner of the township for a distance of five miles, then angled west, and thence north, in order to connect with the northeast corner of the township. This the witness stated was an incorrect method. It is clear from the foregoing testimony that the state has failed to sustain the burden of showing that the true line between its land and the land of the defendant is that claimed by it.
—*State* v. *Vanderkoppel* (Wy.) 19 P 2d 955.

The plaintiff sued in equity alleging ownership and possession of a small tract of land which is described as follows: "Bounded on the south by the lands of Wilk Witten; on the west by the lands of James Le Master; on the north by the lands of S. G. Preston; on the east by the lands of W. W. Brown, and containing 15 acres, more or less." The defendants claim this to be at the head of a hollow while appellants claim it to be further down the same hollow. Its determination rests largely upon the ascertainment of the boundary lines of the old Stambaugh patent. The disputed calls are "N 13° E, 10 poles to two maples and a small oak in a dogwood gap; N 23° E, 68 poles to a beech and maple; N 18° W, 80 poles to a beech on the bank of a branch." These are the 7th, 8th and 9th calls of the patent. A certified copy of the plat of the patent is the same except that the 10 poles is 100 poles. In addition to this a number of witnesses have testified as to the location of the dogwood gap.

We are unable to determine which is correct, as the abutting lands are not described with any degree of certainty. The burden is in the plaintiff and the doubt should be resolved against him.
—*Green* v. *Witten*, 200 Ky. 725.

Ordinarily the burden of proof is on the plaintiff, but defendant has the burden of proof in matters of defense. Where the statute divides the burden of proof equally (as in Louisiana), each side has an equal responsibility. The burden of proving surveys, monuments, agreement, acquiescence or a change of boundary is on the party asserting that fact.

In an action in boundaries, the law requires proof from each of the contiguous owners, and the burden is divided.
—*Russell* v. *Producrs Oil Co.*, 143 La. 217.

Hageman, a civil engineer, presented a plat and stated that he examined the territory in question and found the course of the stream was not as shown on the government plat, but is as described on a plat which he produced which shows the stream flowing east of south instead of southwesterly into the Quillayute River, and that as a result thereof Lot 3, if extended to the actual stream, would contain a little more than 82 acres of land instead of 39.50 acres. It was shown that Hageman made no instrumental survey of the territory of the stream, merely a visual survey without any actual measurement, without determining the exact course, sinuosities, and location of the stream. He had never even made any land surveys, although he was probably professionally competent to do so. What he did in this matter, however, is merely nonexpert, and we cannot consider that this evidence was competent to prove the fact which it was offered to prove, or sufficient to contradict or impeach the official government survey, that fixes the character of the entry and determines whether the possession is adverse.

—*Rue* v. *Oregon and Washington Ry. Co.*, 109 Wash. 436.

When there is possession, especially for long duration, doubt in property location evidence is usually resolved to leave the status quo; proof of true lines must be positive.

Three separate surveys were made by three surveyors and each of said surveyors admitted he was unable to state whether or not the point from which he started his re-survey corresponds with the corner of said block as it was originally surveyed. We think that the appellant has failed in this court to establish where the true boundary line lies. In case of disputed boundary lines, based upon uncertain evidence the courts ought not to disturb boundary lines between lot owners which have been acquiesced in for years.

—*Westgate* v. *Ohlmacher*, 251 Ill. 538.

2-11 Preponderance of Evidence

PRINCIPLE. *In civil cases it is not necessary to prove "beyond a reasonable doubt" as in criminal cases; it is only necessary to prove a "preponderance of evidence."*

The surveyor cannot always prove conclusively "beyond a shadow of doubt" that his monuments are positively in their correct position. If he is to be upheld by the courts, he must be prepared to prove that the preponderance of evidence is in his favor. A second surveyor, in disagreement with him, must be prepared to prove by a preponderance of evidence that the other is wrong.

A preponderance of evidence is not necessarily just more than 50 percent of the weight of the evidence. It has been variously defined as that which inclines "an impartial mind as to one side rather than the other" and that which removes "the cause from the realm of speculation." In a survey, the surveyor should satisfy himself about the preponderance of evidence, giving due weight to presumptions, *prima facie* evidence, and law, that his location is probably correct to the exclusion of other possible methods of monumenting the property.

The surveyor investigates all possibilities, excludes the unlikely or improbable, and monuments in accordance with the most certain.

> The law does not require demonstration; that is, such a degree of proof as, excluding the possibility of error, produces absolute certainty because such proof is rarely possible. Moral certainty only is required, or that degree of proof which produces conviction in an unprejudiced mind.
> —Sec. 1826 CCP Calif.

The question to be determined is whether an oil well has been drilled upon the one side or the other of a line which divides two quarter sections of land in a section and township which has been surveyed and subdivided by the authority of the government of the United States. One might think that it is a very simple question, but it is not so. No one on earth can furnish the information necessary for its decision, save the gentlemen of the civil engineering and surveying profession, and those of them who have testified in that behalf in this case have arrayed themselves on opposing sides. Several surveys were made in order to re-establish the lines and corners originally established by Jones and Moore. Mr. W. E. Martin made his survey in which he located the line 22.2 feet east of the oil well. Mr. H. E. Barnes, on behalf of the defendants, found the line to be 15.7 feet west of the well. A private survey made by Mr. H. A. Jenkins, on behalf of the defendants, located the line 35 feet west of the well. Mr. A. D. Kidder, acting on behalf of the United States government, made a survey for purposes not connected with this litigation, located the line 1.74 east of the well. Mr. Welman Brandford, who was employed by defendants, found the line to be 31.2 feet west of the well. Mssrs. Martin and Williams, surveying under direction and instructions of the trial court, reported the line run under these instructions to be 14.7 feet west of the well. Many other surveyors testified in the case, and the diversity of opinion among these gentlemen is most bewildering to the lay mind. The gentlemen who have been employed or consulted in the matter are all reputable members of the civil engineering and surveying professions; their work seems to have been done with care, and their opinions expressed only after deliberate consideration, and it therefore seems to be impossible to determine and to prove with mathematical and absolute certainty precisely where the corners designated on the sketch belong. That situation may as likely be the result of error in the original field notes as in the resurvey made in connection with this litigation. We shall proceed to establish the limits of the property according to what we consider to be the preponderance of evidence in the case. Of the several surveys made, one stands out most as worthy of consideration of the court. It was not made on behalf of any of the parties in the interest of this litigation, but it was ordered by the United States government, and it was executed under instructions from the general land office, whose stamp of approval has been placed upon it. The engineer under whose personal supervision was done was Mr. Arthur D. Kidder, supervisor of chief of surveys of the general land office. He is the author of a treatise on the improved solar transit and the compiler of tables of azimuth of polaris used by the government in its surveys. In the performance of his work he used two improved solar transits recognized as scientific instruments of great precision, and, with the aid of tables published in the ephemeris of the sun and the north star, checking one instrument with the other to obtain in the field itself an accurate north and south line, and to determine with absolute accuracy the variation of the needle. Neither he nor his assistants, though aware of a contest involving the ownership of an oil well, knew, at the time of making the survey, any of the litigants in the case, nor did they know on which side of the line in dispute the said litigants respectively claimed the oil well to be located. Under these circumstances the recognized ability and competency of Mr. Kidder, the total absence of any possible bias on his part, and the great care he exercised in the performance of his work, and the most modern and scientific methods adopted by him, and the further fact that the results of his work bear the approval of the General Land Office, are, in our opinion, sufficient to establish a preponderance of evidence in favor of the plaintiff, to justify a decree based upon his findings under the law applicable to the case.

Russell v. *Producers Oil Co.*, 143 La.

2-12 Presumptions

DEFINITION: *Presumptions at law are deductions that the law expressly directs to be concluded from certain known facts.*

Presumptions are for the purpose of expediting court cases by reducing the amount of evidence necessary. Technically, presumptions are not evidence, but are considered as substitutes for evidence. A presumption affects only the burden of offering evidence; thus it becomes a "procedural tool" based on considerations of (1) probability, (2), practical convenience, or (3) public policy.

In applying such a rule proof can be offered that a properly stamped and addressed envelope was placed in the mailbox; this will give an inference that is was received; thus the courts will "presume" it was received in the absence of evidence to the contrary.

Conclusive presumptions are those that are irrebuttable, and this by law permits no contradiction. A commonly known conclusive presumption is "everyone is presumed to know the law." Whether a person does or does not in fact know the law is immaterial, since no evidence can upset this conclusive presumption. In the absence of pleadings of fraud or illegality, the truth of the facts recited in a written conveyance is a conclusive presumption as between the parties. But there is an exception in the recital of a consideration.

Disputable or rebuttable presumptions are those that may be proved incorrect by other evidence; but, unless proved otherwise, the jury's findings must be according to the presumptions. Common-law presumptions usually have force in the various states, provided they are not in conflict with statutory presumptions or provisions.

One who grants a thing is also presumed to grant whatever is essential to its use. Thus a conveyance of land includes all existing easements necessary for the use of the land whether said easements are recited or not. The burden of proof is with the person trying to prove that the easements did not pass with the property. Evidence willfully suppressed is presumed to be adverse if produced. If inferior evidence is produced, it is presumed that the higher evidence would be adverse. In many states a letter duly directed and mailed is presumed to be received in the regular course of mail. This is known as the *mailbox rule*. Each of these presumptions may be overcome by acceptable contrary evidence.

Many rules of surveyors are equivalent to presumptions. Normally, the order of importance of conflicting deed elements are senior rights, intentions of the parties, monuments, measurements, and area. Of course, this is not a hard and fast rule but is merely a good disputable presumption to be followed until contrary evidence is developed. The surveyor, in his quest for evidence, sometimes discovers or seeks facts for the purpose of proving a presumption wrong, but until he proves the contrary the presumption governs.

Before a presumption can be assumed the person who wants to seek the benefit must establish the "basic fact" that is the foundation to the "presumed fact." Courts for the most part hold that a presumption is not evidence but is a deduction that must be drawn from evidence.

2-13 Inferences as Evidence

DEFINITION. *An inference is evidence in the form of a logical conclusion from a set of facts without express directions of the law to that effect.*

Surveyors frequently resort to inferences to prove a given property location. Inferences are not based on imagination or supposition; they are based on probabilities, and the drawing of inferences is a matter of discretion left up to the trier of fact. A failure to speak and explain when it is a duty to do so gives rise to an adverse inference. The negative testimony of a person in a position to see a monument in a particular location supports the inference that the monument was not there.

2-14 Extrinsic Evidence, When Used

Once an agreement or deed is reduced to writings, testimony cannot be used to overcome clear, unambiguous, written words. But if the words are not clear and need explanation, *extrinsic evidence*—that is, evidence other than the writing itself—may be sought to explain the words. A deed reading "beginning at the southeast corner of Jones' watermelon patch" may require a lot of extrinsic evidence to explain where the southeast corner was located.

Extrinsic evidence may also be taken to explain a local meaning to a particular word. In Texas and the West, the old Mexican vara, a unit of measurement, was used. Extrinsic evidence is sometimes needed to explain the length of a vara, since it was not the same in all localities.

Courts generally take extrinsic evidence to explain latent but not patent ambiguities in descriptions. A *patent ambiguity* is a defect appearing on the face of the conveyance itself. A deed reading "a house and lot on Main Street" describes nothing in particular and contains a patent ambiguity that cannot be remedied by other evidence. A *latent defect*—that is, a defect not apparent on the face of the instrument but apparent when the instrument is applied to matters outside the instrument—can be cured by extrinsic evidence. If an ambiguity has been raised by extrinsic evidence, it is only logical that the courts would allow the same kind of evidence to explain it.

The rigidity of the original rule defining these two types has been greatly relaxed. The present tendency is to regard a description as valid rather than void and to extend the word *latent* to its most liberal meaning. A patent ambiguity is construed to exist only when persons of competent skills are unable to interpret the deed.

> In general if a competent surveyor can take the deed and locate the land on the ground from the description contained therein, with or without the aid of extrinsic evidence, the description will be held to be sufficient.
> —*Blume* v. *MacGregor*, 148 P 2d 656, Calif. 1944.

> The general rule of law is that possession must start under a claim of color of title which purports to be valid. If the deed under which he claims be void or insufficiently formed to pass title, the possession is not adverse under our statutes. The description reads, "One tract of land lying and being in the county of the foresaid, adjoining the land; John J. Phillips and Pender, containing 20 acres more or less." This description fails to identify or furnish a means of identifying. It gives neither course nor distance of a single line, nor a single point, stake or corner anywhere to begin at.
> —*Dickens* v. *Barnes*, 79 N.C. 440.

> Where land is conveyed by a general description, extrinsic evidence is admissible to ascertain the location of adjoining tracts called for, so as to apply the conveyance to its proper subject-matter. If with the aid of these the land granted can be sufficiently identified, it is all that is necessary.
> —*The Sulphur Mines Co. of Virginia* v. *Thomson's Heirs*, 93 Va. 293.

2-15 Admissibility, Relevancy, and Conclusiveness of Evidence

The purpose of any trial is to determine the truth regarding the issues that are presented by both sides. In court not all evidence is admissible to prove property location. Evidence to be admissible must be relevant to the issues, competent under established rules of law, and material in the sense of having some reasonable tendency to prove or dispute points in issue. The decision about what evidence will be admitted lies exclusively with the judge. Parol evidence, with some exceptions, must be based on direct observations and normally cannot be something the witness had heard someone else say. Lay witnesses generally do not express conclusions from evidence. Thus there are rules about what evidence should be considered. The admissibility, relevancy, or conclusiveness of evidence are discussed in evaluating the various types of evidence. Evidence in order to be offered must have a probative value in that it must tend to prove the fact for which it is offered. If one were to consider the Federal Rules of Evidence, rule 401, it is explained that "relevant evidence means evidence having any tendency to make the existence of any fact that is of consequence to the determination of the action where more probable or less probable than it would be without the evidence."

2-16 Judging the Effect or Value of Evidence

The jury judges the effect or value of evidence. But the judging must be in accordance with the laws of evidence.

> The jury, subject to the control of the court, are the judges of the effect or value of evidence addressed to them, except when it is declared to be conclusive. They are, however, to be instructed by the court on all proper occasions:
> (1) That their power of judging of the effect of evidence is not arbitrary, but one exercised with legal discretion, and in subordination to the rules of evidence;
> (2) That they are not bound to decide in conformity with the declarations of any number of witnesses, which do not produce conviction in their minds, against a less number or against a prescription or other evidence satisfying their minds;
> (3) That a witness false in one part of his testimony is to be distrusted in others;
> (4) That a testimony of an accomplice ought to be viewed with distrust, and the evidence of the oral admissions of a party with caution;
> (5) That in civil cases the affirmative of the issue must be proved, and when the evidence is contradictory the decision must be made according to the preponderance of evidence; that in criminal cases guilt must be established beyond reasonable doubt;
> (6) That evidence is to be estimated not only by its own intrinsic weight, but also according to the evidence which it is in the power of one side to produce and of the other to contradict; and therefore,
> (7) That if weaker and less satisfactory evidence is offered, when it appears that stronger and more satisfactory evidence was within the power of the party, the evidence offered should be viewed with distrust.
>
> —Sec. 2061 CCP, Calif.

2-17 Weight of New Evidence

In a boundary dispute being settled in court, each party presents evidence that tends to prove his side of the case. After all evidence is presented the judge decrees a property position as based on law, facts, and evidence. After a period of time, if the decree is not reversed by a higher court, the decree becomes final. The judge has this advantage: he draws a conclusion from the evidence and facts presented and from his knowledge of law. After a decree is final, discovery of new evidence does not alter the property-line location decreed.

The surveyor, when making a property survey, gathers evidence, makes a finding, and sets his property markers. But the surveyor's findings do not have the finality of a court decree. Another surveyor, uncovering additional evidence, may come to a different conclusion and locate his lines in another position. And the second position might be the correct location. But on the other hand, still another surveyor might be more diligent and uncover further evidence that proves still another location to be proper. Thus it is of the utmost importance that a surveyor seek and find all the evidence in the first instance and this irrespective of costs. Many so-called lost monuments have been found on later surveys. In fact, one of the major causes of disagreement between surveyors relates to the lack of discovery of all available evidence at the time of the initial survey. If every surveyor uncovered all the evidence, differences would be reduced to a minimum, and the surveys would have a finality of location.

BEST AVAILABLE EVIDENCE

2-18 Scope

After a conveyance is made, the location of the land is determined by conclusions drawn from the best-available discovered evidence, all in accordance with the law of evidence. The surveyor's area of necessary knowledge includes what is acceptable evidence, what conclusions can be drawn from evidence, and what evidence must be rejected. The law of evidence, in effect, assigns an order of importance to evidence, and, if two pieces of evidence are in conflict, conclusions are drawn in accordance with the evidence's importance as stressed in the law of evidence.

The written title to land can be partially (rarely, all) extinguished by the acts and behavior of adjoiners, usually with wrongful possession for a period of time. This chapter pertains to written title rights; the transfer of title by acts of possession and behavior of adjoiners is treated in the next chapter. Normally, surveyors locate land in accordance with a written description, but it is important that the surveyor understand the significance of possession rights (usually called *unwritten title*) for reasons of liability as discussed in Chapter 4.

Occasionally, portions of land are sold to two parties; the first buyer (sometimes the first person to record) has the right of ownership. Consideration of this subject is included in this chapter.

2-19 Understanding the Law of Boundaries and Evidence

PRINCIPLE. *The surveyor is presumed to know the law of boundaries and the law of evidence, and, when he agrees to locate a written conveyance on the ground, he agrees to locate it in accordance with the laws governing how written conveyances should be located.*

The surveyor does not take verbal evidence to explain the law. The law is a matter of judicial notice and is gathered from statutes and writings of learned people. Everyone is presumed to know the law, and the surveyor is no exception. This is an irrebuttable presumption that may not be overcome by contrary evidence. If the surveyor agrees to monument a certain written conveyance on the ground, he also agrees to locate the conveyance in accordance with the laws regulating the interpretations of written conveyances. The surveyor is a professional specialist who knows how to read and interpret deed words and how to set monuments in accordance with the words. Understanding and applying the correct law (and that includes the laws of evidence) are certainly part of his duties.

The surveyor may not practice law; he merely obeys the law as it exists. Law is written. It is not the opinion of a lay witness. When performing a property survey it is the prerogative and duty of a surveyor to interpret the meaning of the written words of a conveyance. It is this fact that elevates him above the layman and promotes him to professional stature.

2-20 Conclusive Evidence, Senior Rights, and Third Persons

PRINCIPLE. *The words of a written conveyance are conclusive evidence as between the buyer and seller, but they are not conclusive evidence as against a third party who has senior rights or a person with occupancy rights.*

If a person conveys part of his land, he cannot at a later date convey more than his remainder. The first buyer has what is known as *senior rights*, and the second buyer has *junior* or *remainder rights*. The senior buyer is entitled to all land conveyed to him according to his description; the junior buyer is entitled to all land conveyed to him, provided it does not interfere with the senior rights. If such interference occurs, the junior deed loses. Evidence proving senior rights ranks first in the order of importance of written evidence.

Suppose that Jones sells the easterly 50 feet of his record 100-foot lot and then, at a later date, sells the westerly 50 feet. Prior to unwritten rights ripening into a fee, a proper survey reveals that the lot is only 98 feet wide. Although the second buyer has a title that clearly gives him 50 feet, his written title does not prevail as against the senior rights of the adjoiner. This second buyer gets 48 feet. The written evidence, the deed, is only conclusive evidence as between the parties, not as against a third party's valid claim.

Rarely, the evidence disclosed in a junior deed, written from a survey, locates the monuments called for in a senior deed. In a recital of "thence N 18°01'W 201.03 feet to a found original iron pipe marking the Easterly corner of Jones' land . . . " the bearing and distance serve as evidence to relocate the iron pipe in the event that it is lost. In such event the junior deed is not controlling the senior deed, but it is merely furnishing evidence of the location of the original senior property.

> Where in an ejectment the plaintiff's claim [is made] under one warrant and the defendant under three warrants, and it appears that the returns for all the warrants were made on the same day and that the location for the three surveys under which the defendant claim call for the plaintiff's warrant on the northeast, while the location of the plaintiff's land calls for vacant land on the southwest boundary, the evidence is conclusive that the plaintiff's survey was earlier in time than the defendent's survey, and there is no question of priority to submit to the jury.
> —*Collins* v. *Clough*, 222 Pa. 472.

The date of the conveyance, not the date of recording, determines the seniority of a deed. But in most states, unless a deed is recorded, it does not prevail as against the innocent rights of third parties.

The writings, of course, are not conclusive as against claims of third parties who have rights as a result of other causes such as an occupancy right.

VALUE OF WRITINGS AS EVIDENCE

2-21 Writings as Indispensable Evidence

PRINCIPLE. *A written conveyance is indispensible evidence in the proof of ownership.*

The law, as enforced by the Statute of Frauds, requires all conveyances to be in writing; some form of written proof of title is indispensable evidence. Because of this, surveyors survey from written conveyances in accordance with law. After the written conveyance is located on the ground, the surveyor then notes possession not in agreement with the writings and, as a minimum, informs the client of the possible significance of prolonged possession.

Most important, the surveyor should be sure that he has the correct document describing his client's land. The original description as contained in the original conveyance could have been altered by a court case or by a legal establishment as described in Chapter 7.

2-22 Identification of Property Descriptions

PRINCIPLE. *Descriptions do not identify themselves, and all that is required of a deed is that it furnishes evidence of a means of identification. If the description in a deed is sufficient when the deed is made, no subsequent change in conditions or loss of evidence can render it insufficient.*

Although a description may have been sufficient at the time it was written, subsequent disappearance of evidence may render the certainty of location exceedingly doubtful. But courts declare where property is located irrespective of the poor quality of evidence presented; the best available evidence is accepted even though the evidence may not be admissible in other types of litigation. Courts have held that vagueness in a description will render a deed void or voidable.

> Written descriptions of property are to be interpreted in the light of surrounding facts and circumstances well known in the community; descriptions do not identify themselves, and all that is required of a deed is to furnish a means of identification. If the description in a deed is sufficient when the deed is made, no subsequent change in conditions can render it insufficient.
> —*Blair* v. *Rorer's Administrators*, 135 Va. 1.

> In determining the original government lines and corners, where they are in dispute, course, distance, measurements, plats and field notes must all yield to the actual corner and lines established by the original government surveyor. It must be conceded that these monuments, which were artificial, have now to a large extent

been obliterated; but in determining where they were, we may consider not only the testimony of those who saw and identified them when they were discernible, but evidence of practical location made at a time when they were presumably in existence; acquiescence of the parties concerned in supposed boundary lines; acts of public authorities in the well established boundaries of others contiguous tracts; and reputation and tradition is also to be considered in some cases.
—*Rowell* v. *Weinenann*, 119 Iowa 256.

After 90 years have elapsed and time has destroyed in large measure the evidence left by the original locator, it is then permissible, not only permissible, but of necessity is required, that we resort to any evidence tending to establish the place of the original footsteps which meets the requirement that it is the best evidence of which the case is susceptible.
—*Taylor* v. *Higgens Oil and Fuel Co.*, 2 SW 2d 300.

Where marks left by the original surveyor have disappeared it is permissible and necessary to resort to the best evidence of which the case is susceptible.
—*Reynolds* v. *Bradford*, 233 SW 2d 464.

The purpose of a resurvey is to trace the footsteps of the original surveyor. When the marks of his footsteps are found, they control. When they cannot be found, old use and occupancy, old recognition, must suffice.
—*Ballard* v. *Stanolind Oil & Gas Co.*, 80 F 2d 588.

In the older areas of the United States, especially in rural settings, where considerable loss of evidence exists, reputation assumes a role of greater importance.

2-23 Conclusiveness of the Written Words of a Deed

PRINCPLE. *When the terms of an instrument, deed, or will have been reduced to writing, it is to be considered as containing all those terms, and there can be, between the parties or their representatives or successors in interest, no evidence of the terms of the instrument other than the contents of the writing, except in the following:*

1. *Where a mistake of imperfection of the writing is put in issue by the pleadings.*
2. *Where the validity of the agreement, deed, or will is the fact in dispute.*
3. *To establish illegality or fraud.*
4. *To explain an extrinsic ambiguity.*
5. *To show the circumstances under which the instrument was made for the purpose of properly construing the instrument.*[1]

In the foregoing, the surveyor is not concerned with items *1*, *2*, and *3*. These are matters for title companies, abstractors, and attorneys. The surveyor does

[1] Sec. 1856 CCP Calif.

frequently seek evidence to explain extrinsic ambiguities and sometimes evidence to understand the circumstances under which the deed was written (was the basis of bearing magnetic or astronomic?). In addition to items *4* and *5*, the surveyor is charged with "judicial notice" of the meaning of conveyance words.

> A survey of land and the erection of some monuments, with the view of a subsequent conveyance, cannot be admitted to vary or control the deed afterward made, when that deed calls for none of the monuments so erected and contains no reference to such survey.
>
> All prior negotiations must be taken, so far as the construction of the deed is concerned, to have been merged in that instrument; the conclusive presumption being that the whole engagement of the parties, and the extent and manner of it, were reduced to writings.
>
> The actual intention of the party, or the surveyor, is not admissible to effect the construction of the deed.
>
> —*Wells* v. *Jackson Iron Manufacturing Co.*, 47 N.H. 235.

The conclusiveness of the written words of a deed also applies with the same force to *all* documents called for in the deed. If a certain map is called for, all the writings and monuments called for on the map have just as much conclusive control as though the writings and monuments were called for by the deed itself.

Because of the conclusiveness of the written documents, the writings themselves are often called the *best available evidence.*

> When the description of the premise conveyed in a deed is definite, certain and unambiguous, extrinsic evidence cannot be introduced to show that it was the intentions of the grantor to convey a different tract. Thus, where the land is described by course and distance, parol evidence is inadmissible to prove that the true boundary is a line of marked trees not mentioned in the deed, or that a deed stating a definite thing was intended to express another, or to alter or vary the legal import of monuments the location of which has been ascertained. If the calls in the grant when applied to the land correspond with each other, parol evidence will not be admitted to vary them to show that in point of fact they were not the calls of the survey actually made (but not called for).
>
> —*Blair* v. *Rorer's Administrators*, 135 Va. 1.

After a surveyor has completed a plat and has certified to the same that he has surveyed it, his testimony would be incompetent to prove that he did not in fact make a survey. The evidence on the plat, being writings, is not to be impeached by verbal testimony.

> The appellants contend the surveyor whose certificate appears on said plat admitted, while testifying as a witness in the cause, he had not, in fact, made the survey. The plat was completed, certified by the surveyor, acknowledged by those claiming to own the land, and duly recorded, in compliance with . . . law. Said . . . law provides that plats "may be used in evidence to the same extent and with like effect in the case of deeds." The surveyor was not competent as a witness to impeach his certificate (writings).
>
> —*Almendinger* v. *McHie*, 189 Ill. 308.

As with many rules of law, exceptions occur in a few states as discussed in Section 2-57 ("Uncalled for monuments").

2-24 Rules for Interpretation of Writings

In the interpretation of the meaning of various words and phrases contained in written evidence, the following rules explaining the significance of evidence are generally applicable in most states.

> The language of a writing is to be interpreted according to the meaning it bears in the place of its execution, unless the parties have reference to a different place.
>
> In the construction of a instrument, the office of the judge is simply to ascertain and declare what is in terms or in substance contained therein, not to insert what has been omitted, or to omit what has been inserted; and where there are several provisions or particulars, such a construction is, if possible, to be adopted as will give effect to all.
>
> In the construction of the instrument the intentions of the parties as expressed by the writings, is to be pursued, if possible; and when a general and particular provision is inconsistent, the latter is paramount to the former. So a particular intent will control a general one that is inconsistent with it.
>
> For the proper construction of an instrument the circumstances under which it was made, including the situation of the subject of the instrument, and of the parties to it, may also be shown, so that the judge be placed in the position of those whose language he is to interpret.
>
> The terms of a writing are presumed to have been used in their primary and general acceptation, but evidence is nevertheless admissible that they have a local, technical or otherwise peculiar signification, and were so used and understood in the particular instance, in which case the agreement must be construed accordingly.
>
> When the characters in which an instrument is written are difficult to be deciphered, or the language of the instrument is not understood by the court, the evidence of persons skilled in deciphering the characters, or who understand the language, is admissible to declare the characters or the meaning of the language.
> —Secs. 1857 to 1863 CCP Calif.

Words expressed in a deed or conveyance reflect the intent and meaning of words of the parties at the time of preparation, not necessarily the present meaning. Most frequently, the need for explanations of words arise from local customs such as a local name for a tree. What does the term "spotted tree" mean in Maine? Is it a blazed, hacked, or marked tree? What is meant by "plug" (a 2 by 2 inch wooden stake)?

2-25 Deeds Construed as a Whole, No Parts Rejected if Possible

PRINCIPLE. *Each deed is construed as a whole; parts are read in the light of other parts; nothing stands alone and nothing is rejected if it is possible to give a part meaning.*

Although they may be rejected if it is absolutely impossible to give them meaning, parts are rarely rejected. The following cases represent standard thinking throughout the United States.

> A deed must be construed as a whole, and a meaning given to every part thereof.
> —*Lyford* v. *City of Laconia*, 75 N.H. 220.

> A particular will control a general description of boundaries.
> —*Herrick* v. *Hopkins*, 23 Me. 217.

> Where one part of a description in a deed is false and impossible, but, by rejecting that, a perfect description remains, the false part should be rejected, and the deed held good.
> —*Chandler* v. *Cline*, 99 N.H. 202.

> Any particular of a description may be rejected, if it is manifestly erroneous, and enough remains to identify the land intended to be conveyed.
> —*Lane* v. *Thompson*, 43 N.H. 320.

2-26 Reference Calls for Writings or Plats

PRINCIPLE. *"Writings or plats, referred to in a conveyance for the purposes of identifying land, are to be regarded as a part of the conveyance itself, as much as if incorporated into it."*
—*Ferris* v. *Coover*, 10 C 590.

The evidence of writing (including the writings of plats) always includes in scope the document and all writings referred to by the document. Further, the document or plat referred to may also call for additional writings that may also serve as evidence of the original intent.

In this chapter on evidence these remarks are equally applicable to all writings, including all words and symbols on plats.

> When another deed is referred to for a description of the premises conveyed, the deed referred to is regarded as of the same effect as though it had been copied into the deed itself, and whatever is described in it will pass.
> —*Basso* v. *Veysey*, 110 A 2d 706, 118 Vt. 401.

> Where an actual survey is made, and monuments marked or erected, and a plan afterwards made, intended to delineate such survey, and there is a variation between the plan and the survey, the survey must govern; but such is not the rule where the survey was made subsequent to the plan.
> —*Thomas* v. *Patten*, 13 Me. 329.

> A deceased surveyor's ancient plans, minutes, or field books, clearly describing and identifying the lots and bounds in controversy, are admissible in evidence.
> —*Morse* v. *Emery*, 49 N.H. 239 and
> *Smith* v. *Forrest*, 49 N.H. 230.

In *Morse* v. *Emery* above, wherein a deceased surveyor's written records are admissible evidence, the principle is not applicable in all states, especially where records are commonly filed. In the majority of states, only publicly stored field notes, as, for example, government-sectionalized land surveys, are admissible for the reason that a land owner should not be held to know what is contained in private files. In states where extensive obliteration of surveys has occurred and the location of the descriptions of monuments are unknown, the strict rules for boundary locations are often relaxed toward allowing reputation and inferences to be of importance, especially in older states.

The usual rule is that someone must testify as to the accuracy of field notes of private surveyors, and, even if the notes are admissible, they may not be used to prove the clearly stated words in a conveyance are in error.

2-27 Contrasting Plats and Writings as Evidence

Plats have certain advantages as evidence over written descriptions in that the symbols used on a plat convey definite meanings without usage of words. In Fig. 2-1 the north arrow indicates direction, and the scale given indicates relative size. Without words the picture conveys the idea of direction of all lines (north and south or east and west), length of lines (100 × 200 feet as scaled), and a closed figure (rectangle). Some of the early maps contained no more information than this. Evidence of this type is graphic evidence, is peculiar to maps or plats, and is regulated in value by the law of evidence.

The major objective of filing maps, plats, and field notes is to avoid the necessity of encumbering deeds or descriptions with lengthy wording.

2-28 Field Notes as Written Evidence

PRINCIPLE. *Where the law requires field notes as a part of a survey and the notes are stored in a public place, reference to the survey includes reference to all the field notes of the survey.*

Evidence of the field notes of a survey, if filed in a public place, is easily introduced in court. Private field notes must, in general, be identified by the person taking them, and in the absence of such evidence, the notes are rarely

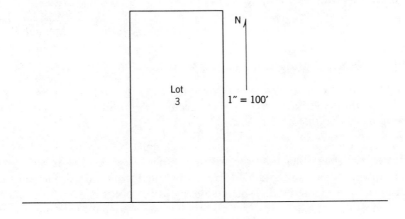

Figure 2-1.

admissible. At times, private notes prepared by surveyors who act in an official or semiofficial capacity may be considered as *prima facie* evidence according to county or state statutes. Even in the event that private field notes are introduced to explain an error or omission, others cannot be held to know the secret information of the surveyor. For this reason private survey notes, kept in private files, have little effect on the outcome of litigation involving the interpretation of written documents.

Field notes kept by public officials, as required by law, have a similar effect as does recording a deed; the public is charged with knowledge of information contained within them. In the survey of the public domain, a call for Section 19, Township 12 South, Range 2 East, San Bernardino Meridian, automatically includes all field notes required by law. Further, the contents of the field notes are usually more certain than the plat itself; the field notes represent what the surveyor did, whereas the plat is a copy of what he did.

Normally, the field notes augment the plat; they explain what was omitted from the plat. In the event of conflict between the two, the courts' interpretation of which controls has varied. Where the parties acted by the plat without considering the field notes, the courts have generally held that the plat controls. In the normal situation, neither refutes the other—in which circumstance both would control.

> The object of the bill is to have the court go behind Irwin's [the surveyor's] mapped plat and investigate his field notes, which field notes, it is claimed, properly extended, would have made the plat, if properly drawn, show the forty acres to be in fact east of the creek. The court below refused to allow this in this proceeding, and so do we. Martin's conveyance was by the recorded plat, not by the field notes. If they were incorrectly mapped, the conveyance was none the less by the plat, and this cannot be affected by any mistake in them. It is questionable if any man buying a recorded map would bother about field notes. Without the plat Haley would have nothing. He is bound to claim under it, but wants to claim under it as incorrect, and to have the error corrected by the field notes. It is not possible that he can claim under the field notes, because the conveyance was not by them, but by the plat.
> —*Haley* v. *Martin*, 85 Miss. 698, 1904.

2-29 Extrinsic Ambiguities of the Writings

PRINCIPLE. *Extrinsic evidence may be sought (1) to explain the meaning of words existing within a written conveyance and (2) to explain the conditions existing as of the date of the deed.*

Extrinsic ambiguities are ambiguities that must be explained from some source of evidence other than the writings. Once a conveyance is reduced to writings, words may not be added to or subtracted from a deed, but extrinsic evidence may be gathered to explain existing words of the deed or to explain a latent ambiguity. If a deed is exact and explicit, no amount of verbal testimony can change it. A deed reading "thence to a stone mound" goes to the stone mound regardless of verbal protests. But is must be proved by evidence (1) that the stone mound is the one referred to and (2) that the stone mound has not been disturbed. Parol evidence may be taken to prove or explain these two points, but it may never by taken to refute the fact that the line goes to the spot occupied by the stone mound as of the date of the deed.

Extrinsic evidence is often necessary to understand a deed. If a deed is written "my house and lot," it can be a perfectly valid description, provided verbal evidence or written evidence can prove that the seller had one and only one house and lot. If it can be shown that the seller had two houses and lots, verbal evidence cannot be taken to differentiate between the two houses. The deed would be invalid because of uncertainty. Words cannot be added to identify between two houses, but an explanation of where one house exists is permissible.

Most deeds have words or conflicts that need explaining, but such explanations are only proper as long as the laws of evidence are not violated. "Beginning at Jim Brown's hen house" certainly needs explanation about where the hen house was located as of the date of the deed. The fact that the deed started at Brown's hen house is indisputable; it is written in the deed. But the location of the hen house may take real detective work. If the hen house is now gone, old nails or a few fence posts may indicate where it used to be. Again, Farmer Brown may be called on to testify about the location of the old hen house. One thing is certain, if the location of the hen house as of the date of the deed can be established in accordance with rules of evidence, it will control the starting point. Frequently, extrinsic evidence must be used to identify the physical objects called for.

> Parol evidence is admissible to locate and identify monuments and calls in a description of land.
> —*Proprietors of Claremont* v. *Carlton*, 2 N.H. 369.

> When grants and conveyances of lands are made, the usual mode of describing the premises conveyed is by reference to natural or artificial monuments as boundaries, and by this means the premises may be found and distinguished from other tracts or parcels of land. The same object is also attained by describing the premises conveyed by a specific name, but in either case the location is not always determined alone by the description in the conveyance, independent of extrinsic evidence. The deed describes the objects bounding the premises, but parol evidence is usually resorted to, for the purpose of identifying the objects themselves. It is well settled that all monuments, and things referred to in a deed for the purpose of locating the land, may be established and identified by extrinsic evidence. In most instances, however perfect the description employed by the conveyance, the premise could not be located and identified without reference to extrinsic evidence, wither more or less proximate, and for the purpose of sustaining a grant, extrinsic evidence may always be used to identify, explain or establish the objects of the call in a deed.
> —*Williams* v. *Warren*, 21 Ill. 549.

> The trial court admitted extrinsic evidence showing that by "due north" the parties to the boundary agreements meant that the call in the deeds should be

surveyed only by a basis of bearings method. Richfield contends that this evidence requires reversal of the judgment, invoking another aspect of the parol evidence rule: that evidence of the circumstances surrounding execution of a written instrument is admissible to determine its meaning. (Code Civ. Proc., § 1860.) Carpenter and Henderson and defendants, however, were not parties to the boundary agreements, and it is not shown that they had actual notice that Richfield and Norris Oil Company attached a particular meaning to the words "due north." The conduct of Carpenter and Henderson indicates that they believed that the "due north" description in the March 15th quitclaim deed was interchangeable with the government section description in the 1947 sublease. Although evidence of the negotiations preceding execution of the boundary agreements may be admissible as between the parties thereto . . . , a special interpretation of the quitclaim deed could not be enforced against a party relying upon the words of the instrument and without knowledge of the meaning attached thereto by the other party to that deed . . . Thus, in the present case the expert testimony of surveyors and engineers showing the proper method of surveying the calls in the deed was admissible as against both parties to the deed, since they must be deemed to know that "technical words are to be interpreted as usually understood by parties in the profession or business to which they relate" (Civ. Code, § 1645), but evidence of the negotiations of the parties to the boundary agreements could not be admitted to the detriment of defendants, for it would be manifestly unjust to charge them with the secret interpretation of other parties after defendants had relied upon the ordinary meaning of the words of the instrument.

—*Richfield Oil Co. v. Crawford*, 39C, 2d 729, Oct. 1952.

2-30 Practical Location

PRINCIPLE. *Evidence of practical location, when appropriate, may be received to clarify ambiguity of writings.*

On occasion deeds may not be definite in defining the limits of a particular thing. For example, a deed reading "all of lot 1 and a 20-foot wide road easement across Lot 2" lacks an exact location for the easement mentioned. But if the owner of Lot 1 uses a particular place for a period of time, that particular place becomes the correct easement line by practical location. The reasoning is simple; since that location was used, there must have been an agreement, either verbal or implied, between the parties. Seeming ambiguities may not be ambiguities at all; a simple inspection of the land for evidence of usage sometimes provides the answer.

Evidence of practical location may only be received to clarify an ambiguity; it may never be used to overcome clear, concise, unambiguous words in a deed; it may not be used to create an ambiguity. If the easement above were defined as to location in the deed, evidence of usage would be incompetent to overcome the written words unless such usage ripened into a possession right.

2-31 Patent Ambiguities

PRINCIPLE. *Ambiguous deeds or descriptions that cannot be cured by extrinsic evidence may be considered as void or voidable by the courts.*

"Ten acres in Section 21" is not locatable and is probably void. "My house and lot" is patently ambiguous if the party owns two houses and lots. A latent ambiguity—that is, an ambiguity that is not apparent until it is being located on the ground, can be explained by extrinsic evidence. A patent ambiguity—that is, an ambiguity that appears on the face of the document—cannot be explained by extrinsic evidence.

2-32 Ancient Survey Plats and Documents

PRINCIPLE. *An ancient survey, recorded or accepted as a public document, made by a competent authority, and produced from proper custody, is generally admissible as evidence to prove the location of a boundary line.*

A private survey may be admissible on proof of its correctness by the party making it, whereas an ancient survey, by proper authority, and recorded as a public record, may be accepted without further verification. Private surveys, in general, are not accorded the same standing as publicly recorded surveys.

> An ancient map of the public roads of a county, purporting to have been made by authority, and coming from the proper custody, is competent evidence to show the existence and location of the public roads of the county at the time it was made; and in a contest between coterminous land owners, where a road delineated on the map is claimed to be a boundary, such map is relevant to the question of the boundary. The theory of which such ancient maps are received is that where the matter in controversy is ancient and not susceptible of better evidence, traditionary reputation of matters of public general interest is competent evidence of the matters in which it relates. The admissibility in evidence of an ancient map of matters of a public general interest is not to be confounded with a map which a land owner causes to be made of his premise. It was doubted that the rule admitting a map 30 years old as an ancient document applied to private maps. An unofficial survey is admissible in evidence when it has been proved to be correct.
> —*Bunder* v. *Grimm*, 142 Ga. 448.

Each state usually has a statute declaring how old a document must be before it can be classified as ancient, and it is often 30 years. Such items as deeds, wills, and other documents used in conveyances are most often applicable to surveying evidence. The following court cases indicate what is found in the law library.

> Ancient deeds, more than thirty years old, are admissible to prove ancient possession.
> —*Mentz* v. *Town of Greenwich*, 118 Conn. 137.

> By reason of statute, it would seem that in New York a period of 20 years satisfies the age requirements to bring a map or survey within the "ancient document" rule.
> —*New York* v. *Wilson & Co.*, 278 N.Y. 86.

> A will, executed more than 30 years before it was offered in evidence, when produced from the proper custody, and otherwise free from suspicion, is entitled to the benefit of the same presumptions as an ancient deed.
> —*Appeal of Jarboe*, 91 Conn. 265.

EVIDENCE TO PROVE INTENT OF CONVEYANCES

2-33 Interpreting Conveyances

The meaning and intent of words are normally the subject of judicial notice; that is, the meaning is sought from evidence found in books or authoritative sources but not from the opinions of the parties to the deed. It is only when words have a particular meaning for a particular locality or a particular profession that the courts resort to evidence to determine the meaning of a term. Normally in the matter of surveys, the surveyor is the expert who interprets the meaning of unusual words as applied to his profession. Under certain circumstances, it may be necessary to seek verbal explanations of a word from others. A deed reading "commencing at an oak tree" certainly needs verbal explanation about where the oak tree is located, but it needs no explanation of the fact that it is an oak tree.

Courts rely on *Black's Law Dictionary* and *Webster's Dictionary* for definitions of words. West Publishing Company publishes a set of books (100 volumes, more or less) entitled *Words and Phrases*.

> The legal interpretation of the words, "so-called," in a deed is, not what the parties, but what the public generally, says about the premises.
> —*Madden v. Tucker*, 46 Me. 367.

2-34 Evidence That Determines the Intent of a Conveyance

PRINCIPLE. *The intentions of the parties to a conveyance, as expressed by the evidence of the writings, are the paramount considerations of the court in interpreting the meaning of a deed, and that intent is gathered exclusively from the written words of the deed, except where the written words have extrinsic ambiguities or where explanations of conditions existing as of the date of the deed are necessary.*

Surveyors do not ask a person what he intended; the surveyor reads what was signed, takes into account explanations of extrinsic ambiguities and the legal and survey meaning of words and phrases, then decides for himself what the written intent was. The written deed is conclusive evidence; verbal testimony is of no avail except to explain an extrinsic (latent) ambiguity or to explain a condition existing as of the date of a deed.

Circumstances existing as of the date of the deed may be inquired into to explain the situation or meaning of words as of that date. A deed written in the Gold Rush days that failed to specify the basis of bearings was interpreted to be on a magnetic basis. Witness evidence and other documentary evidence revealed that this was the custom at the time. Allowing witness evidence to reverse the usual interpretation placed on bearings is not altering the meaning and intent of the deed's words; it is merely explaining the meaning and intent of the words as of the deed's date.

2-35 Evidence of Intent of Senior, Equal, or Junior Rights

PRINCIPLE. *Depending on the evidence discovered, a conveyance is classified with respect to the adjoiner as being senior in rights, as being equal or simultaneous in rights, or as being junior in rights. An intent, gathered from the written evidence, cannot alter senior rights.*

The surveyor, if he is to correctly survey a given parcel, must usually classify the seniority of that parcel with respect to the adjoiner, and such classification is sometimes difficult or impossible to derive from discovered evidence (see Section 8-12 for the surveyor's duties with respect to research).

If a person has conveyed part of his property to another, he cannot at a later date convey it to someone else, irrespective of his written intent in a new conveyance. The first deed (senior) has a right to all the land that is called for, and the seller (junior) owns the remainder. If a person owns a remainder, no excess or deficiency exists. A remainder does have a definite size, but it is "more or less" in character until measured. A person may have more or less land than he expects, but, as long as he has a remainder, the unexpected quantity of land, be it large or small, is all his. It is not divided among several owners.

If parcels are *created in sequence* with a lapse of *time* between them, senior rights exist. Metes and bounds descriptions are usually created with a lapse of time between each creation and hence are created in sequence; whereas, all lots on a subdivision map are normally created at the same moment of time (when the map is filed or accepted), even though the lots are sold in sequence. Lots created simultaneously will have equal rights.

Examples of simultaneous descriptions appear in the following: (1) wills and gifts, wherein none of the heirs or benefactors are designated to receive a remainder; (2) lots in subdivision, wherein a map is filed with a governing body and no lot is sold prior to filing the map; (3) lots in any legal subdivision, wherein it is impossible to distinguish an intent to give senior rights to buyers in sequence; (4) court proceedings in partition, wherein each litigant is given a proportionate share of the whole, and no one is designated to receive the remainder; (5) metes and bounds descriptions that are created simultaneously, and no one is designated to receive a remainder.

VALUE OF EVIDENCE OF A SURVEY

2-36 Evidence of a Survey

PRINCIPLE. *For a survey to be a consideration of a conveyance, it must be called for by the conveyance, or it must be required by law as a part of the conveyance proceedings, or, in a few states, if a survey is made soon after the conveyance proceedings by the parties of the deed and approved by them, the survey becomes controlling.*

Although most conveyances calling for a survey are of the map or plat type and most conveyances not calling for a survey are of the metes and bounds type, either can recite a call for a survey. In Texas the state required a survey as a condition for obtaining a patent to state-owned land. East Texas lands, in general, although not always, were surveyed as the need arose—that is, land was not laid out in lots and blocks prior to needs. As a result, many of the Texas patents calling for a survey are sequence conveyances having junior and senior considerations. In the sectionalized land states, where surveys are called for, most patents are of the lot and block type (called *sections, townships,* and *ranges*). A call for a survey is not then limited to one type of conveyance; it may be inserted in either a sequence or simultaneous conveyance.

If the evidence of a survey is to have force, it must be called for in the writings or it must be presumed by law. A survey not called for by the written conveyance cannot be considered a part of the writings; words cannot be added to a conveyance. But where the law requires a survey to be made as a part of the conveyance, it is presumed that the law was obeyed. The United States required, by law, surveys of sectionalized lands; Texas, since 1879, required surveys of patents; hence, whether or not a survey was called for in the writings for Texas or U.S. patents, it is presumed that one was made. Contrary proof—that is, proof that no survey was made—will overcome this presumption, but the proof must be real, not merely surmised.

> Our statute required that anyone laying out an addition should cause to be made out an accurate map or plat thereof, particularly setting forth and describing all lots for sale, by numbers, and their precise length and width. There are two strong reasons for finding that the lots and streets and their location were accurately surveyed and marked on the ground as a basis for the plat. The first reason is that the law presumes that such survey and markings were made. The plat bears on its face facts which, taken in connection with conceded extrinsic facts, show that the plat was based on a survey, not only of the boundaries of the addition but of the interior lot lines. The plat shows that the course of the railroad is not straight, but it curves to the eastward as it goes southward. The lots bordering on the right-of-way of the railroad are irregular in shape and the lengths of their boundaries are marked accordingly. It would have been difficult if not impossible to correctly indicate such distances without such survey. We have a right to take notice of the historical fact that on the invention of wire fences, hedges were no longer planted in this state, and that the hedge fence at or near the southwest corner of the addition was there when the addition was platted. We have a right to draw the inference that such fence was adopted by the maker of the plat as the west boundary; in other words as the center of the section. It was right on the line called for by the course and distance.
> —*Dolphin* v. *Klann*, 246 Mo. 477.

> A deed which does not refer to a partiular survey, and which is unambiguous in its terms, cannot be enlarged, or in any manner modified by the introduction of parol evidence tending to prove that one of parties to it understood in descriptive words to have reference to a particular private survey.
> —*Rowland* v. *McCown*, 20 Ore. 538.

> A resurvey not shown to have been based upon the original survey is inclusive in determining boundaries, and will ordinarily yield to a resurvey based upon known monuments and boundaries of the original survey.
> —*Pallas* v. *Daily*, 100 N.W. 2d 197, Neb. 1960.

Exceptions to the foregoing general rule are to be found in some of the original states, especially where widespread loss of evidence makes it necessary to rely on reputation and hearsay. Evidence of a survey that cannot be proved to have been made before or after a conveyance may be accepted merely because it has the reputation of being correct. Here are two cases involving surveys made after a conveyance. In the first the conveyance calls for certain monuments that did not exist at the time of the conveyance but were set some months later by a surveyor. The second case pertains to a survey made after the conveyance and approved by the acts and behavior of the adjoiners. The second case should be discussed under the subject of unwritten rights, but it is presented here to emphasize differences.

> Where land has been conveyed by deed, and the description of the land in the deed has reference to monuments, not actually in existence at the time, but to be erected by the parties at a subsequent period: when the parties have once been upon the land and deliberately erected the monuments, they will be as much bound by them, as if they had been erected before the deed was made. In this case, there was a reference in the deed to monuments not actually existing at the time, but the parties soon after went upon the land with a surveyor, run it out, erected monuments, and built their fences accordingly; and this is not all. They respectively occupied the land according to the line thus established, for nearly ten years. And there is now no evidence in this case of any mistake or misapprehension in establishing the line.
> —*Lerned* v. *Morrill*, 2 N.H. 197.

In the foregoing case it should be noted that the monuments *were called for* by the deed but not set until some 18 months later. In New Hampshire when such monuments are set by the parties, they become a controlling consideration.

In the following Maine case the property line was run immediately *after* the conveyance, and the parties then built a fence. Several years later it was discovered that the fence did not agree with the writings; the surveyor was wrong. The court was of the opinion that the acts of the parties following the deed indicated their intentions and were controlling. Whether this principle was applicable to surveys made by strangers to the original transaction or was applicable to surveys made many years after the original conveyance was not in issue. Because of the importances of this case in the northeast, it is quoted in its entirety. It represents a variation of "practical location" discussed elsewhere.

> DICKERSON, J. WRIT OF ENTRY. Both parties claim title through the same grantor, Henry Smith, who, in the first instance, conveyed "parts of lots numbered 9 and 10, on the east side of Sandy river," to the defendant. After reciting the other boundaries, the description in the deed continues as follows, "thence easterly by a

line parallel with the north line of lot No. 9 to the county road," the grantee taking the land north of the line now in dispute, and the grantor retaining the land south of it. The line was run and marked by a surveyor immediately after the conveyance, and the parties then built a fence on it, intending it for a division fence, Smith occupying to the fence on the south, and the defendant on the north side of the fence, for some six years, when Smith conveyed his remaining parcel to the plaintiff's grantor, describing the line in controversy as follows, "to land supposed to be owned by George Toothaker, thence easterly on said Toothaker's south line to the county road." About eight months afterwards, the grantee conveyed the last named premises to the plaintiff, describing it as "the same she purchased of Henry Smith." The plaintiff claims to hold to the line described as running "easterly by a line parallel with the north line of said lot No. 9 to the county road," in Smith's deed to the defendant, which is several rods northerly of the fence, and the defendant claims to hold to the divisional line made by the fence; and the question is, which is the true line between the parties?

The presiding judge ruled that the words, "on said Toothaker's south line," would limit the plaintiff's land to the line established by Toothaker and Smith, on which the division fence was built, and that she could not hold beyond this line, even if she could satisfy the jury that it did not conform to the original lot line; thereupon the parties agreed to submit the question to the law court, judgment to be rendered for the defendant if the ruling is correct; if not, the action is to stand for trial.

But for the acts of the parties in interest, in running, marking, and locating the line, building a fence upon it immediately after the conveyance, and occupying up to it down to the commencement of this suit, the line on the course described in the deed, if it could be ascertained, would be the line between the two parcels. Did these acts fix and establish the divisional line as the true line?

It was early held that where a deed refers to a monument, not actually existing at the time, but which is subsequently placed there by the parties for the purpose of conforming to the deed, the monument so placed will govern the extent of the land, though it does not entirely coincide with the line described in the deed. *Makepeace v. Bancroft*, 12 Mass. 469 (1815); *Kennebec Purchase v. Tiffany*, 1 Greenl. 211 (1821); *Lerned v. Morrill*, 2 N.H. 197 (1820).

Again it was held in *Moody v. Nichols*, 16 Maine, 23 (1839), that when parties agree upon a boundary line, and hold possession in accordance with it, so as to give title by disseisin, such boundary will not be distrubed, although found to have been erroneously established. In that case the call in the deed was "a line extended west, so as to include" a certain number of acres, the boundaries upon the other three sides having been accurately described. The parties to the deed agreed upon and marked that line, erected a fence upon it, and held possession according to it for thirty years.

The same doctrine was held by the supreme court of the United States, in giving construction to a line described in the deed as "running a due east course" from a given point. *Missouri v. Iowa*, 6 How. 660.

So the court in Massachusetts, in giving effect to a deed, describing a line as "running a due west course" from a given point, held that the line located, laid out, assented to, and adopted by the parties, was the true line, though it varied several degrees from "a due west course." *Kellogg v. Smith*, 7 Cush. 382 (1851).

In *Emery v. Fowler*, 38 Maine, 102 (1854), the call in the deed was a line from a given point, "on such a course . . . as shall contain exactly one and a half acres."

The lots to be conveyed were located upon the face of the earth by fixed monuments, erected by referees mutually agreed upon; and the parties to the several conveyances assented to and adopted the location before the deeds were given. Deeds intended to conform to the location thus made were then executed by the parties. The respective grantees entered under the deeds, built fences, and occupied in conformity with the location for fifteen years, when, it being found that more land was contained within the limits of the actual location upon the face of the earth than was embraced within the calls of the deed, a dispute arose. The court held that the monuments thus erected before the deed was given, must control, thus extending the rule adopted in *Moody* v. *Nichols* to cases where the possession had not been long enough to give title by disseisin. That decision also makes the rule of construction the same, whether the location is first marked and established, and the deed is subsequently executed, intended to conform to such location, or whether monuments, not existing at the time, but referred to in the deed, are subsequently erected by the parties with like intention.

In construing a deed, the first inquiry is, what was the intention of the parties? This is to be ascertained primarily from the language of the deed. If this description is so clear, unambiguous, and certain that it may be readily traced upon the face of the earth from the monuments mentioned, it must govern; but when, from the courses, distances, or quantity of land given in a deed, it is uncertain precisely where a particular line is located upon the face of the earth, the contemporaneous acts of the parties in anticipation of a deed to be made in conformity therewith, or in delineating and establishing a line given in a deed, are admissible to show what land was intended to be embraced in the deed. It is the tendency of recent decisions to give increased weight to such acts, both on the ground that they are the direct index of the intention of the parties in such cases, and, on the score of public policy, to quiet titles. The ordinary variation of the compass, local attraction, imperfection of the instruments used in surveying, or unskillfulness in their use, inequalities of surface, and various other causes, oftentimes render it impracticable to trace the course in a deed with entire accuracy. If to these considerations we add, what is too often apparent, the ignorance or carelessness of the scrivener in expressing the meaning of the parties, we shall find that the acts of the parties in running, marking, and locating a line, building a fence upon it, and occupying up to it, are more likely to disclose their intention as to where the line was intended to be, when the deed was given, than the course put down on paper, if there is a conflict between the two.

Hence, the rule of law now is, that when, in a deed or grant, a line is described as running from a given point, and this line is afterwards run out and located, and marked upon the face of the earth *by the parties in interest* [emphasis added], and is afterwards recognized and acted on as the true line, the line thus actually marked out and acted on is conclusive, and must be adhered to, though it may be subsequently ascertained that it varies from the course given in the deed or grant.

The acts of the defendant and Smith, through whom the plaintiff claims, in surveying and marking the line in dispute upon the face of the earth by stakes and stones and spotted trees, building a fence thereon, intending it to be the line between them, and occupying up to it, make and establish such line as the divisional line between the two lots.

The ruling of the presiding judge was in accordance with this construction of the deeds, and there must be judgement for the defendant.

—*Knowles* v. *Toothaker*, 58 Me. 174.

A careful reading of the case just cited indicates that where the parties to a deed in Maine, soon after its consumation, make a practical location of the deed on the ground, by survey, the parties are bound by it whether it agrees with the writings or not. The theory being, the acts of the parties disclose their intentions. In a later case, unrelated to the *Knowles* case just cited, two adjoiners, not a part of their original deeds as of the time and the land was originally divided (*Bemis* v. *Bradley*, 126 Me. 462), built a fence along a line that they believed to the the true dividing line. The court decreed that since the parties were trying to mark the true deed line, they were not bound by the fence. In other words, strangers to the original transaction of dividing the land, when mutually marking a line that they believe to be the true line, are not bound by the line unless it is in fact the true line.

Proof of where the surveyor marked his lines is usually inferred from the evidence of monuments; hence, proof will be discussed under that heading.

2-37 Ancient Private Surveys

In most states the records of private surveys of a former generation are not admissible in evidence for the reason that landowners cannot be held to unavailable records. In a few states where evidence of surveys is not recorded and there is a widespread loss of evidence, courts have accepted ancient private survey information. Unless a court case exists within a state approving old private survey records, the surveyor should assume that such records cannot be used in court. One such court case occurred in New Hampshire as follows.

> A deceased surveyor's ancient plans, minutes, or field books, clearly describing and identifying the lots and bounds in controversy, are admissible in evidence.
> —*Morse* v. *Emery*, 49 N.H. 239 and
> *Smith* v. *Forrest*, 49 N.H. 230.

2-38 Intent of a Survey

PRINCIPLE. *The evidence of intent of a survey as called for in a conveyance is to be interpreted from the map of the survey, the field notes of the survey, and the acts of the surveyor, but not from the unwritten, unexpressed intent of the surveyor.*

A survey called for cannot be interpreted in the light of the secret or hidden intentions of the surveyor. What the surveyor did and what he recorded in the evidence of writings count. The surveyor's testimony that he intended to include all the lands up to the adjoiner cannot enlarge a survey to include omitted lands.

> The question seems to have been whether or not the calls for course and distance or those for lines of older surveys should prevail. Upon this question, we are of the opinion that the testimony of the surveyor stating his intention in making the survey was not admissible. In determining the location of the land in such cases, the courts seek to ascertain the true intention of the parties concerned in the survey; but the intention referred to is not that *which exists only in the mind of the surveyor.* When

reference is made in the decisions to the intentions of the surveyor, the purpose deduced from what he did in making the survey and description of the land is meant, and not one which has not found expression in his acts. Hence, if the intention of the surveyor appears from his field notes and his acts done in making the survey, his evidence to prove his intention is superfluous, while if it does not so appear, it cannot control or affect the grant.

—*Blackwell v. Coleman County*, 94 Tex. 216.

The surveyor can, of course, testify about things he actually did but not about his intent. A statement by the surveyor that a certain found monument was the one that he set would have force. But a statement to refute writings would more than likely be rejected, since parol evidence is inferior to written evidence.

VALUE OF MONUMENTS AS EVIDENCE

2-39 Evidence of Monuments

PRINCIPLE. *Generally for a monument to be a controlling consideration as evidence, the monument must be called for in the written evidence proving conveyancing or it must have been required by law. A call for a specific survey also calls for monuments set by that survey. The call for a monument is a call for the spot occupied by the monument as of the date of the written conveyance.*

A monument to control the intent of a deed must be called for either directly, indirectly by reference, or required by law. A deed may call for an oak tree in the writings, or the deed may call for a map which in turn calls for an oak tree, or the deed may call for a survey by Jones, and Jones' field notes may call for an oak tree. If the law requires a survey and set monuments, extrinsic evidence may be taken to explain what monuments were set as required by law. One very important fact that is sometimes overlooked is that a call for a monument is in actuality a call for the particular spot occupied by the monument as of the date of the deed. The monument itself is merely a symbol or object to mark the spot. A found monument that is uncalled for or is not referred to has no weight in substantiating that survey unless it can be shown by other evidence that it is occupying the spot of the original monument.

If there is a call for a monument, that monument, if discovered undistrubed and uncontradicted by the remainder of the writings, is conclusive. A deed that calls for bearing and distance but does not call for a monument either directly, indirectly, or by reference, and is not required by law cannot be altered by giving control to a monument found in the vicinity of the bearing and distance termination.

> Witness Dewey testified that he saw a rock pile in 1872, and for some years subsequent thereto, at the north end of the line as run by Holman. It was the evident purpose to have the jury believe that this rock pile was made at said place for the northeast corner of the Lampasas County School land. If it was made there for that

purpose, it was wholly irrelevant and immaterial, unless it was placed there by the surveyor who made the original survey, or by someone who knew that it was at the corner of said survey. There is no evidence as to how the rock pile got there. It was not called for by the original field notes.

—*Runkle* v. *Smith*, 133 SW (Tex.) 745.

In surveying terminology the phrase "original monument" is applied to the monument or monuments called for, either directly or indirectly, by the deed. Other monuments, so far as a particular deed is concerned, are not original monuments. With the possible exception of monuments called for in a senior deed, original monuments control a conveyance location.

As noted, the spot occupied by the original monument, as of the date of the deed or as of the date of a survey called for by the deed, waters excepted, is the controlling consideration. All monument evidence sought is to explain where that particular spot exists on the ground. Discovery of the original monument itself is not a necessity, since many types of evidence can be resorted to that will suffice as proof of the original location. A disturbed monument may be of no value; the original spot occupied by the monument may not be identifiable. An obliterated monument—that is, one lost from view—may be restored to its former position by competent witness evidence. Evidence is to prove where it was as of the date of the deed, not where the measurements say it ought to have been set.

> In resurveying a tract of land according to a former plat or survey, the surveyor's only function or right is to relocate, upon the best available evidence, the corners and lines at the same place originally located. Any departure from such purpose and effort is unprofessional, and, so far as any effect is claimed for it, unlawful.
>
> —*Pereles* v. *Gross*, 126 Wis. 122.

> In all cases of disputed lines the following rules shall be respected and followed; natural landmarks, being less liable to change and not capable of counterfeit, shall be the most conclusive evidence; ancient or genuine landmarks, such as corner station or marked trees, shall control the course and distances called for by the survey. If the corners are established, and the lines not marked, a straight line, as required by the plat, shall be run, but an established marked line, though crooked, shall not be over-ruled; course and distance shall be resorted to in the absence of higher evidence.
>
> —Georgia Statute 85–1601.

As pointed out under Section 2-36, if parties to a deed, soon after its formation, erect monuments to indicate their intent by a practical location, they are bound by it in a few states. This was the finding in the following case wherein monuments were called for but not set until after the deed's date.

> Where a monument does not exist at the time a deed is made, and the parties afterwards erect such a monument, with intent to conform to the deed, such monument will control.
>
> —*Lerned* v. *Morrell*, 2 N.H. 197.

In most states the mutual designation of a property line by parties to the original conveyance is considered as an unwritten conveyance and is called a *practical location*. Further discussion of this subject appears in Section 3-13.

2-40 Control of Each Original Monument

PRINCIPLE. *No one corner or monument recited in a description has any greater dignity than any other corner or monument recited.*

Each corner monument called for has just as much control as any other monument called for; all are to be given control, if possible—that is, of course, if the monument or corner is properly identified, is undisturbed, and corresponds to the one called for.

2-41 Monuments as Indispensable Evidence

PRINCIPLE. *Every survey of a conveyance must start from evidence that proves the postion of at least two monuments somehow related to the written record.*

When locating a conveyance, the locations of at least two monument positions are indispensable evidence. All measurements commence from a monument and go in a direction determined by another monument (star, magnetic pole, north pole, or physical object).

Normally, in the interpretations of the intent of a deed, all monuments called for by the writings are given preference over conflicting calls of distance, directions, or area.

In locating written deed lines, not all discovered monuments are of value as evidence; some are accepted; others are rejected in accordance with the law of evidence. In general, for a monument to be controlling as evidence, it must be called for by the writings, either directly or by reference. A call for an oak tree is a direct call for a monument. A call for a survey by a particular surveyor is a call for any monuments set by him. Many of the old maps have statements on them that read "surveyed by John Doe." Although on the face of the map there is no mention of monuments set, it is always necessary and proper to seek an explanation of what was meant by "surveyed by John Doe." If monuments were set by this named person, they can be accepted. This is a question of proof by evidence.

In some areas of the United States early surveys were made, but it is unknown whether they were made as a consideration of a conveyance or not, because of widespread loss of original evidence. Where there is long-standing acquiescence in a boundary that could have been marked by an original surveyor and there is no evidence to the contrary, there is no reason to disturb the status quo. Most important, though, is that the surveyor must be absolutely certain better evidence proving another location does not exist.

2-42 Sufficiency, Amount, and Kind of Evidence to Prove Monuments

The amount, kind, and quality of evidence necessary to prove the correctness of a monument is variable, depending on the circumstances. The law of evidence is not an exact law but is a relative thing stating general principles that may have flexibility, depending on the circumstances. In questions of civil litigation, under which land disputes fall, the courts accept the premise of the "preponderance of evidence," which is not the same as "beyond a reasonable doubt." Proving a monument, or rather proving the position as occupied by a monument as of the date of the deed, is done by the following evidence:

1. The physical characteristics of the monument itself.
2. Probability or possibility of being disturbed or moved.
3. Public records and sometimes private records, such as maps, notes, and documents, that prove historical sequence.
4. Witness evidence.
5. Hearsay evidence.
6. Common report.
7. Measurements to prove proximity to record measurements.
8. Witness monuments, old fences, and old lines of possession.

The monuments set by the original deputy United States surveyors for the west section corners must control as to the proper location of those corners. The question where they were located, if destroyed, is one of fact, and not of law, for the jury to determine under all the evidence.

—*Goltermann* v. *Schiermeyer*, 111 Mo. 404.

2-43 Physical Characteristics of Monuments

Some monuments are conclusively identifiable by the evidence of their physical characteristics alone. No trouble should be encountered when identifying the Great Lakes or a particular river. To be sure, the exact spot that locates the average water mark on the shore or the exact location of the thread of the stream may be difficult to determine, but there should be no trouble identifying the monument called for. An oak tree with a particular type of blaze mark should not present identity troubles, but an oak tree without a blaze mark and located in an oak grove could give substantial troubles. In the first case, if a blazed oak tree is found, witness evidence would be incompetent to overcome the location of the blazed tree. But if two blazed trees are found, witnesses can identify between the two trees. In the case of no identifying marks, called-for measurements from other known monuments are the best evidence to distinguish between the various trees. However, if measurements are not certain and conclusive, witness evidence may be resorted to. Where the monument is certain of identification from the physical characteristics of the monument itself, witness evidence is incompetent to prove any other monument. But if the wording of the writings is ambiguous so that alternate monuments are possible, less conclusive witness evidence may be used to distinguish between the monuments.

Evidence concerning the physical characteristics of a monument may or may not be described in the writings. If in a subdivision the original surveyor stated on his map that he set a stake at a particular corner in the year 1900, the stake must be identified as to type and markings. The fact that the surveyor set a stake is not disputable; the only question is what kind of stake? Each surveyor has peculiarities and habits, and many old-time surveyors had certain specific and characteristic habits. In the San Diego vicinity surveyors usually used redwood hubs or so-called plugs. In one area 4 × 4 redwood posts were used; another used 3 × 3 redwood posts. See Figs. 2-2, 2-3, 2-4, and 2-5. In sectionalized land surveys rock mounds were common. Knowledge of what former surveyors did, what their habits were, and the customs of surveyors at the particular date is all evidence. In the case of a stake set in 1900 in San Diego, if anything other than a redwood stake and of the type commonly observed were found, the surveyor would be put on notice to look further. An old Model T Ford axle found for a corner certainly could not be an original corner of an 1880

Figure 2–2 A scribed 3½ × 3½-inch redwood post set for the corner of quarter Section 42 of Rancho de la Nacion in 1868 and replaced in 1958. Paint protected the area below ground, and weathering deteriorated the portion above ground.

Figure 2–3 A scarfed redwood post 2½ × 2½ × 18 inch set for the sideline of Salinas Avenue, Del Mar Heights, Del Mar, California, in 1887. The post was stenciled with black paint, and weathering has embossed the letters so that they stand out. The paint is now gone.

Figure 2-4 A scarfed redwood post set for lot 22. Weathering has removed much of the upper portion as compared to that below ground.

Figure 2-5 A witness stake to a concrete monument. White paint on the numbers protected the wood; whereas the unpainted wood was eroded away by the elements.

subdivision; Fords were not manufactured then. These minor, seemingly unimportant, facts of evidence are often the difference between a successful resurvey and an erroneous survey.

One thing that may make the older experienced surveyor more valuable than the new surveyor is the older surveyor's knowledge of past history. The older surveyor knows what is correct; after all, he was there when it happened.

2-44 Evidence of Being Disturbed

PRINCIPLE. *An original monument to be of value must be located in the same spot occupied as of the date of the deed, waters subject to riparian rights excepted.*

Evidence, either real or parol, may be sought to prove this point. Often the condition of the monument itself will prove whether it has been moved. A monument found in a cut bank is doubtful without further proof. A monument not in the measured proximity of where it ought to be casts doubt and suggests seeking further evidence.

Water is an exception to the rule that the spot occupied by the original monument as of the date of the deed in the controlling consideration. Whenever a deed calls for naturally occurring waters (rivers, lakes, ocean, etc., but not

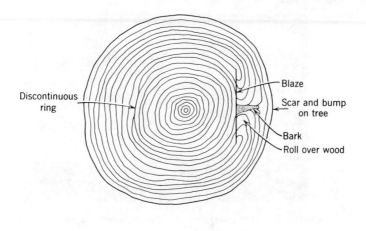

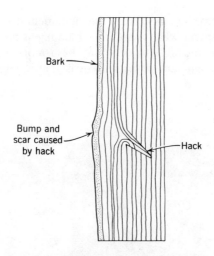

Figure 2-6 Annular rings.

artificial sources of water such as dams or canals), the location of the governing line of water at any particular moment is controlling. Since the subject of waters is quite complex, better treatment of the subject calls for a separate chapter including evidence and location procedures.

2-45 Evidence of Witness Objects

PRINCIPLE. *Witness objects called for by writings, if found undisturbed, are evidence proving original corner locations and have the same weight and dignity as the object itself.*

When alienating land, the sovereign often required a survey, monumentation, and measurements to nearby identifiable witness objects. When found, witness

48 / EVIDENCE

Figure 2–7 Tree overgrowing wire that was wrapped around it.

objects are considered equivalent in value to the monument itself and constitute proof of where the monument was. If a corner monument is easily movable, witness trees and other immovable objects may be a more certain means of identifying the original monument position than is the monument itself. See Figs. 2-8 and 2-9.

Figure 2–8 Bearing tree with scribing exposed.

2-46 Evidence Used to Identify Trees

In forested areas, marked trees were the most commonly used witness objects. Identification of witness trees is a specialized science, and those expert in that field sometimes find their services in demand for land litigation.

Surveyors use the hack, notch, cross, and blaze to single out a particular tree as a monument or line identifier.

A *hack* is a single horizontal axe cut deep enough to penetrate beyond the bark. Three hack marks on each side of a tree was one of the usual symbols to indicate a line tree. When hack marks heal, scars are noticeable on the bark years later. In healing over the hack, wood grows as shown in Fig. 2-6, and the age of the hack can be determined by ring count.

Notches are made by cutting in two directions so as to leave a horizontal notch extending into solid wood. Notches are larger than hacks and require longer periods of healing. Because of this the danger of wood rot increases, especially so if the notch extends through the sapwood into the heartwood. Notches are not recommended for line trees, since hacks will serve the purpose equally well.

Blazes are larger blocked-out areas extending below the bark and growth layer. The danger of wood rot, due to exposure to bacteria and virus, is greatly increased, and the possibility of tree death is potential. Blazes should never extend below the sapwood into the heartwood. Blazes have been and are being used for marking (1) tree lines, (2) witness trees, (3) witness trees to a line, and (4) the monument itself. Marked line trees have a blaze on each side and usually have hack marks above and below each blaze. Trees are and have been blazed for many reasons, but the hack marks distinguished the surveyor's trademark. Trees directly on line were said to be fore-and-aft blazed. Side-line trees, that is, trees within about 3 feet of the line, were customarily marked with three hacks facing the line but sometimes by a blaze facing the line. Witness trees are blazed on the side facing the corner. If the corner itself happened to be a tree, the tree was blazed and often scribed to assist certainty of identification.

In Texas and some other areas, crosses rather than blazes were made on witness trees. This practice has merit since the danger of wood rot is lessened and a positive identifiable mark exists.

In identifying trees the following considerations are taken into account:

1. Species of tree and the tree's characters (life expectancy, regrowth habits, susceptibility to disease, etc.).
2. Particular original markings called for.
3. Stumps.
4. Ring count.

Trees and Their Characters. Identification of a particular witness tree may only require identification by species, especially so where only one tree of that species is found. But usually when one species is found, growth conditions are favorable and many other trees of the same species will exist. Particular markings, size, or age are then used to identify different trees. Loblolly pine rarely exceeds 100 years of age; hence any survey over 100 years old and calling for a young loblolly pine as a witness tree cannot very well be identified by the witness tree. A 50-year-old survey calling for a large loblolly pine may not be identifiable by the witness tree, since the tree may have been more than 50 years old when it was marked. Redwood or cedar trees may exist for more than a thousand years, and, regardless of the size of the tree called for, the chances of existence are probable for another few hundred years.

Certain trees, such as redwood (*sequoia sempervirens*), black oak, and sycamore, have the ability to regrow from roots. Fire, disease, or humankind may fell the main trunk, but regrowth from the roots will preserve the original tree position, minus, of course, any particular original markings.

Tree rot attacks some trees much more readily than others. Trees susceptible to rot may be made open to attack by axe marks. Although the rot that enters may not destroy the whole tree, the markings will be unidentifiable. In blazing trees it is recommended that the blaze be painted. Certain trees, such as redwood and cedar are almost immune to rot and are ideal for blazing.

Many forested areas have been logged by man. Where only one witness tree is called for, identification of a stump is seldom a certainty. If the tree happened to be the only one of a species in the vicinity, an unlikely fact, identification of the wood stump by microscopic examination may be possible. If three witness trees are called for, and even sometimes when only two are called for, the relationship of bearing and distance between existing stumps may serve for identification. The size of a stump, minus its estimated growth from the time of the original survey, may be sufficient for identification.

Tree Growth. Trees grow in and out from a paper-thin cambium layer. Bark growth is on the outside, and wood growth is on the inside. The inner part of a tree is composed of live sapwood and dead heartwood. Both sapwood and heartwood are incapable of reproducing cells, and any injury to the wood can only be repaired by fill-in grown from the cambium layer. Once an injury occurs to a tree, by an axe or other means, the cambium layer is stimulated into more rapid growth at the point of injury. Often the stimulated growth produces a hump or bump at the point of former injury. As the cambium layer grows outward to fill in the gap, the characteristic appearance has given rise to the term *roll-over wood* (see Fig. 2-6).

Since sapwood cannot regrow, any injury caused by cutting into the tree leaves a permanent mark always identifiable. A nail driven into a tree or wire attached to a tree does not move up in elevation with tree growth; it stays in the same place and the wood grows around it (see Fig. 2-7). Injuries to wood are just as permanent as the wood itself. Paint applied to a blaze and overgrown by wood may be exposed years later.

Trees of the same species do not have the same growth rate. One tree may have better food than the other; one tree may have a genetic makeup that insures greater size than another. The size of a tree comparative to that originally reported is a poor means of age identification.

Ring Count. Counting rings to determine the age of a tree is not an absolute indicator but may be accurate within a small margin of error. Differences in the rate of growth of wood for seasons of the year cause rings of greater density to deposit annually. Sometimes, as a result of various causes, annual rings are discontinuous as shown in Fig. 2-6. Loss of a limb, destruction of a root, or an injury may temporarily stop ring growth on one side of a tree and not hinder it on another. Ring count based on a segment or portion of a tree may not be absolutely accurate. In addition, in the first few years of a tree the ring growth is probably below where the sample was taken. Ring count does not give the absolute age of a tree but is an indicator of within 2 to 5 years. In surveying, the ring count from a blaze outward is usually wanted, and it can usually be determined more accurately than the tree's absolute age.

In some varieties of trees the difference in the density of wood from different seasons is so slight that ring count is difficult to impossible. Members of the palm family have no annual rings, and the age of palms cannot be determined accurately.

In the 1903 edition of the *Manual of Surveying Instructions for the Survey of the Public Lands of the United States* there is found the following:

> The marking of trees and brush along lines was required by law as positively as the erection of monuments, by the act of 1796, which is still in force. The old rules therefore are unchanged.
>
> All lines on which are to be established the legal corner boundaries will be marked after this method, viz: Those trees which may be intersected by the line, will have two chops or notches cut on the sides facing the line, without any other marks whatever. These are called sight trees or line trees. A sufficient number of other trees standing within 50 links of the line, on either side of it, will be blazed on two sides diagonally or quartering toward the line, in order to render the line conspicuous, and readily to be traced in either direction, the blazes to be opposite each other, coinciding in direction with the line where the trees stand very near it, and to approach nearer each other toward the line, the farther the line passes from the blazed trees. In early surveys, an opposite practice prevailed.
>
> Due care will ever be taken to have the lines so well marked as to be readily followed, and to cut the blazes deep enough to leave recognizable scars as long as the trees stand. This can be attained only by blazing through the bark to the wood. Trees marked less thoroughly will not be considered sufficiently blazed. Where trees two inches or more in diameter occur along a line, the required blazes will not be omitted.
>
> Lines are also to be marked by curring away enough of the undergrowth of bushes or other vegetation to facilitate correct sighting of instruments. Where lines cross deep wooded valleys, by sighting over the tops, the usual blazing of trees in the low ground when accessible will be performed, that settlers may find their proper limits of land and timber without resurvey.

The practice of blazing a random line to a point some distance away from an objective corner, and leaving through timber a marked line which is not the true boundary, is unlawful, and no such surveys are acceptable. The decisions of some state courts make the marked trees valid evidence of the place of the legal boundary, even if such line is crooked, and has the quarter-section corner far off the blazed line.

On trial or random lines, therefore, the trees, will not be blazed, unless occasionally, from indispensable necessity, and then it will be done so guardedly as to prevent the possibility of confounding the marks of the trial line with the true. but bushes and limbs of trees may be lopped, and stakes set on the trial or random line, at every ten chains, to enable the surveyor on his return to follow and correct the trial line and establish therefrom the true line. To prevent confusion, the temporary stakes set on the trial or random line will be removed when the surveyor returns to establish the true line.

Surveyors must be careful about tree names in field notes and in land descriptions, because colloquial names abound. Most trees have several different local names, some of which are very misleading, for example, larch or tamarack being called "juniper," hemlock being called "spruce pine," and the like.

Sometimes trees or their blazes are scribed with corner information. Even when grown over, if the tree is in good enough condition, this can be recovered by cutting into the tree and removing the overlaying layers of wood and bark. By taking out a section the original marks can be uncovered and a mirror image exists on the removed section.

Remains of trees can often be identified by anatomy. Stumps, even rotten wood, and charcoal retain their internal structure that can be identified by an expert. This, combined with other evidence, can aid in the identification of the true monument or corner.

All blaze marks on trees should be treated with a certain amount of skepticism. As A. C. Mulford in his book *Boundaries and Landmarks* said, "It must be remembered that a small boy with a hatchet can mark up more trees in one Saturday afternoon than a dozen surveyors can in a year."

Three examples of scribed trees are shown in Figs. 2-8, 2-9, and 2-10.

2-47 Chain of History of Monuments

PRINCIPLE. *The best evidence of a monument's original position is a continuous chain of history by acceptable records, usually written, back to the time of the original monumentation.*

Deeds have a chain of title back to their inception, and the validity and correctness of a deed is based on this chain of title. Similarly, monuments should have a continuous chain of history. The original surveyor set a stone mound for the section corner. Surveyor number two finds the stone mound and sets a 2-inch iron pipe. Surveyor number three finds the 2-inch pipe and sets reference points 30 feet on each side of a new proposed road. Surveyor number

2-47 CHAIN OF HISTORY OF MONUMNETS / 53

Figure 2-9 A pine tree with scribing exposed by cutting in with a chain saw. "S 25" is visiable in the lower portion. "R 9 E" is visible above.

four finds the reference monuments and resets the true section corner in the centerline of the new road. Surveyor number five finds the new monument in the center line and wants to prove its identity and the correctness of its position. How can surveyor number five accomplish his goal without a continuous record of what each previous surveyor did? It is because of the need for continuous records that California has a law making it mandatory to file a record of survey under certain circumstances.

History or chain of record for monument position is valuable evidence, but all too often there is an interruption in the history, and a continuous chain of records cannot be proved. In such an event a different type of evidence must be resorted to.

54 / EVIDENCE

Figure 2–10 Bark-scribed beech tree in Alabama.

2-48 Surveyors' Records on Monument Location

PRINCIPLE. *In general but not always, the county surveyor's records (or city engineer's) are* prima facie *evidence, whereas that of private surveyors are not.*

Proof of monument positions is frequently dependent on surveyors' records and surveyors' testimony in court. But not all records of surveyors can be admitted in evidence. With regard to admissibility of evidence, surveyors can be divided into two classes: (1) public and (2) private surveyors. A regularly written record of a public officeholder or public employee, whose records were written as part of his job, is usually acceptable in evidence; but the records of a private surveyor, without the testimony of the private surveyor, are difficult to use as evidence in court since they would be hearsay.

Public surveyors and deputy surveyors have an official duty to run lines, establish boundaries, and make and file reports of their results. When such reports are publicly filed, no question exists about their admissibility as evidence. Sectionalized land field notes are difficult or almost impossible to impeach despite the fact that numerous instances of definite fraudulent surveys are known.

> The original surveys made by the United States are not to be taken as conclusive presumptions of law; they may be rebutted and impeached as to their correctness; but, prima facie, they are to be presumed to be correct until properly impeached.
> —*Brayton* v. *Merriman*, 6 Wis. 14.

Some states make public survey returns *prima facie* evidence; no proof need be presented to prove a monument set by public surveyors, but others may disprove the monuments so set. Private surveyors do not enjoy a standing of this type. The whole value of a private surveyor's results must usually come from his personal testimony, except in (1) case of death or (2) common repute that is discussed later.

The statute declaring that,"No survey or resurvey, hereafter made by any person, except that of the County Surveyor or his deputy, shall be considered legal evidence in any court in this State, except such surveys as are made by the authority of the United States or by mutual consent of the parties," does not disqualify any surveyor, private or official, from testifying concerning surveys made by him. It simply makes a survey by the County Surveyor prima facie evidence of its correctness, but withholds that quality from surveys made by private or other public surveyors, but nevertheless permits them to be received in evidence where they are first shown by competent evidence to be correct. A survey made by the County Surveyor, unless its face shows it was not made as the statute requires, is entitled to be admitted as evidence without further proof, but it may be shown by competent evidence that it was not made as the statute directs, and upon such showing or when a defect appears on its face, it is no longer official; whereas surveys made by any other surveyor must first be shown by competent evidence to be correct before they can be admitted in evidence, but when that is shown they are competent evidence.
—Sec. 11,301, R.S. 1909 or 60.150 R.S. 1949 Mo.

The court also instructed the jury that the survey made by the County Surveyor for Mrs. Coffman was *prima facie* evidence of the correct line so far as it appeared from the survey. This instruction was erroneous. Our statute provides that the County Surveyor shall keep a record of every survey made by him under the statutes and that a certified copy of this record under the hand of the surveyor shall be deemed *prima facie* evidence in any court of record.

In the present case the official record of the County Surveyor was not placed in evidence nor was it shown that notice that his survey would be made was given as provided by the statute. The statute is precise in prescribing that it is only a certified copy of the record of the County Surveyor which shall be admitted as *prima facie* evidence. The oral evidence of the County Surveyor gave his acts no more validity than the acts of any other surveyor.
—*Sherrin* v. *Coffman*, 143 Ark. 8.

Decisions about the admissibility of evidence is looked on in terms of how relevant it is to the question at hand and is determined by the judge.

2-49 Witness Evidence to Prove Monuments

PRINCIPLE. *Witness evidence may be used to locate the former position of a monument. But witness evidence cannot refute a found, undisturbed monument called for by the written conveyance.*

Those who have personal knowledge may testify about where the monument was, and such evidence, if by a reliable witness and undisputed, is conclusive. In California the law allows the surveyor to administer oaths and take verbal evidence concerning monument position. It might be pointed out that this oath and evidence are of no value in court as long as the person giving the evidence is alive, since the witness will be called into court and be required to testify directly. In the event of death the oath may be of value, and it also serves as a deterrent to a change in testimony.

> Appellant contends, and rightfully, that it is to be presumed that the government field notes are correct, and that where such notes place the quarter corner on a direct line between the section corners it will be presumed that the corner, as actually located, was on that line, and that it will take clear and satisfactory proof that it was located elsewhere to justify a court in finding that it was located other than on such line. No witness swore that he had ever seen a government mound, or what he thought to be a government mound, upon the direct line between the section corners; but there were four witnesses other than plaintiff who were positive that there was formerly a mound and pits answering the description of government mounds and pits, located at a point some distance south of the straight line running between the section corners. One of these witnesses had known of this mound ever since he was a boy. He testified positively to seeing the mound in 1909; that this mound was in a field which a party was then breaking; and that it was so located that, in breaking such field, the mound would be and was destroyed. He testified to its location in reference to a road that formerly ran from section corner to section corner, which passed very close to the mound. The other witness also testified to its location in relation to the old road. We are satisfied that there was sufficient evidence from which the trial court could find, and from which it was clearly bound to find, that the original government corner had become obliterated, that such corner had been seen by those witnesses. It was therefore not a lost corner. It certainly cannot be contended, merely because a quarter corner is become obliterated, and the exact location thereof cannot be fixed, the courts must treat the same as though it were a lost corner, and locate it on a direct line between the section corners.
> —*Kohlmorgan* v. *Roswell Township*, 41 S.D. 124.

Evidence has only as much value as would be accredited to it by the court. Obviously, the person giving the testimony must have had an opportunity to have observed the corner prior to its destruction; hence he must have personal knowledge, he must have lived, or resided, or been in the area at the right time. If a person has too much financial interest in the corner's location—that is, if he stands to gain—his testimony should be viewed with suspicion. Hearsay evidence, that which someone hears someone else say, is generally not admissible evidence, except in the case of what a deceased person has said or to prove a contradiction.

Parol evidence may not refute written evidence; hence the undisturbed position of an original monument called for by the writings may not be altered by witnesses.

> The lines of a grant must be established by the calls in the field notes. If these calls are inconsistent, then certain rules of construction and even parol evidence may be resorted to in order to resolve the doubt and to establish the line which was actually run by the surveyor. It is but a case of a latent ambiguity in a written instrument. A writing, unambiguous upon its face, may become doubtful when applied to the subject matter of the description. On the other hand, if there be no conflict in the

calls found in the field notes of a survey, there is no room for construction, and the calls must speak for themselves. To permit the introduction of parol evidence to vary the calls would be to violate the familiar rule, that extraneous evidence is not permissible to vary a written instrument.
—*Thompson* v. *Landgon*, 87 Tex. 254.

2-50 Summary of Competent Parol Evidence

Verbal evidence that is not competent under rules of law should not be received by the surveyor. From the foregoing it is obvious that a lay witness cannot (1) testify about what the laws pertaining to boundaries are, or (2) testify to alter written words of deeds, or (3) express opinions.

The general rule is that testimony may not contradict, vary, or modify writings. Testimony in general is limited to the following items, although there may be others. (1) Testimony may be taken to explain a latent ambiguity that is not a question of law. (2) Testimony may be taken about the former location of a monument, about the identity of a monument if the writings are not clear, about whether the monument has been moved, but testimony may not be taken about the location of a monument identifiable from the clear, unambiguous written words of the conveyance. (3) Testimony may be taken about the usual customs and meaning of words as of the date of the deed. (4) Testimony, if needed, may be taken about the general reputation of a monument. (5) Testimony may be taken about the surrounding circumstances as of the date of the deed.

This boils down to the fact that a lay witness may testify to facts within his own perception but not about conclusions, opinions, inferences, or facts contrary to the writings.

> We find no merit in the contention that the parol evidence above referred to was improperly admitted. Contrary to appellant's claim, the evidence neither altered nor varied the terms of the written instrument. It was offered by the respondent, and received by the court below, not for the purpose of changing or adding to the deed, but for the sole purpose of explaining the language therein contained. That parol evidence is admissible for such a purpose cannot be denied. If the word "block," as used by the common grantor in the subdivision map and several deeds executed pursuant thereto, was intended by him to have a particular and peculiar meaning, it was competent for the court below to permit the introduction of parol evidence to establish that meaning.
> —*Ferris* v. *Emmons*, 6 P 2d 950, Calif., 1931.

> Parol evidence was admissible to show that at the date of the deed M Street was straight, but that its course was subsequently changed to a curve.
> —*Abbot* v. *Frazier*, 240 Mass. 586.

2-51 Passing Calls

PRINCIPLE. *Passing calls for objects do not have a high rank.*

Passing calls are calls for found objects along a line. A passing call does not determine the start or terminus of a line and is not given the stature of locative monuments. This type of call is more frequently found in surveyor's field notes, particularly in sectionalized land areas and in Texas surveys. Notes reading "thence north 3 chains cross a creek, 15 chains on a ridge, 40 chains set a stone" have two passing calls of a creek and a ridge. The locative call, "set a stone," is controlling, but, in the absence of the locative call, the passing calls may have probative force to indicate the surveyor's footsteps.

> Passing calls mybe resorted to for the purpose of ascertaining a located corner where the locative calls have all disappeared, or cannot be identified, and there are no means, other than the incidental calls, of ascertaining the place where the located monuments for the corner was placed by the surveyor.
> —*Davenport* v. *Bass*, 153 SW 2d 471.

During the time the sectionalized land area was being surveyed numerous calls for passing objects were given in the field notes, and in some areas such calls are given strong probative value in the absence of the original monuments. Initially, these were used to show topography features on maps. In the following California court case the court adopted the topography calls to determine the location of a line and rejected the proportional method.

> On the west line of Section 6 Larson went north from the southwest corner thereof to the last recognizable call in the government field notes beyond the Garcia River, then there being no further ascertainable calls except that of "north," he continued north to the township line. On the east line of Section 6 Larson followed the calls from the southeast corner of said section north to the Garcia River. He could find no recognizable calls beyond that, so he followed the only call he could be sure of, and that was north to the township line. Cummins apparently made no attempt to follow the calls, even as far as the Garcia River. He started right out at the southeast corner of Section 6 and went on a straight line about 40 degrees off of a true north direction, ran many more chains than the notes called for, and in a different direction, and came out on a township line a mile westerly from where he would have had he gone north all the way from the corner he started from.
>
> "There can be no doubt, and I understand all parties to agree, that the government field notes beyond the Garcia River are grossly incorrect, as to distances, and consequently as to courses. It does not seem possible that such errors could occur by one in the field. One can guess that parts of said lines were not actually run, but only estimated from afar (not good estimates either). It is the duty of a later surveyor to try and retrace steps of the original surveyor where such is possible. This, Larson attempted to do. Kenneth Cummins "found no relation between the line I ran and the calls in the field notes." Larson followed the call north, Cummins did not. It is pointed out his closing point was between two streams and a spur, which description is found in some of the field notes with reference to other political subdivision. However, this is far from being conclusive when we ascertain that in order to get there we must go along a line which leads us to a point a mile away from where it would be if the line went straight north from the southeast corner to the northeast corner there of said Section 6. And where the distance is so far off from the distance calls in the field notes and where it is not shown there were

other places along the township line that could not qualify as coming within the same description there is not much left to go on in that regard.

"All the original government plats in evidence show the east and west lines of Section 6 to be straight lines north and south. The Larson line is a straight line north, the Cummins line is not. such plats may be considered in the over all picture. Indeed, Russell Cummins testified, 'The original surveyor may have intended to run on a true line north until it intersected the boundary of the township next north.' If he made a correct survey, that is what he did. The Court is forced to the conclusion that the great weight of the evidence shows that was what he did do. It would certainly be a peculiarly shaped section if the Cummins line was adopted as the east line thereof. Even so, if we were satisfied he had thereby traced the line of the original surveyor, that is the way it would have to be. But the evidence, including much of the testimony of Kenneth Cummins himself, shows that he did not do that. He ignored the calls of the field notes from the southeast corner northerly altogether and started out on a line by the single proportional method in a case where it was not authorized. Such a line cannot be adopted."

—*Hanes* v. *Hollow Tree Lumber Co.*, 12 Calif. Rptr. 713.

The following case took 340 trial days, 36,302 transcript pages, 844 exhibits and 1072 pages of briefs. One surveyor, William Wattles, was cross-examined for 46 days. The final decision was made on passing calls (topography called for).

Obliteration of a monument does not justify adoption of the proportional method of locating a common corner as a lost corner where the surveyor's fieldnotes refer to certain natural objects that can be found along the line mentioned so as to approximately locate it.

The task of the court, when confronted with an obliterated corner, uncertain boundary location or the like, is to decide from the data appearing in evidence its approximate position when the exact spot cannot be found, and fix the place at a point where it will best accord with the natural objects described in the fieldnotes as being about it, and found to exist on the ground, and where it will be least inconsistent with the distances mentioned in the notes and plat.

—*Chandler* v. *Hibberd*, 332 P 2d 133, Calif. 1958.

2-52 Possession Evidence

PRINCIPLE *Possession representing the location of original survey lines may be used to prove original survey lines.*

Possession that represents the original location of original monumented lines is distinctly different from unwritten title lines. In many instances, after all original monuments have disappeared from view, the best available evidence about where the original lines were is evidence of old fences built soon after the original stakes were set.

In the *City of Racine* case, the city resurvey was made by proportionate measurements from distant points. The comments on this and possession were the following:

In determining whether a fence which has been maintained for more than 40 years is upon the true line of the street or not, the question is as to the location of such line according to the original plat made prior to the erection of the fence, and not according to a resurvey by the city authorities nearly 40 years later, by which after fixing the line of one street from one of the original monuments, the distances were apportioned between the several blocks and the streets changed accordingly.

Although according to the resurvey the fence in question was more than two feet within the street, evidence that it was built according to stakes set by the surveyor who made the original plat, that it is on a line with other fences and buildings erected according to stakes set at the time the original plat was made, and that all said fences have been maintained on substantially the same line for more than forty years, is held to show that the fence in question was built on the true line.
—*City of Racine* v. *Emerson*, 85 Wis. 80.

The city tie points were valueless in this case. In California a similar situation with the same conclusion was recorded in *Perich* versus *Maurer* (29 Calif. App. 293) in Sacramento.

For possession to represent the original lines of the original surveyor, the following five facts must pertain. (1) There was an early survey that, if located, is controlling the line between the adjoiners. (2) The lines of possession are along the lines surveyed or presumed to have been surveyed by the surveyor. (3) Usually, but not always, a series of possessions, in agreement with one another, substantiate one another. (4) Possession is an ancient matter of a former generation (if it is of a present generation someone can testify about its origin). (5) Possession has the reputation of being correct survey lines.

The fact that the line between the parties, being the $\frac{1}{16}$ line, was not at the time marked by original monuments, deprives it of the right accorded ancient fences on the presumption that such fences were located upon the original stakes then visible.
—*Wollman* v. *Ruehle*, 104 Wis. 606.

Not all possession represents controlling survey lines. Although some lines of occupancy may become title lines by the process of unwritten agreement, adverse rights, or other unwritten means, such lines should not be confused with original survey lines. The surveyor relates possession to his survey lines, and he tries to gather evidence explaining the origin of possession. If possession came about merely for the purpose of a cattle enclosure and the person erecting the fence had no idea of ownership lines, or if possession follows a line that was not originally surveyed, obviously possession cannot represent original survey lines.

Where possession could be an original survey line, it is one of the duties of the surveyor to seek evidence to explain its origin.

2-53 Common Report, Reputation, and Hearsay

PRINCIPLE. *By common report or by reputation, monuments or former monument positions are sometimes proved. But title to land can never be proved by reputation.*

Although there is a general rule that hearsay evidence—that is, evidence of what you heard someone else say—is not admissible in court, many exceptions to the rule exist. In a court trial over land boundaries, the court is charged with making a location of the boundaries based on the best-available evidence, and if hearsay evidence is the best available, it can be used. The reputation of a monument as being correct is mere hearsay, and if better evidence of a monument's stature is not available, the reputation may be sufficient to prove its authenticity.

Although differences of opinion do exist about the application of the principle of reputation, none exists regarding the legal force of the principle. The rule is, of course, one of last resort.

Once a boundary or line is run and the survey is called for, the line is fixed in position although it cannot be heard, seen, or felt. All living things die, decay, and return to the earth that once nourished them. Stones, mounds, and physical objects disappear. Finally, all original markings of lines are gone. The original position of the ground remains the same, but how can it be monumented after the locative objects are gone? Can the certainty of title vanish with the objects, or can the land be located from the best available evidence?

> These boundaries over 30 years old are ancient boundaries. Reputation and hearsay are admissible to prove ancient private boundaries. Such testimony was admitted and there is no escape from the conclusion that dating from the time when the original boundaries were freshly and plainly marked on the ground, the southern boundary of claim Number 6 has been considered and established as coincident with the northern boundary of claim Number 7 and was so recognized by those who were familiar with the stakes and other boundary markings.
> —*Rickert* v. *Thompson*, 8 Alaska 398.

> The actual location of a deed may be satisfactorily established, not only by the natural objects found on the ground, but by the fact that all parties that knew the facts and were interested in the land, located the deed in a certain way or acquiesced therein. Time obscures all things, and facts that might by clearly shown 50 years ago may be incapable of proof now, when all the men of that generation have passed away.
> —*Wilson* v. *Commonwealth*, 243 Ky. 333.

After all original monuments in a subdivision have been lost along with the chain of records proving the new monuments to be replacements of the originals, no method exists, other than common report, to prove the stature of some monuments. All surveyors at times accept monuments and use monuments that cannot possibly be proved by direct evidence or chain of history evidence to be in their original positions. Reputation evidence is important to prove monuments that are not originals but are accepted as replacements of the originals.

After a monument has been used by numerous surveyors, the proof of error of location must be conclusive, not just surmised. The mere fact that all surveyors use a monument, without additional proof, does not and will not make it correct by continued use; the monument must be initially correct. Thus, in a superior court case in Alpine, California, it was shown that at an early date the State Highway surveyors tied in a fence corner and for some unexplainable reason described it as a section corner. A later surveyor in 1928 accepted the fence corner and set numerous corners from the accepted section corner. Up until 1950 some 10 or 15 surveyors filed maps and accepted the old fence corner as correct. When surveying an old holding dating back to 1900, another surveyor found that fences did not fit the proclaimed section corner. In a routine check it was discovered that the original government field notes stated, "Set a rock mound 3 feet south of a 12-foot-high boulder." Not only was the 12-foot-high boulder found but also a witness testified that in 1898 he had seen a stone mound just south of the boulder. All the expert testimony, reputation, and recorded plats could not overcome the fact that the true corner was 70 feet east of the accepted fence corner. The best available evidence was the written government field notes, and it prevailed. Reputation evidence does not overcome contrary proof, but the contrary must be proved, not just surmised. As a side light on that Alpine case, those with substantial enclosures were awarded title as based on unwritten occupancy rights, and said occupancy was described from the old original location of the section corner.

Reputation evidence is hearsay evidence and is an exception to the hearsay rule. Reputation is resorted to only when other means of proof are lost because of a long time lapse. The necessity of such evidence can only arise from the lack of better evidence. Certain safeguards have been set up by the courts so that the usage of reputation will not be abused. First, the reputation must be of ancient matters such as an old fence of unknown origin and reputed to be a property line. Recent surveys are excluded. Second, the reputation is of a former generation. What has happened in the present generation is provable by other evidence. Third, the reputation must predate the boundary litigation; otherwise the reputation will merely be a contention by one party. Fourth, if the reputation is based on the statements of an individual, such as a surveyor, the individual must be shown to be now dead and that he was disinterested at the time of his statements. Fifth, the statements of the individual are generally, although not always, required to be in reference to some monument or be supported by occupation. The reason is that a witness can remember a certain fence or monument but has no way of remembering isolated spots.

> It is also well settled that ancient boundaries may be proven by evidence of common reputation.
>
> —*Cockrell* v. *Works*, 94 SW 2d 784.

> The true location of the point where the gate post stood will settle the dividing line in dispute. The gate post has long since disappeared. It was shown in evidence that the adjoiners had a dispute as to the location of the gate post corner about 20 years ago. Appellant and her husband then applied to the county court for the appoint-

ment of processioners to establish and remark the obliterated corner. They were appointed and with the county surveyor met upon the ground. They had called a number of old people living in the neighborhood, who were requested, and probably sworn, to locate the old gate post. Each of the witnesses so called stuck a stick down at the point where he remembered the post to have been. They differed by some few yards. The surveyor placed his Jacob's staff in the center of the point so selected, and from thence ran the line to the beech which was then pointed out as the next corner. The county surveyor who was present at the time made a memorandum of what had been done, and signed it. But it was not signed by the processioners, nor was it returned to the county clerk, as was required by statute. On the trial of this case the surveyor who then ran the line was called as a witness, and testified to the foregoing facts, and produced the certificate which he had given at the time. His testimony was objected to. The old citizens who were called upon by the processioners, and upon whose statements the corners were established, are now all dead. It is competent to prove the location of the corner or line of public survey by reputation. In the nature of the thing those who marked the original corner, and knew personally of its location, will in time pass away, and so in some instances will the corners themselves. Such matters of common knowledge are discussed in the neighborhood, and are accepted and treated by those interested as being of a certain nature, so that their reputation becomes established and known of all in the community. After the death of the original witnesses and the destruction by time of the monuments marking the corner, the only thing left by which its location might be identified is the reputation established and made notorious when both witnesses and corner were in existence. It is, therefore, that the law receives the evidence of the reputation in proof of the fact to the location of such corners and lines as the best evidence obtainable in the nature of the case. The surveyor's certificate was not receivable because it did not conform to the statute, but what was said then to the surveyor by the persons who were then before him was evidence of reputation of the location of the original monument and corner. This evidence outweighs, in our opinion, the conflicting and unsubstantiated statement of appellee's witness that the gate post was north of the point located by Wood.
—*Phillips* v. *Steward*, 133 Ky. 134.

Where for more than 40 years the southeast corner of a block had been recognized as being at a certain place, and lots, blocks and streets located, and buildings built with reference thereto, such universal usage and acquiescence outweighs indefinite notes of the surveyor who many years before replatted the block, and under which it is claimed the corner is located three feet farther south.
—*Crandall* v. *Mary*, 67 Ore. 18, 1913.

In some areas—particularly along the eastern seaboard, where there has been extensive loss of original records and destruction of monuments—the courts are inclined to accept ancient evidence of old boundaries in preference to modern measurements. This in no way means that a surveyor should accept all ancient boundaries; the surveyor must seek an explanation of the boundary. Many times old fences have been proved to be mere barriers of convenience. Accepting a location as based on reputation is a rule of last resort.

2-54 Evidence of Fences to Prove Boundary Locations

Frequent allusions have been and will be made to the value of fences in determining the correct location of boundaries. One of the most difficult tasks of the surveyor is to determine when he should or should not use existing fences as proof of original survey location.

The first wire fence was erected about 1816, with most of the barbed wire patents, as now known, coming along in the 1880s. Fences called for in early descriptions may have been board fences, stump fences, or stone walls. Wire may have been erected later, however, and the fence thus perpetuated.

It is difficult to say when a fence was erected, since the fence may have been put up years after the wire was purchased. However, knowing when a particular type of fence was invented (there are about 1000 different patents) one can be sure when the fence was not there.[2] Sometimes a probable age of a fence can be determined. Ring comparison of fence posts with living trees can illustrate approximately when the post was cut, and a determination can be made of when wire was attached to a tree by examining the rings.

Fences are not always erected on boundary lines. There are, of course, all sorts of interior fences on farms for various purposes. Sometimes fences were put up to keep animals from straying into the woods or going into a brook or swampy area, with no intention of marking any boundaries. Even when fences were erected at or near the boundary they would not be put up according to any survey so that the boundary indicated by them can only be close at best. Farmers would set posts where they could dig a hole without hitting a big rock, or they would tack the fence to a convenient tree. If intending to be relative to the line, the fence builder would put the fence where he *thought* the line was or often slightly inside the true line to avoid ancroaching on his neighbor. Today, fences are sometimes accepted as the best available evidence, or the only evidence remaining to indicate the line.

> Ancient fences used by a surveyor in his attempt to reproduce an old survey are strong evidence of the location of the original lines and, if they have been standing for many years, should be taken as indicating such lines as against the evidence of a survey which ignores such fences and is based upon an assumed starting point.
> —*James* v. *Hitchcock*, 309 SW 2d 909, Tex.

> A long established fence is better evidence of actual boundaries settled by practical location than any survey made after the monuments of the original survey have disappeared.
> —*Diehl* v. *Zanger*, 39 Mich. 601

> Evidence of ancient fences and improvements is competent to prove boundary, where monuments and lines of original survey cannot be shown.
> —*Cay* v. *Stenger et ux.*, 274 P 112.

---

[2]See the discussion of barbs and prongs in Robert T. Clifton, *Points Prickers and Stickers*, University of Oklahoma Press, Norman, 1973.

2-55 Evidence of Deceased Persons' Sayings to Prove Monument Reputation

PRINCIPLE. *What a person heard a deceased person say regarding a boundary line may be admissible evidence provided (1) better evidence is not obtainable, (2) the deceased person had peculiar means of knowlege of the boundary, (3) the deceased person was disinterested at the time of declaration, and (4) in some jurisdictions, the declarations of the deceased person was part of the* res gestae.

The sayings of deceased property owners, adjoining owners, surveyors, and tenants are the ones commonly coming under this rule. Surveyors as a group are usually considered disinterested. The sayings of deceased owners are sometimes excluded because of interest, but deceased owners' sayings with regard to things against their interest are always admitted.

In general, but not in all jurisdictions, to be admissible, the saying must be part of the *res gestae*; that is, the saying must have been in connection with some act. The sayings of a surveyor in connection with his professional duties, such as telling the owner the stake being driven is his property corner, would meet this requirement. The declarations of a deceased owner could be made in connection with pointing out boundaries.

> The rule that declarations of a deceased owner with respect to boundaries are competent evidence only when made on the ground, applies also to declarations of a deceased surveyor. The field notes of a deceased surveyor are admissible as declarations contemporaneous with the work done on the ground, provided they are authenticated in some other way than by the mere declarations of the surveyor himself.
>
> —*Collins* v. *Clough*, 222 Pa. 472.

> Objection is made to certain evidence dealing with the location of monuments and the claims of former owners because it is hearsay.
>
> J. Abrams and M. Abrams testified that Bob Blake and Gene Clements, men dead when the testimony was given, told them where the bed of the old creek lay, and showed it to them on the ground. Ephraim Bowler, who had himself worked on the dyke, testified to the line pointed out to him by Col. Carter. There is other testimony of the same general character.
>
> These statements were made by men now dead, by former owners of these properties, by their tenants and by neighbors who were particularly qualified to speak on the subject. They are admissible, and although hearsay, evidence of this character is often the only evidence that it is possible to obtain. It is a rule of necessity and gives us the best evidence available. Monuments perish and streams fill up. At times, if they cannot thus be relocated, they cannot be relocated at all.
>
> That boundaries may be proved by hearsay testimony is a rule well settled; and the necessity and propriety of which is not now questioned. Some differences of opinion may exist as to the application of this rule, but there can be none as to its legal force.
>
> There was introduced a plat of a Panpatike farm. Mr. S.S. Robinson, the son of the surveyor and himself a surveyor, was asked to explain what certain lines of the

plat meant. He might well have been permitted to answer, for in cases of this character the courts should be liberal in their reception of evidence.
—*Woody* v. *Abrams*, 160 Va. 693.

2-56 Weight of Evidence Accorded City Engineers' Monuments

PRINCIPLE. *Monuments set by public agencies, such as the city engineer or the county surveyor, in pursuance of their official duties, are presumed to be correct.*

The burden of proof of error lies in the person disproving them. In some states monuments set by the city engineer are declared to be *prima facie* evidence; that is, unless the monuments are proved to be in error, the monument itself will establish the fact of location of the point that it purports to mark.

Discovery of a city engineer's monument does not relieve a surveyor of the obligation to look further. The city engineer's monument is only proof in the event superior evidence cannot be discovered. If the city engineer's monuments were set as replacements of original monuments, the search would cease with the discovery of the city monuments. But if the city engineer's monuments are set by measurement, the discovery of an original monument or proof of where the original was, will overcome the presumed control or *prima facie* evidence of the city engineer's monuments. The surveyor must seek all other evidence and use the city engineer's monuments as though they were the last resort.

2-57 Uncalled for Monuments

The rule is that "for a monument to be legally controlling, it must be called for either directly or indirectly in the writings conveying the property." Court interpretation says the *position* occupied by the original munument is the controlling consideration. Any monument occupying the spot of the original monument can be controlling, provided it can be identified as occupying the original monument's position. Thus it is not always necessary to find the exact type and kind of monument described in the original writings. Here are some examples.

About 1870 Horton's map of the city of San Diego was filed with a note that it was surveyed and lots were 50 × 100 feet. No mention was made concerning what or where monuments were set. At the present time monuments exist in the sidewalk at 7 × 7-foot offsets from block corners. Since sidewalks did not exist at the time of the original subdivision, the present offset monuments can in no way be considered original monuments. Furthermore, no one can testify about what the original monuments were or where they were set. These undescribed monuments are accepted merely because they are the best available evidence explaining where the original survey was located.

This is different from the following. In the city of Hartford, Connecticut, Mrs. Parsons agreed to sell to Mr. Kashman a house and the land enclosed by a fence. The deed described the property as being 60 × 147 feet on the corner of Main and Canton without mention of the fence. A later survey disclosed that the fence only occupied 57 feet at the rear. In the litigation that followed (39 A 179) the court observed, "Fences not referred to in a deed cannot control the distance stated in the deed." What is the difference between this case and the offset monuments found in the city of San Diego? Horton's map called for a survey, and it was up to the surveyor to determine the best evidence of where that survey was. In Mrs. Parson's case no survey or fence was called for; the only thing called for was 60 × 147 feet and it held.

In Maine a case involving an uncalled-for monument was *Bemis* versus *Bradley*, 139 A. 593. Fifty acres off the east end of Lot 5 was conveyed in 1843. In 1929 Bradley owned the eastern part and Bemis owned the western part. At the time the land was sold to Bemis an old stump was pointed out to both parties as one corner, and the line was said to run south from it. A fence was built. Later Bemis ordered a survey, and it was discovered that the stump was 15.6 rods too far west. Since the stump was not mentioned in the deed, Bradley claimed it to be an agreed corner and the fence an agreed line. The court ruled that the fence and stump were agreed on because both parties *thought it to be the true line*, and since it was not the true line, either party could claim to the true line. The uncalled-for monument (the stump) did not control. The interesting part of the case was the testimony of the surveyor who ran the lines for Bemis. The surveyor found an early survey line that enclosed "slightly" more than 50 acres (51 plus). "On this line there is evidence of spotted trees which indicates that at some former time a division line was run on this location." The evidence consisted of a spotted post or stump on the south line on Lot 5, the spot being described as very old: an old spotted beech stump about 12 rods northerly therefrom; another spotted beech stump about 10 feet further north; a spotted beech about 30 rods still further north, located 4 feet east of line; and a spotted hemlock tree at the end of the line. This line was accepted by the court, even though it contained more than 50 acres as called for. In most states exactly 50 acres would be laid off; no survey was called for. In this case Bemis in the pleadings asked for the survey line, not the 50-acre line; it is not known what would have been the outcome if Bemis had asked for the line that marked off exactly 50 acres.

In summary, the surveyor may accept a found monument that the evidence indicates to be occupying a position which legally controls the location of the writings. The controlling consideration is a position on the face of the earth, not necessarily the monument occupying that position as of the date of the deed. Any monument that can be shown by acceptable evidence to be occupying a legally controlling deed position may be acceptable. The surveyor is looking for controlling monument positions called for in the writings, and, if it can be shown that an undescribed monument is occupying that position, the monument is acceptable.

2-58 Fraudulent surveys

The Federal Government contracted for the survey of the public domain. Whereas, in general, the work actually was performed in most areas, a number of fraudulent situations have been proved, the mosts notorious being that of the Benson Syndicate, so called. John Adelbert Benson was a deputy surveyor of California, Oregon, and probably Washington in the years 1873 to 1885, and he and his associates or employees contracted for most of the work during these years. Several of the associates listed were fictitious, and those that were not often just signed the necessary documents. After "surveying" about a million acres Benson was convicted and sent to jail; Benson died a year after release of a heart attack and left an estate of $431. For details of the case and for the areas surveyed see *Chaining the Land* by Francois D. Uzes, Landmark Enterprises, Sacramento.

Since considerable areas of California land were sold in accordance with the fraudulent surveys, and those are the only "surveys" to go by, California surveyors should be acquainted with the fraudulent townships in their area. In general, part of a township was actually surveyed, and the rest was faked in. If the township plat is within the dates cited previously and if signed by any of the following, the "survey" is suspect: G. W. Baker, P. Y. Baker, W. F. Benson, N. L. Berdam, John A. Berson, Theo. Binge, James Branham, S. W. Brunt, H. E. Buckley, Geo. S. Collins, J. W. Fitzpatrick, J. L. Glover, L. B. Gorham, J. D. Hall, A. P. Hanson, S. A. Hanson, John Haughn, D. M. Hill, Chs. Holcomb, John L. McCoy, William Magee, Henry Meyrick, jr., William Minto, W. H. Myrick, W. H. Norway, James O'Brien, Geo. H. Perron, Geo. W. Person, C. F. Putnam, C. F. Ragsdale, M. F. Riley, Milton Santee and Jas. E. Woods. Prior to teaming up with Benson, Minto did do some acceptable surveys.

MEASUREMENTS

2-59 Measurement Evidence to Prove the Proximity of Monuments

PRINCIPLE. *Measurements may be used to prove the validity of monuments. Such monuments to be acceptable should be within reasonable proximity of the record measurements.*

There are many circumstances that can vary the meaning of the word "reasonable." Some deeds are written without the benefit of a survey; hence measurements may be mere estimates. Surveyors have been known to make gross errors. No definite rule exists that identifies exactly how close a monument should come to the record measurement. If an iron pipe shows signs of having been disturbed by cultivation, it is only right that it should be bent back a fraction of a foot to its measured position. However, if an oak tree is identified as the one called for, a substantial mesurement error probably would not disprove it. An error of 100 feet in going to the ocean may not cause the surveyor anguish, but an error of 100 feet measured to an easily movable stone mound would present serious doubts. The amount of measurement error allowed is dependent on the movability of the monument and the certainty of identification of the spot occupied by the monument as of the date of the deed.

2-60 Evidence of Measurement

In the order of importance of items that may prove the intent of a deed, evidence of measurements ranks below senior rights and monuments. A surveyor who needs his ego deflated needs only to reflect on the fact that the courts in the matter of interpreting deeds have placed the least importance on measurements.

PRINCIPLE. *Unless proved otherwise, measurements of distances are presumed to be horizontal. In the colonies early measurements were often surface distances, and, when proved as such, surface measurements may control.*

In some areas of the United States, especially along the eastern seaboard, early linear measurements were made on the ground without corrections for slope. When it can be shown that the original surveyor did use surface measurements as a local custom, the presumption of horizontal distance is overcome. The following cases illustrate this point.

In *Justice* v. *McCoy* (1960 Ky., 80 ALR 2d 1208) in Kentucky two streets differed in elevation by about 40 feet and were about 100 feet apart. McCoy's lot called for 60 feet "up the hill." If 60 feet were measured horizontally the measurement would be about 70 feet on the surface and would include parts of the houses of the upper lot. The plaintiffs contended that there was an absolute rule requiring the use of horizontal measurements; the court's opinion was that surface measurements were proper when such was the custom of the locality or was dictated by the circumstances of the case. Other cases reported were as follows:

> While the surface and not the level or horizontal mode of measurement is generally adopted in surveys, and the general presumption is that a survey of the surface was contemplated by the parties to a deed, yet that presumption prevails only where it appears feasible and reasonable to have pursued that course.
> Where a line of survey crossed a perpendicular cliff at a place where it could not be climbed, and to give the quantity of land called for by the survey and to take the line to a boundary shown to have been marked in an old survey, it was necessary to exclude the distance up the face of the cliff, it was not error to instruct the jury to exclude it in determining the boundary.
> —*Stack* v. *Pepper*, 119 N.C. 434, 1896.

> Surface measurement is the only kind in practice by the district surveyors of this state. But in making that, it is usual for the surveyor in chaining over an uneven surface to make allowance by elevating the chain.
> There was nothing to show that this was not the practice when the original survey was executed; whatever it was, would regulate the measurement now; and we see no error, therefore, in the answer of the learned judge to the defendant's point, that they were entitled to the measurement adopted by the surveyor who located the warrants. In other words, to the ordinary measurement of official surveyors.
> —*Boynton* v. *Urian*, 55 Pa. 142, 1867.

As can be observed from the foregoing cases, it is a matter of proof by evidence about whether the original surveyor made his measurements on the ground surface or horizontally. That is, it is a matter of proof in those states that permit it. In most of the states west of the Appalachian mountains, the standard of measuring horizontally cannot be refuted. In Tennessee in 1860 a U.S. court rejected surface measurements even though the original surveyor testified using it as in the following.

> The surveyor who made the survey on which grant No. 22,261 is founded, deposed at the trial, "that no actual survey was made in 1838 of said land, except the first line from A to H. That the other three lines of the grant were not run, but merely platted. That the proper mode of making surveys was by horizontal measurement, but that he had not been in the habit of making them in that way; that in making the line from A to H, in this survey, he had measured the surface; that the custom of the country was to adopt surface measure; and that he had made the survey in accordance with such custom."
>
> The grantee was bound to abide by the marked line from A to H; but the other lines must be governed by a legal rule, which a local custom cannot change.
>
> In ascertaining the southwest corner of the tract at 894 poles from the poplar corner, the mode of measuring will be to level the chain, as is usual with chain-carriers when measuring up and down mountain sides.
>
> —*McEwen* v. *Den*, 1860, 24 Howard (U.S.) 242.

All private surveys today are, of course, relative to the level standard foot, and in the near future will be relative to the meter. The Federal Government uses the chain as the official unit of measurement as prescribed by federal law. The foregoing cases are only cited to explain some of the unusual customs of the past. In the sectionalized land system commenced in 1785, horizontal measurement was specified and has always been used. In most states the assumption that horizontal measurements were made cannot be refuted. Certain areas, such as Miami, Florida, are having the metric system imposed by virtue of an influx of foreign-trained surveyors.

PRINCIPLE. *In most states the presumption is that bearings are relative to the astronomic north, unless otherwise specified. In a few states, especially along the eastern seaboard, the presumption for older surveys is that the magnetic meridian as of the date of the measurement was used.*

Basic presumptions can be overcome by contrary evidence. In the western mining areas where it was shown that the custom was to survey on the basis of magnetic north, the courts have upheld such basis of bearings. In the original 13 colonies, until about 1800, and in certain places after that, it was the custom to survey by the compass without correcting for declination. Which method was used, magnetic north or true north, is a question to be proved by evidence, but if contrary evidence is not presented, the presumption will prevail.

In M'Iver's *Lessee* v. *Walker* tried in a U.S. court in 1815 (13 U.S. 173) Chief Justice John Marshall noted that in North Carolina it was undoubtedly the practice of surveyors to express on their plats and certificates of survey the courses that were designated by the needle, and if nothing exists to control the call for course, the land must be bounded by the courses according to the magnetic meridian.

An 1815 case in Georgia (*Riley* v. *Griffin*, 16 Ga. 141) held that if nothing exists to control the call for course and distance, the land must be bounded by the courses and distance of the grant, according to the magnetic meridian. A similar result was found in *Vance* v. *Marshall* in 1813 (6 Ky. 148), wherein the judge observed that calls for course was usually and generally understood to be according to the magnetic meridian. Furthermore, instruments were not then available to determine the true meridian. To this the judge noted that the magnetic meridian to be used was as of the date of the original survey, not as of the date of the resurvey.

In the *Wells* v. *Jackson Iron Mfg. Co.* case occurring in 1866 (47 N.H. 235) the judge found that the term "due north" did not mean by the true meridian; as a matter of common knowledge boundaries had almost uniformly been run out according to the magnetic meridian. It was common law in the state of New Hampshire that the course in deeds of private lands are to be run by the magnetic meridian when no other is specifically designated.

In 1953 in North Carolina (*Goodwin* v. *Greene*, 237 N.C. 244) the court ordered a new trial because the judge had failed to instruct the jury to take into consideration the change in magnetic north since 1898. In 1858 in Ohio (*McKinney* v. *McKinney*, 8 Ohio St. 423) the court ruled that the custom in private boundaries was to use the magnetic meridian, and this was done even though sectionalized lands were done on the true meridian.

In the act of Congress of May 18, 1796, all sectionalized lands were to be surveyed as based on true north (astronomical north). Although the compass was to be used, the direction of magnetic north was to be corrected to true north.

From the foregoing it is readily apparent that the burden of proving what the basis of bearing was for a deed is the responsibility of the surveyor. After it has been decided that the original deed was based on a magnetic bearing, then comes the task of determining what the magnetic variation was as of the date of the survey or deed.

The fact that magnetic north and the direction the needle points vary from time to time and from hour to hour in the day is well established. Basically, three causes of magnetic change are present: (1) daily or diurnal changes, (2) annual changes, and (3) magnetic storm changes. The magnitude of daily changes is about 7 minutes and is brought about by the effect of the sun on the earth. The annual change is gradual and can vary from 1 minute up to several minutes. Although the change for 1 year may not be significant, the change over several years can become very significant. The change may be in one direction for several years and then change to the opposite direction. Although charts and books are available predicting variation and annual rates of change, these are nothing more than predictions and will not reflect the actual numbers. Surveyors cannot predict with certainty what magnetic changes will be, but they can tell, within limits, what the changes were. Magnetic storms that occur at infrequent irregular intervals may cause as much as 7° variation of the needle during the time they are in progress; thus there is a shadow of doubt about what occurred in the past.

Local attraction can, of course, affect the needle. If there is a problem in relocation of an original compass-surveyed line, the best solution is to find two monuments that existed as of the time of the survey and were referred to, then make an actual determination of the present magnetic direction on that line. The differences between the two readings will give the change in magnetic deflection. In the absence of monuments to make a direct determination, tables published by the U.S. government must be resorted to. The U.S. Coast and Geodetic Survey (USC and GS)—now National Geodetic Survey, NGS—published tables and maps showing the magnetic declination as it existed in the United States at various times.[3]

PRINCIPLE. *If there is a conflict within a deed and a choice must be made between bearing and distance regarding which controls, no uniform rule has been laid down by the courts. Variations between states do occur.*

In the sectionalized land system of the Federal Government, in replacing lost corners in accordance with BLM rules, distances are always kept in their original proportions and bearings yield. In Texas (*Stafford* v. *King*, 30 Tex. 257) bearing is given definite preference to distance. Since it takes both bearing and distance to determine a line, it would be a rare instance to give one control over the other. In the usual situation, the surveyor gives control to as many items in the deed as is possible, rejecting as few as possible. In this situation, N 20°10′ E, 100 feet to the south line of Palm Avenue, the bearing would be given force and the 100 feet would yield. In this situation, N 80°10′ E along Palm Avenue, the distance would hold and the 100 feet would go along Palm Avenue regardless of the bearing. The custom is to reject as few terms in a deed as is possible.

In Fig. 2-11 is shown a block with regular lots of 50 × 115 feet each. A resurvey discloses that the block frontage measures 300.60 feet instead of 300.00 as originally recorded. The rear of the lots measures 301.20 feet instead of 300.00. Assuming that only block corners were originally set, in most states the excess would be evenly divided between all the lots (frontage is 50.10 feet; the rear, 50.20 feet). In a state where direction is more important than distance, the angle would be prorated as shown in Fig. 2-11. Although the two methods theoretically yield differences in lot dimensions, the differences are too small to measure; both methods are comparable.

In Kentucky the rule is definite; give direction (course) control over distance as stated in this 1801 case:

> When a departure from either course or distance becomes necessary, reason as well as law seems to suggest that the distances, taken in our mode of mensuration, ought

---

[3]National Geophysical and Solar-Terrestrial Data Center, EDS/NOAA, Boulder, Colorado.

2-60 EVIDENCE OF MEASUREMENT / 73

	300.60						
	50.10	50.10	50.10	50.10	50.10	50.10	
	90°00'	90°03'	90°06'	90°09'	90°12'	90°15'	90°18'
115.00	1	2	3	4	5	6	115.00
	50.20	50.20	50.20	50.20	50.20	50.20	
			301.20				

Figure 2–11 In some states direction is considered more important than distance. In the above subdivision excess exists. If the excess is apportioned by distance, the results are almost identical to apportionment by direction.

to yield, as being much the most uncertain of the two. Indeed, it never has been held, that in laying off vacant land purchased from government, that a line established by a surveyor can be altered on account of its being longer or shorter than the distance specified in the plat and certificate he has returned.

These considerations seem to dictate an answer to the question which has been stated, to wit: From one of the adjacent corners which remain, the courses and distances of the lost lines ought to be run, as called for in the plat and certificate of survey, and if they close with the other adjacent corner which remains, the true situations of the lost corners, and consequently the true situations of the lost lines, will be satisfactorily ascertained. But, if the courses and distances thus run do not close the survey, it must be accomplished by running the same courses, and either lengthening or shortening the distances, as each case may require, and in proportion to the length of each line, as called for in the plat and certificate of each survey. And if the survey cannot be made to close by this means, then, and not otherwise, a deviation from the courses called for must also aid in accomplishing the purpose. For example, where all the corners of a survey are lost but one, lost corners ought to be ascertained by running comformably to both the courses and distances specified in the plat and certificate of survey. Where there is but one corner of a survey lost, the courses called for in the plat and certificate of survey ought only to be regarded; and nothing more is necessary than to extend those courses from the adjacent corners which remain, until they intersect each other, and the place of intersection will be the situation of the corner to be ascertained.
—*Beckley* v. *Bryan*, Ky. Nov. 12, 1801.

Compass courses called for in a deed prevail over measurements
—*U.S.* v. *Murray*, 41 F. 862.

When a departure from either course or distance becomes necessary, the distance should yield.
—*Curtis* v. *Aaronson*, 7 A 886, 49 N.J. Law 68.

PRINCIPLE. *In a metes and bounds description each course is normally run in the order described in the deed; no one course is superior.*

This principle is a presumption that can be overcome by contrary evidence. Usually, the determinate factor, bearing or distance, is determined by case law within a particular jurisdiction or state. The majority opinion is expressed by the following court cases.

> Whether established by proof or admission, it being in evidence that the line was reversed, when by running forward, a different result would have been attained, it was error to refuse to instruct the jury that the location made by running, as the deed was originally run from a known corner or one established by proof, was to be adopted rather than one ascertained by running in the opposite direction. It is a fact of which the courts must take and have taken notice that the measurements of boundary lines in making the original surveys for deeds and grants is often, if not always inaccurate.
>
> It is therefore a well known fact that, owing to inaccuracies in measurement, different results will follow from adopting one of the other of the two methods of surveying where many of the old monuments have perished or been removed. In determining which is correct the courts proceed upon the idea that the object of legal investigation and inquiry is to find the lines, corners and monuments which were agreed upon by the parties to the original conveyance, and that in order to attain that object the lines should be run in the direction and order adopted by them. . . . There are some exceptional instances in which it is manifest that reversing a line is a more certain means of ascertaining the location of a prior line than the description of such prior line given in the deed, but such cases are the rare exceptions to a well established general rule. . . . The general rule is an established law of evidence adopted as best calculated to ascertain what was intended to be conveyed, and it is incumbent on a party asking the courts to depart from it to show facts which bring the particular case within the exception to the rule.
>
> —*Duncan* v. *Hall*, 117 N.C. 443, 1895.

> In order to determine what land was intended to be described in case of ambiguity in deed, the description may be read by reversing the courses.
> —*Blume* v. *MacGregor*, 148 P 2d 656, Calif., 1944.

The principle of running each course in the sequence in which it occurs places the error of closing in the last course. In Kentucky where course (direction) is preferred to distance, the error of closure would not fall in the last course in the following situation shown in Fig. 2-12. The last two courses read, "West 100 rods; thence south 100 rods to the point of beginning." A found monument controls the beginning of the course "west 100 rods." After running west 100 rods, the point of beginning is found to be S 3° E, 110 rods away. Do you run west to a point that is north of the point of beginning, or do you run the full 100 rods and let the last course be S 3° E 110 rods? In Kentucky you would run west to a point where you could run south to the point of beginning. In other words you would run west 94.76 rods and then south 109.85 rods to the point of beginning. In states where the error is placed in the last course, you would run west 100 rods and let the last course fall where it may (S 3° E, 110 rods). The *Manual of Instructions for the Survey of the Public Domain* recommends

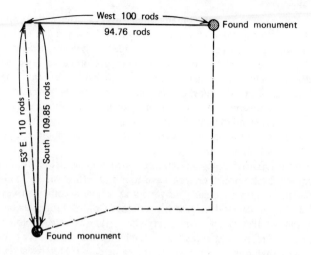

Figure 2-12 Direction controls over distance in some states.

using the compass rule to distribute the error of closure in metes and bounds descriptions.

In *Curtis* v. *Aaronson* (New Jersey) the supreme court objected to the instructions given to the jury in a case involving where to put the error of closure as follows.

> Courses govern distances in a survey; and it is error to instruct the jury that a certain line must stop at the distance given in a description in a deed from the corner from which it was run, when, on any theory of the location of the grant, the length of some line must be increased, and on that of one party (which there was evidence to support) this particular line must be lengthened, or the course of another changed, in order to reach the monument called for.
>
> —*Curtis* v. *Aaronson*, 7 A 886.

2-61 Evidence of Office Surveys

In Texas, in the U.S. public domain, and in other states, the law required a survey prior to or simultaneous with the disposal of public lands. Occasionally, the surveyor, under contract, failed to perform an actual survey on the ground but did turn in notes and plats purporting to be the results of his survey. Such fraudulent surveys have been nicknamed by surveyors as "office surveys," "paper surveys," "chimney surveys," "chamber surveys" and "tent surveys."

Technically speaking, fraudulent office surveys cannot legally be the foundation for conveyances; but, since an officer of the state, the surveyor, perpetrated the fraud, most states have passed acts validating the illegal surveys or have accepted them by court decree.

In West Texas there are a great many office surveys, often called chimney corner surveys. That is to say there was never any actual survey made on the ground. It would seem that these surveys would be invalid, but the state legislature validated these by legislative act. It is not difficult to detect office surveys. Sometimes a working sketch will have to be made of an entire block to be certain. If the field notes call only for stakes, stakes and mounds, and earth mounds, you may rest assured that the locating surveyor was never on the ground. The call for a stake, and for a stake and mound seemed to be a conventional sign, so to speak, for no corner. These calls were used by all the early surveyors. Why the earth mounds? I am unable to say, unless it was to break the monotony.

Wherever the name of Dolores Gonzales appears on the field notes of a survey as a chain carrier, look no further, you have found an office survey. This may have been an arrangement among early surveyors to denote an office survey, or it may have added a bit of humor. It is possible too, that Dolores Gonzales became a sort of legend and he, like Kilroy, was everywhere. At any rate, Dolores Gonzales' name appears on more field notes than any chain carrier who ever lived. He even served on more than one survey party on the same day. If Dolores served on all the survey parties that the field notes say he did, he was the "walkinest" man in all Texas.[4]

In California in one township the southern tiers of sections were located with no difficulty, but no corners were ever found in the northern tiers. At that time cedar posts cut in the mountains and transported to the desert were used. Discovery of all remaining unset, properly marked monuments in a gully explained the reason. Why the surveyor stopped in the middle of his job, however, is a mystery.

Despite the fact that many instances of office surveys are suspected, the fact must be proved by acceptable evidence, not merely surmised. The evidence of writings, by those authorized to do original surveys, is in general conclusive until proved otherwise.

2-62 Variable Interpretations of the Meaning of Evidence

By now it should be apparent that the significance assigned to evidence varies between states and the Federal Government. Although states tend to adopt like rulings, such does not always occur. To illustrate contrary thinking, the restoration of a lost section corner is used. It is known that the original survey was made under authority of the United States. The surveyor general and the federal courts say that when you restore a lost section corner on the interior of a township you use double proportionate measurement. In Missouri by statute law 60-290 you use the intersection of straight lines connecting the nearest found corner in each direction. Needless to say, in most instances, the restored spots are in different locations. In a court case that followed (*Simpson* v. *Steward*, 281 Mo. 228) the state court observed that neither method would

---

[4]Hugh L. George, Registered Public Surveyor, Texas, "The Identification of Original Survey Corners in West Texas," *Fifth Annual Texas Surveyors Association Short Course*, 1956.

result in the exact location of the original corner; either method was equally valid, and in Missouri the intersection of lines would be used. In Missouri the only time double-proportion measurement would be used would be for lands that never passed to the state's jurisdiction.

In California the courts have made it abundantly clear that proportional measurement is a rule of last resort (*Weaver* v. *Howatt*, 161 Calif. 77; and *McKenzie* v. *Nichelini*, 43 Calif. App. 194); any surveyor appearing and testifying that he made his location in accordance with topography will probably prevail as opposed to one using proportionate measurement. In agreement with the Missouri theory, the Federal Government rules of evidence do not control what the states will accept as evidence. The state court said in *McKenzie*:

> The rule (proportional measurement) as laid down by the surveyor general of the United States is for the guidance of United States deputy surveyors in running lines in which the government is interested. It cannot control a state court in its choice of means for establishing a fact, to wit, the point where the government surveyor originally placed a section or quarter corner.
> —*McKenzie* v. *Nichelini, 43 Calif. App. 194, 1919.*

Since there are 50 states and the Federal Government, each with final jurisdiction in some circumstances, differences between jurisdictions are bound to occur. It is the responsibility of each surveyor to know the laws of his state.

2-63 Duties of Surveyors in Finding Evidence Needed to Locate Deed Writings

Deeds valid when formed are not made invalid because of a loss of evidence of location; the surveyor must hunt, seek, and find the best available evidence that determines a deed's location on the ground. In those areas where there has been widespread obliteration and loss of evidence, it may become necessary to accept evidence of an inferior type, such as hearsay and reputation, but whatever is accepted, it must be the best of that found after an extensive and complete search of the record, the ground, and adjoiners.

The fact that a *valid* written deed has a location on the ground is not to be disputed; the surveyor is charged with finding the ground location in conformity with the law of evidence.

Ownership of land can be obtained by either of two methods: (1) a lawful written title and (2) an unwritten possession right (called an *unwritten title*) as discussed in Chapter 3. An unwritten right to land ownership can never be transformed into a written right without either a court decree or a written agreement between adjoiners. In title insurance offices, written titles are classified as marketable, and unwritten titles, as nonmarketable. Lending agencies will not loan money on the basis of an unwritten title. In other words, land claimed by some unwritten right to title has a lower market value than land claimed by a written title and possession. For this reason, surveyors are obligated to distinguish between each type of ownership right. Details of unwritten rights are given in Chapter 3, along with further explanations of the surveyors' duties with respect to unwritten rights.

Evidence is an essential part of every property-line survey. For a property surveyor to gain prominence and for him to keep himself free of liability for negligence, he must have an intimate acquaintance with the laws of boundaries and the laws of evidence to prove boundaries.

Surveyors rarely fail to come to an agreement about the distance existing between two known monuments. Disagreements more often result from differences of opinion over which of two known monuments should be used. Perhaps the worst disagreements arise from a failure of one surveyor to uncover all available evidence. Two surveyors having the same evidence, if equally educated and equally intelligent, should come to the same conclusions. Unfortunately, all surveyors are not equally diligent in their search. The one with all the evidence usually comes to the correct conclusion, whereas the one with partial evidence makes faulty locations.

PRINCIPLE. *All surveyors are presumed to know the laws of evidence as pertaining to the location of land boundaries described by writings, and they are charged with the responsibility of knowing how to apply the laws of evidence when they locate deed boundaries.*

It is the responsibility of the surveyor to obtain all the evidence available to make a correct property-line location in accordance with controlling writings; failure to find all the necessary evidence to make a correct location is not an excuse for an incorrect survey.

The surveyor should correctly locate the written title lines and also take into account unwritten title rights as discussed in Chapters 3 and 4. Evidence needed to prove the location of water boundaries was not included in this chapter; it is included in Chapter 9.

REFERENCES

Brown, Curtis M., *Boundary Control and Legal Principles*, 2d ed., John Wiley & Sons, New York, 1969.

"Boundaries," *Corpus Juris Secumdum*, Vol. 11, American Law Book Co., Brooklyn, 1938, p. 535.

Kaplan, J. and Waltz, J. R., *Evidence*, Gilbert Law Summaries, Gardena, Calif., 1979.

Sipe, F. Henry, *Compass Land Surveying*, private printing, 1404 Harrison Ave., Elkins, W. Va., 1979.

Tharp, B. C., "Trees and Land Lines,"*Fifth Texas Surveyors Short Course*, Texas Surveyors Association, Austin, 1952.

Tracy, John Evarts, *Handbook of the Law of Evidence*, Prentice-Hall, Englewood Cliffs, N.J., 1960.

Directions for Magnetic Measurements, Serial No. 166, U.S. Government Printing Office, Washington, D.C., 1947.

CHAPTER 3

Unwritten Transfers of Land Ownership

3-1 Introduction

The Statute of Frauds, as originally established in England, required titles to land to be transferred in writing. As the courts soon found, enforcement of the statute sometimes created a fraud. For this reason, certain exceptions to the law evolved and have been, ever since, applied in England and the United States. Under the exceptions, where an occupant of land is entitled to go to court and obtain a written title, that person is said to have an *unwritten title*. Although the term "unwritten title" may not be the best epitaph for the exceptions to the Statute of Frauds, it has been used sufficiently to warrant its continuation. The term "unwritten title" should not and cannot be equated to a written marketable title; rather, it identifies rights obtained by law.

Although an unwritten right to the ownership of land is in theory good against the written title, it is only so, provided *it can be proved in court and then subsequently adjudicated and then granted.* Since one cannot be certain of the outcome of a court case, especially when parol evidence is required, and the attorney costs, plus court costs, often exceed the value of the land in question, it cannot be said that the value of land claimed by occupancy alone is equal to the value of the same land claimed by both written title and occupancy. One duty of the surveyor should be to differentiate between boundaries claimed by the different modes of rights to ownership. This would include encroachments on the client's written deed and the client's encroachments on an adjoiner's written deed.

An unwritten ownership right can come into being by either of two routes: (1) by a peaceful settlement of a boundary by agreement or acquiescence or (2) by hostile relations (adverse rights). The second method usually involves theft of land, and it is not looked on with favor by the courts, whereas an agreement to fix a disputed boundary often has the opposite effect.

The fact that an unwritten title can extinguish a written title is well established. Land ownership lawfully gained by unwritten means supersedes written title; all the evidence in the world to prove where written titles are located is of no avail against a legal unwritten transfer of title. In the order of importance of elements proving where boundary lines lawfully exist, evidence to prove unwritten conveyances ranks first.

> On the trial a motion was granted striking out the evidence of the surveyor as to the location of the measured line. If there was an agreed line, the location of the measured line was immaterial, and there was no error in striking out the evidence.
> —*Doyle v. Bradshaw*, 183 P 185.

Determining when possession on the ground, which is in disagreement with the written deed, has ripened into an ownership right is complex and may often depend on oral testimony. Few states delegate the land surveyor the authority to administer oaths and take oral testimony, and none delegates the land surveyor the power to compel witnesses to talk. When oral testimony is an indispensable part of determining ownership rights to land, it is often impossible for a surveyor to decide which party has the superior right. The surveyor must first locate the client's written rights. In the event the client encroaches on an adjoiner or the adjoiner encroaches on the client, the surveyor must fully inform the client of the significance of unwritten rights in such a manner as to protect himself from liability as discussed in Chapters 4 and 5.

3-2 General Concept of Unwritten Conveyances or Rights

Five general means by which a person can gain or lose land areas, or rights to land areas, without writings are these:

1. Those involving agreement, either expressed or implied; this would include agreement, practical location, silent recognition and acquiescence, and even the doctrine of estoppel, where both agreement and dishonesty enter.
2. Adverse relationships (hostile) either expressed or implied.
3. Acts of nature (accretions, erosion, and relictions).
4. Statutory proceedings (Chapter 7).
5. Prescription ripening into easements.

Unwritten Agreements. Adjoining owners verbally agreeing on the location of an unknown disputed line may be bound, at law, to the agreement. They are not, in the eyes of the court, transferring land but are merely agreeing on the location of that which is already owned and is uncertain. Technically, a parol agreement is not an unwritten conveyance, since it is merely giving definiteness to that which is already conveyed. Usually, the end effect of a parol agreement is the alteration of the written words of a conveyance; therefore, a parol agreement is often considered as an unwritten conveyance, although, in the eyes of the court, it may not be regarded as such.

In many jurisdictions unwritten agreement must be consummated by occupancy, by practical location, or by recognition and acquiescence. In law literature practical location and recognition and acquiescence are sometimes treated as definite types of unwritten conveyances. A more convenient concept is one that considers these as variations of unwritten agreements.

The best way to settle a dispute is to have the disputing parties come to an agreement. Once adjoining owners peaccably settle their boundary problem by a verbal agreement, the court should not interfere by declaring some other line as being correct. In many instances, where the court can act appropriately without violating the Statute of Frauds, verbal agreements are recognized as being valid. In some jurisdictions the right of the court to recognize verbal agreements is affirmed by statute law.

> The principle is well established, that the owners of adjoining tracts may, by parol agreement, settle and establish permanently a boundary line between their lands, which, when followed by possession according to the line so agreed upon, is binding and conclusive upon them and their grantees. The line is established, not by transfer of title of either to the other, because that can only be done by deed properly executed, but such settlement determines the location of the existing estate of each, and, when followed by possession and occupancy, binds them, not by way of passing title, but as determining the true location of the boundary lines between their lands. Having agreed upon the line, or agreed upon a mode by which it shall be determined, and having accepted and acquiesced in it by the unequivocal act of taking possession according to the line, they and their privies are estopped from afterwards disputing it.
> —*Berghoefer* v. *Frazier*, 150 Ill. 577.

> Where there is no dispute between adjoining landowners as to the true location of the dividing line between their tracts, and there is no doubt as to the superiority of the title of one of the owners where it conflicts with that of the other, a parol agreement between the respective owners establishing a new boundary line in such a position that one is given more land than he previously had, there being no marking or establishment of a distinct line, and no consideration for the cession of the land by one owner to the other, is within the provisions of the Statute of Frauds requiring contract conveying land to be in writing and signed by the party to be charged.
> —*Howard* v. *Howard*, 271 Ky. 773.

Agreement Deed and Unwritten Agreement. An agreement deed is in writing, whereas an unwritten agreement is not. The usual object of an agreement deed is to change the location of that which is known. Legal unwritten agreement fixes in position that which is unknown or uncertain; it may not be used to alter that which is known.

Hostile Relationships. Acquiring land by adverse rights always has the idea of hostilities (either direct or implied) and can be effective only because of the operation of the statute of limitations. It is an actual taking of one's rights and giving them to another. If a person has been lax and has not asserted his rights for a statutory period of time, it is presumed that he has given up his rights of ownership. But the law does not look with favor on such thefts of land and has provided many safeguards for the benefit of the rightful owners. In some states the acquiring of land by adverse occupancy is prohibited by statute law.

Acts of Nature. A person may lose land by erosion or gain land by reliction or accretions. The initial right to follow receding or encroaching waters is defined or implied by a written conveyance, but the actual land area of a person's holdings is increased or decreased without a change in the writings. The subject matter of erosion, accretions, reliction, and other riparian rights is developed in Chapter 9.

Statutory Proceedings. A person may have his land forceably taken from him without his written consent because of his failure to pay taxes or assessments. In some states, the law provides that certain officials (surveyors, processioners, county surveyors, etc.) may establish boundary lines between adjoining owners, and, if the lines are uncontested for a definite period of time, their location may become conclusive or may become *prima facie* evidence. The establishment of boundaries under the Massachusetts Land Court System and under the Torrens system are conclusive after the final decree. The taking away of land for failure to pay taxes or assessments is not discussed, but the establishment of lines by officials is included in Chapter 7.

Prescription Ripening into Easements. Long usage of land does not necessarily ripen into a fee title; at times the court awards an easement. Usage of land for a traveled way or for a pipeline seldom results in a fee right.

3-3 Unwritten Conveyances and Government

In general, although there are exceptions, unwritten conveyances (the gaining of rights in land without writings) cannot operate against the federal, state, or local governments. In most states, irrespective of how long a person occupies a road, occupancy gives no rights against the public.

In Texas most lands were disposed of by the state in sequence by a series of surveys. Needless to say, vacancies between grants did occur. When discovered these vacancies belonged to the state, not the person occupying them, and under Texas law (Article 5421c) others could claim them. In the oil fields vacancy hunters often profited handsomely. One judge humorously referred to a vacancy as "a strip of land completely surrounded by oil and gas wells." New legislation now enables the adjoiner in possession to claim the land, and the discoverer of the vacancy can no longer claim title. This condition also exhibits itself in other of our nonpublic domain states: Tennessee, North Carolina, and Georgia.

Lands held by the state in a private manner (road maintenance yards, etc.) may be subject to the same rules of unwritten rights as are private lands. Lands held in trust for the public, except where a statute declares otherwise, are not subject to alienation by unwritten means.

3-4 Unwritten Title Supersedes Written Title

PRINCIPLE. *Land lawfully gained by unwritten means extinguishes written title rights.*

The fact that an unwritten right can extinguish a written right is well established, but it is not until after a written right has been identified on the ground that it becomes apparent that a unwritten right might exist. What the surveyor should do about unwritten rights is discussed at the end of this chapter.

3-5 Written Title

PRINCIPLE. *Written title must precede unwritten title.*

Although lands in some instances in the past were acquired by squatter sovereignty, now it is always true that before acquiring land by prescriptive means, some form of prior written title must exist. The law with respect to unwritten conveyances operates to enlarge or decrease a person's right but not to give him a right where none existed. Normally an unwritten right does not extinguish all of a person's rights, it merely alters his written title rights.

ESTOPPEL

3-6 Introduction

The doctrine of estoppel, a legal doctrine and not a type of unwritten agreement, is a principle at law that sometimes makes unwritten agreements effective. *Estoppel* is a bar or preclusion to one's alleging or denying a fact because of his own previous actions. Estoppel by deed precludes a party to a deed from denying, to the detriment of the other party, anything written therein. The written words of a deed are conclusive. Estoppel by record precludes denial of a final judgment of a court. Estoppel by conduct may contribute to unwritten transfers of landrights.

3-7 Estoppel by Conduct

PRINCIPLE. *The doctrine of estoppel by conduct is based on the theory that, where a party by his acts, declarations, conduct, admissions, or omissions misleads another so that the other is induced to injure himself to the benefit of the person misleading him, the party causing the wrong must suffer the loss and is estopped from revealing otherwise.*

The doctrine of estoppel is not limited to land boundary disputes; it has many applications in other situations at law. Often the principle of estoppel is sanctioned by statute as in the California CCC, but in most jurisdictions it applies at common law.

> Wherever a party has, by his own declaration, act or omission, intentionally and deliberately led another to believe a particular thing true, and to act upon such belief, he cannot, in any litigation arising out of such declaration, act or omission, be permitted to falsify it.[1]

Four things are essential for the doctrine of estoppel by conduct to become effective in the classical cases, although these have been liberalized in recent thinking:

1. The party being estopped must know the facts.

[1] Sec. 1962, CCP Calif. pt 3.

2. The party being estopped must intend, or at least act so the other has a right to believe that he intended, that his conduct shall be acted on.
3. The person claiming estoppel must be ignorant of the facts.
4. The person claiming estoppel must have relied on the conduct of the person being estopped and must have been damaged.

All the preceding points must be proved conclusively to have the doctrine of estoppel apply. A hypothetical example of estoppel pertaining to land occurs if Jones points out to Smith a line that he knows to be 20 feet in error. Smith, not knowing where the true line is, accepts the line pointed out and grades and paves a drive for his benefit. Jones promptly claims the drive as his own, but he is estopped at law from doing so. If the law were otherwise, it would induce people to falsify to derive benefits.

Estoppel can arise from long acquiescence in a line, from failure to speak when a person ought to, and from inferences. As an example of an actual occurrence, a surveyor made an error of 3 feet in locating a lot line for a branch bank. The bank, relying on the survey, built a structure encroaching on the adjoiner. Although these actions were being conducted by both parties, the neighboring landowner sat on a porch and witnessed all of the events. When the bank was completed, the neighboring landowner approached the bank and talked settlement. The results of the court case rested on the fact that the owner knew of the correct location of the line and permitted the acts. Thus the court held that the owner was estopped from claiming otherwise, for there was reliance on his silence.

A case of acquiescence and agreement by parol occurred in Kentucky as follows.

> Where adjoining owners have established a dividing line by parol agreement, and have executed the agreement by taking actual possession of the parts allotted to them, respectively, up to the agreed line, there is an estoppel to claim otherwise, and the agreement is not within the Statute of Frauds, although in establishing the line each has given up some part of his land to the other.
>
> —*Howard* v. *Howard*, 271 Ky. 773.

UNWRITTEN AGREEMENT

3-8 Unwritten Agreement

PRINCIPLE. *If a boundary line existing between adjoining owners is unknown and in dispute and if the adjoining owners verbally agree on a line and take possession to that line, the boundary agreed on may become the true boundary regardless of the location of the written property line.*

The foregoing principle was stated as "may become" because of the wide variation of opinion about what constitutes an unwritten agreement. The rights of innocent third parties (mortgagee, etc.) are protected, and the agreement is only binding between the parties of the agreement and their privies (successors). The requirements of possession vary widely from state to state. Any general rule as that just cited is merely a statement of what is accepted in most states.

> A verbal agreement, accompanied by possession and improvement, is founded upon good consideration, is not contrary to the Statute of Frauds, and will bind the parties and their privies. The agreements between these parties then did not have the effect of conveying from one to another any portion of the land, but merely of determining the boundaries to the land already owned by them respectively. It follows that conveyances by the parties bound by the agreements and their privies giving the same description under which they obtained title, and held the possession, would pass a title according to the agreed boundaries.
> —*Smith* v. *McCorkle*, 105 Mo. 135.

> Where the true dividing line between two tracts of land is in doubt, and there is a dispute between the adjoining owners as to the exact location of the line, which depends upon variable facts or circumstances not susceptible of certain determination, a parol agreement between the adjoining owners establishing a line as the true dividing line is not an exchange of lands, and is not within the Statute of Frauds.
> —*Howard* v. *Howard*, 271 Ky. 773.

Within the state of New Hampshire, Chapter 472, Section 1 of statutes requires that parties may establish an agreement in the manner given in Sections 2, 3, and 4 of the law and not otherwise. Section 4 says the agreement must be in writing. Further details on unwritten agreements are given next.

3-9 Elements of Unwritten Agreement

Boundary lines agreed upon orally and by contiguous owners may become ownership lines if:

1. The true line as described by the writings is unknown.
2. The property line between the contiguous owners is in dispute.
3. The contiguous owners agree to accept the boundary line they establish.
4. In some jurisdictions, possession or acquiscence follows, and the line agreed on is marked.

Although the first three items are uniformly required in all states, the necessity of possession and the amount and kind of possession vary substantially from state to state. The states, in general, may be divided into three classes: those deeming acquiescence or possession unnecessary, those requiring proof of possession or acquiescence for a period at least equal to that required by the statute of limitations, and those requiring that acquiescence or possession for some time (less than that required by the statute of limitations) must be shown. Case reports of some of the states treat agreement, recognitions and acquiescence, and practical location as one subject, all being necessary to establish an unwritten right, whereas other state reports recognize all three ways as being different means of attaining unwritten titles.

> The rule is that whenever the boundary line between adjoining lands is unascertained or in dispute, it may be established by parol agreement and possession in pursuance thereof, and the line so established will be binding on the parties and their privies. The effect of such agreement is not to pass title by parol, for such cannot be done, but to fix the location of an unascertained or disputed boundary line. There are but two conditions under which the rule in relation to establishment of boundary lines by agreement applies, one is, where the line is in dispute, and the other, where it has not been ascertained.
> —*Wright* v. *Hendricks*, 388 Ill. 431 (Opinion by Justice Stone).

3-10 Described Line Known

PRINCIPLE. *If the location of the line described in the written conveyance is known to the parties or one of the two parties, another line agreed on by parol is not binding.*

The Statute of Frauds states that deeds must be in writing. If the true line is known by the parties, a properly written conveyance, based on a consideration and containing words of conveyancing, must be used.

> The fact that a line is capable of being located by a competent survey does not void an unwritten agreement; the line merely need be unknown to the adjoiners and in dispute.
> —*Aldrich* v. *Brownell*, 45 R.I. 142.

3-11 Property Line in Dispute

PRINCIPLE. *Without a dispute no cause for agreement arises; hence many authorities cite the necessity of a dispute preceding an agreement.*

Agreement by both parties is the essential part of conveying title by unwritten agreement. Without a dispute over a division line, adjoining owners have no cause to arrive at an agreement; for this reason, several states consider that a dispute is a necessary part of an unwritten agreement, but not all jurisdictions maintain that an outward dispute is necesary. In some states an uncertainty only needs to exist in the minds of the adjoiners; the fact that the true line could be located is not material. If there is an overlap of adjoiner's papers, followed by an unwritten agreement, the courts almost always will approve the agreement, but it is not always necessary that there be an overlap.

Numerous authorities have been called to our attention involving disputed boundaries. To settle such disputed boundaries an agreement resting on parol must be entered into fixing the line, followed by the erection of a fence on the agreed line and acquiescence in such line thereafter. But these cases are not applicable here. Not only is there a total absence of evidence of any agreement to fix a line before the original fence was constructed, but plaintiff's own testimony shows that this fence was in existence long before the two pieces were separated and when both were owned by the same person. One of the plaintiff's witnesses testifies unequivocally that the fence has been there "50 years anyway." The line of authorities cited is not applicable unless there has been an actual dispute as to where the line is and an agreement for its adjustment.
—*Wilhelm* v. *Herron*, 211 Mich. 339

3–12 Adjoining Owners Must Agree

PRINCIPLE. *Following a dispute over an unknown line, both adjoiners must agree on a location or a mode of fixing a location.*

To agree on a location, that location must be relative to some monument on the ground: a fence, a tree, a hedge, or some similar thing. Such agreement may be a fixed distance from a given object such as an agreed distance from a surveyor's stake.

If the coterminous owners are attempting to establish the true property lines and they err in establishing the true lines, they are not agreeing to a line but merely establishing what they believe to be the true line. In such cases the owners are not denied the right of claiming to their true lines. A mistake in assuming a line to be the true line does not prevent a person from claiming the true line when discovered. Agreeing to accept a particular line, whether it is correct or not, is the thing necessary.

> The building of a division fence by agreement between adjoining owners, in the mistaken belief that they were placing said fence on the true boundary line between their land, there being no dispute as to the line or contention for a different line, does not conclude one of the parties from insisting upon the true line when ascertained.
> —*Henderson* v. *Dennis*, 177 Ill. 547.

> In an agreement fixing a boundary line under the belief that it is the true line, when in fact it is not, is not binding and may be set aside by either party when the mistake is discovered unless some principle of estoppel prevents.
> —*Bemis* v. *Bradley*, 126 Me. 462.

> A landowner is estopped to assert title to his true boundary, where he and those under whom he claims, have, for over 30 years, recognized, by mistake, another line—claimed as a true one by the adjoining owner—by making partition to it, calling for it in deeds, and pointing it out to others; and the adjoining owners, upon the faith of such recognition, have purchased, cleared, improved up to that line. And

it is not material that the landowners were mutually mistaken as to the true line, and could have discovered it, at any time, by survey.
—*Galbraigh* v. *Lunsford*, 87 Tenn. 89.

The element of time, especially if continued beyond the statute of limitations, may change the general rule as it did in the *Galbraigh* case; recognition and acquiescence became controlling.

If a line is capable of being established by a survey, the parties agree to have the true line established by a survey, and the surveyor errs in the establishment of the true line, the adjoining owners can claim to the true line, since no other line than the true line was agreed on. However, if after the error were discovered there was acquiescence for the statutory time, or if there were title by adverse possession, the survey lines may be binding. Where a line is in dispute and unknown and the adjoiners orally agree on a survey and agree that they will be bound by the results of the survey, they are generally bound by the line run.

> Where the owners of land order a survey to determine the lines, and the surveyor makes and returns an erroneous location, upon which they afterwards act without knowledge of the error, the owners are not bound by the line as surveyed.
> —*Inhabitants of Westey* v. *Sargent*, 38 Me. 315.

In the event that one party misrepresents the lines, or deceives the other, the agreed lines are not binding except on the person misrepresenting the lines. Fraud cannot be the foundation for an agreement. The element of estoppel may enter here if the person being deceived is induced to harm himself.

> If either party is deceived by misrepresentations by the other, whether fraudulent or not, the agreed line is not binding.
> —*Bailey* v. *Jones*, 14 Ga. 384.

Owners of adjoining land, or those having vested interests therein, are the only ones who can establish the disputed line by agreement; innocent third parties are not bound. An agreement between owners, but not including the mortgage holder, is not binding on the mortgagee.

Merely erecting a fence with no intention of having it represent a boundary line does not prevent a person from claiming a portion of land fenced out, for a person is not compelled to fence in all his land.

> In the present case there is no testimony to show that the parties made any agreement about the boundary line, or that such agreement, if made, was executed. Mrs. Coffman's testimony only goes to the extent of showing that she had a survey made and that the adjoining proprietor afterwards recognized its correctness and asked permission to move a house situated on the disputed strip. This testimony falls short of showing that the parties made an oral agreement establishing a boundary line which had been in dispute and that the possession of the disputed tracts was taken by Mrs. Coffman by virtue of such agreement. According to her own testimony she took possession of the disputed tract because the survey which

she caused to be made by the County Surveyor showed that it belonged to the tract of land owned by herself and her husband. The testimony does not establish any agreement between herself and Latham as to the establishment of the boundary, or that she took possession of the strip in controversy under any such agreement.
—*Sherrin* v. *Coffman*, 143 Ark. 8.

How soon an unwritten agreement becomes binding varies somewhat in each state. Many say the agreement is binding the moment the agreement is reached; others say it must be followed by acts of dominion, such as erecting a fence. In Maine two time limits are set. If an unwritten agreement is followed by 20 years' possession, it becomes effective; also, if the *parties to the conveyance*, soon after the conveyance is formed, mutually establish an unknown line (by survey or other means), the line so established becomes binding as indicated in the following case.

A boundary line may, under certain circumstances, be permanently and irrevocably established by parol agreement of adjoining owners, and a line so agreed upon by the parties in interest and occupied to for more than twenty years is conclusive. When the principle of estoppel applies a shorter period may be sufficient. A line established by agreement of parties, at or near the time of making the conveyance, may be conclusive, although the occupation be for less than twenty years, as proving the intent of the parties to the conveyance. An agreement fixing a boundary line under the belief that it is the true line, when in fact it is not, is not binding and may be set aside by either party when the mistake is discovered unless some principle of estoppel prevents.
—*Bemis* v. *Bradley*, 126, Me. 462, Dec. 1927.

In legal writings the mutual designation of a line is frequently called a *practical location* (discussed in the next section). The court in Maine theorized that a practical location at about or near the time of the conveyance indicates what the intentions of the parties were. This is a strange extension of the meaning of the "intentions of the parties," since in most jurisdictions the intent can only be determined from the writings themselves and the circumstances at the time; subsequent events do not count. Application of this extended meaning is not recommended except where the court has specifically approved it.

3-13   Practical Location

DEFINITION. *Practical location by adjoining owners is an actual mutual designation of the location of their dividing line on the ground.*

Practical location is an act of the parties and is in itself not a means of causing an unwritten conveyance, but a practical location, along with certain other elements, may be sufficient for an unwritten conveyance of rights.

The difference between practical location of a line and an unwritten agreement on a line is this: practical location is a mutual designation by both parties about what they believe to be the true division line, whereas unwritten agreement on a line is the agreeing to accept the line designated whether it is the true line or not. Agreement is a contract; practical location is an act that may induce a presumption of an agreement. Agreement may be with respect to a mode of establishing a line, whereas practical location includes the act of establishing a line such as erecting a fence.

Practical location calls for cooperation of both parties; the action of one party without the approval of the other has no binding effect. A survey caused by one party, along with notification to the adjoiner, is insufficient to establish practical location, because practical location is dependent on mutual action and mutual agreement to the survey.

In Pennsylvania the term *consentable line* is used to mean where parties have agreed to a line by words or action such as practical location.

One party may be estopped from claiming he was a part of a practical location if his conduct or behavior has misled the adjoiner to the adjoiner's harm.

If a deed given to the buyer is ambiguous and can be interpreted in two or more ways and if the buyer and seller have mutually erected a common fence, the location of the fence is often considered as proof of the parties' intent and is a practical location.

> Where two tracts of land owned by adjoining owners interfere, the boundary called for by the title papers of one overlapping that of the other, there being a dispute and controversy as to the superiority of the titles, and consequently as to the true dividing line, an agreement between the owners to establish a dividing line between them in pursuance of which such line is established and plainly marked, and thereafter recognized by both parties for a considerable time as to the true line, is not within the Statute of Frauds, and will be upheld.
> —*Howard* v. *Howard*, 271 Ky. 773.

> The doctrine of practical location was originally derived from a long acquiescence by the parties in a line known and understood between them, for such a period of time as to be identical with "time immemorial" or "time out of memory." Practical location of a boundary line to be effectual must be an act of the parties either expressed or implied; and it must be mutual, so that both parties are equally affected by it. It must be definitely and equally known, understood and settled. If unknown, uncertain, or disputed, it cannot be a line practically located. Where land is unimproved and uncultivated, mere running of the line through the woods, exparte, by one of the owners, so long as such line is not settled upon and mutually adopted by the adjoining owners as a division line, is an immaterial fact. In such a case, until the adjoining owner shows his assent to it, it would amount to a mere expression of the individual opinion of the owner who ran the line. To constitute a practical location of a line or a lot requires the mutual act and acquiescences of the parties.
> —*Adams* v. *Warner*, 204 N.Y.S. 613.

3-14 Possession Following Agreement

The amount, kind, and extent of possession necessary in conjunction with an unwritten agreement are exceedingly variable and usually are left up to the courts. In some states possession must follow for the statutory period; in others, an agreement can become binding just as soon as a legal agreement is consummated, regardless of possession. In some states a practical location and an agreement to accept the practical location may be binding.

> Having agreed upon a line, or agreed upon a mode by which it shall be determined, and having accepted and acquiesced in it by the unequivocal act of taking possession according to the line, they and their privies are estopped from afterwards disputing it. The estoppel arises from the act of the parties in taking possession and occupying their respective tracts to the line thus agreed upon and determined. In the case at bar the defendant at once repudiated the line as fixed, and retained possession of the strip of land in controversy. It is therefore clear that there has been no practical location of the line by which the parties are estopped. The agreement of the parties is ... insufficient to authorize recovery.
> —*Berghoefer* v. *Frazier*, 150 Ill. 577.

It is generally agreed that possession along a line for a period equivalent to the statute of limitations brings about an implied agreement between adjoiners; that is, in the absence of proof of agreement and in the absence of contrary proof (nonagreement) mere possession for the duration of the statute of limitations is sufficient proof (an implied proof) of agreement. Courts have said that lines acquiesced in for a long period of time may be better evidence of the original lines of the original surveyor than are measurements from distant points (see Section 2–52).

Acquiescence between the public and an adjoiner is not possible, since no public official is authorized to acquiesce.

Acquiescence in a line without agreement (expressed or implied) does not alter written title lines unless such acquiescence is sufficiently long to give adverse rights. Various states have different periods of time requirements for the establishment of lines by acquiescence, said time being 5, 7, 10, 15, or 20 years.

> An unascertained or disputed boundary line between coterminous proprietors may be established; (1) by oral agreement, if the agreement be accompanied by actual possession to the agreed line, or is otherwise duly executed; or (2) by acquiescence for seven years by the acts or declarations of the owners of adjoining land as provided in section 85-407 of the civil code which states: "Adverse possession for seven years gives title, when—adverse possession of lands, under written evidence of title, for seven years, shall give a like title by prescription; but if such written title is forged or fraudulent, and notice thereof is brought home to the claimant before or at the time of the commencement of his possession no prescription shall be based thereon."
> —*Childers* v. *Dedman*, 157 Ga. 632.

> The building of a division fence upon the assumption by both parties that it is upon the true line, and occupation up to it by both parties for less than 20 years (20 years is the statute of limitations in Wisconsin), do not constitute a binding location of the line, where there was no dispute or uncertainty as to the correct line, no

express agreement that the fence should be regarded as such line, and no circumstances such as should estop either party from insisting upon the true boundary.

—*Peters* v. *Reichenbank*, 114 Wis. 209.

3-15 Recognition and Acquiescence

PRINCIPLE. *Recognition and acquiescence over a period of time, especially for the period of time required by the statute of limitations, may imply an agreement.*

Recognition and acquiescence often imply an agreement, although none may have existed. Recognition and acquiescence depend upon the silence of a party when he should speak, on the declarations of parties, or on the inference to be drawn from the conduct of the parties. If the element of estoppel enters, recognition and acquiescence may be effective.

Essentially, establishment of a line by acquiescence is approved by two theories. First, some courts consider acquiescence as evidence of proof of the true line. Second, other courts consider acquiescence as sufficient to give rise to a presumption of a previous agreement between adjacent owners. Often the theory cited is dependent on the peculiar circumstances of the case being adjudicated.

A long period of acquiescence often raises a presumption that the fence is on the true line and often only by contrary evidence can the presumption be overcome. A fence built for mere convenience and admitted by both parties to be not necessarily on the true line, can never become a true line by acquiescence. A person acquiescing by mistake in a false boundary that he believes to be on the true line is not necessarily denied the right to claim to the true line, although his conduct may estop him from so doing. Acquiescing in a false line is treated as a mistake, and, in the absence of an agreement, estoppel, or possession short of the statutory period, a person is not estopped from claiming the true line.

> It seems now well settled by the law in this state that where two adjoining landowners mutually acquiesce in a line marked by a fence for 10 years [statute of limitations is 10 years in Iowa] as a division line between their properties, that the line thus recognized becomes the true line between the properties. Our last expression on the subject is found in Minear.*v.* Keith Furnace Co. (213 Iowa 663), where it is said: "It is the settled law of this state that a line between adjoining tracts, definitely marked by a fence which has been acquiesced in and recognized by the owners of the tracts as a division line for more than 10 years, becomes, as between the parties, the true line between such tracts, although a subsequent survey may show otherwise, and although neither of the parties intended to claim more than his deed calls for."
>
> —*Mullahey* v. *Serra*, 220 Iowa 1177.

Acquiescence for a long period in a line must be with respect to a monument (fence, hedge, or the like) as a boundary, not mere acquiescence in a barrier known to be off from the true boundary. Acquiescence to a monument as a boundary can be deduced from the statements, conduct, or actions of the parties.

The majority of states now require payment of taxes as a necessary part of the proof for an adverse right claim; in California the statute requires 10 years of continuous tax payment. Whether this statute was a bar to obtaining an unwritten agreement right was in issue in *Duncan v. Peterson* (1970).

> In 1966 the defendants learned by accident that the line of the fence might not truly reflect the north-south center-section line. They had a survey made and confirmed that the true center line was 104 feet east of the fence.
>
> Only two findings were made by the trial court: (1) that all taxes on the subject property had been paid by defendants and their predecessors; and (2) that no taxes had been paid thereon by plaintiffs or their predecessors; also the subject was particularly described by the court.
>
> Section 325 of the Code of Civil Procedure provides, as an element requisite to establishment of title by adverse possession, that the party seeking adjudication of such title prove that they "have paid all the taxes, State, county or municipal, which have been levied and assessed upon such land." Throughout the years, all county taxes on these properties were assessed and paid under assessments designating the lands of each party by section references. Assessments had followed the true section and mid-section lines. That meant defendant had paid the taxes on the subject property. But, of course, the parties had no realization of this until 1966.
>
> Elements NOT necessary are: a written agreement, or an unascertainable true location. Payment of taxes is not material (for an unwritten agreement).
>
> —*Duncan V. Peterson*, 83 Calif. Rpt. 744 (1970).

In California, where there is proof of acquiescence over a long period of time, there is an inference of an agreed boundary.

> Where the boundary was established long before plaintiffs and defendant acquired their adjoining parcels, acquiescence for many years in such boundary line is substantial evidence that the fence was built to settle an uncertainty that existed in the minds of the then owners of the two parcels.
>
> The record is clear that the boundary had been acquiesced in for over 40 years before plaintiffs and defendant took title to their respective parcels of property. In the absence of contrary evidence, we believe the law compels a finding that there was uncertainty as to the true boundary at the time the original grantor and grantee fixed the property lines by an implied agreement.
>
> —*Vella v. Ratto*, App., 95 Calif. Rpt. 72.

3-16 Limitation Title

DEFINITION. *A lawful right to land acquired by possession for the period of time required by the statute of limitations is called a "limitation title."*

In some areas of the United States it has been estimated that 50 percent or more of the titles are limitation titles, whereas, in other areas, the incident of such titles is negligible. Limitation titles can be the result of agreements, recognition and acquiescence, estoppel, or adverse possession.

UNWRITTEN DEDICATION

3-17 Dedication—General

Dedication, the giving of land or rights in land to the public, can be either written or unwritten. For dedication to be effective there must be a voluntary offer, either expressed or implied, by a donor and an acceptance. Many dedications, made by unwritten means, present difficult problems to the surveyor, since it is customarily the surveyor's habit to mark off those portions free of public road dedications. The decision of whether a person has properly made an unwritten offer to dedicate and whether the public has accepted the offer (by usage) is often outside the qualification of the surveyor.

3-18 Dedication and Easement

DEFINITION. *Dedication is the giving of land, either as a fee title or as an easement right, by its owner for public, charitable, semipublic, or utility use.*

An easement is an interest in land created by grant or agreement that confers a right on owners (private or public) to some profit, benefit, dominion, or lawful use of the estate of another.

Dedication is the act of giving a right; an easement is the result of one type of giving; it is a right created. A dedication properly consummated may result in either an easement or a fee title, said easement or fee title being held in trust for a particular use.

3-19 Elements of Common-Law Dedications

PRINCIPLE. *For a common-law dedication to have force the following elements must be present:*

1. *A donor capable of dedicating land or dedicating rights in land.*
2. *An identifiable parcel of land to be dedicated.*
3. *An intention on the part of the donor to dedicate (offer, formal or implied).*
4. *Acceptance of the dedication (formal or implied).*

A consideration is not necessary, since a dedication is a gift, and the advantages gained by public use is sufficient. The existence of an individual grantee capable of taking title is not a necessity, since a dedication is intended for the public or a group as a whole. In most states a dedication cannot be made to an individual; it must be made to the public, semipublic, churches, or groups. A railroad is usually a private corporation and ordinarily cannot be the receiver of a dedication.

3-20 Statutory Dedication

Statutory dedication, unlike common-law dedication, must comply with the letter of the law. Statutory dedication is regulated by written law and usually requires that the dedicator sign and acknowledge the offer of dedication before a designated officer. Often, designated governing-body officers must sign an acceptance, and the map must be filed in a particular place. The requirements for the dimensioning and designating areas to be dedicated must be complied with.

In some states the statute specifically provides that an offer to dedicate land is an offer to pass fee title to the land offered for dedication, and in such event fee title passes. In the absence of a statute specifically granting fee title, an easement is granted.

If there is a failure to comply with all the requirements of the law in a statutory dedication, the possibility of common-law dedication is not precluded, since statutory dedication will not operate to void common-law dedication.

Statutory dedications vary considerably from state to state, and it is not the intent to compile all the state statutory laws here.

3-21 Donor of Dedication

Offers of dedication are only binding on those with the power to alienate property. As in written conveyances, the mortgagee cannot be bound by the acts of the mortgagor. Land owned by several owners can only be dedicated by all. A person who has sold part of his land or who has contracted to sell part may not dedicate any part sold. The administrator cannot dedicate the estate of others. Corporations may dedicate by proper resolutions of those empowered to act.

3-22 Location and Description of Dedication

The land being dedicated must be capable of location; otherwise, the dedication would be void from uncertainty. At common law the location can sometimes be inferred from the acts of the parties—that is, a practical location—and need not be an unambiguous, written document. It is only an intent to dedicate that makes a common-law offer of dedication valid; therefore, the land being dedicated need not be described in writing but must be identifiable. A statutory dedication must be described with as much detail as is required for other written conveyances.

3-23 Expressed Intent to Dedicate

At common law it is only necessary for the owner to intend a dedication, and that intent may be by words, acts, or conduct. It is not necessary that the intent be in writing as is required in statutory dedications. A dedication is a gift, and for this reason the offer to dedicate must be clear, although the offer to dedicate may be inferred from conduct.

Since it is only the intent to dedicate that makes an offer valid, it is not necessary that there be a formal offer. The offer of intent to dedicate may emanate from an oral statement, a writing, or a single or series of acts. No definite rule exists about what constitutes intent to dedicate.

The act of platting land into blocks and streets presents evidence of intent by inference, but it does not in itself create a right of the public in the streets or alleys. Acceptance must follow. On the selling of any one lot in the plat, a private easement is created for the benefit of that and all lot owners.

The writing of the word "wharf" on a plat is not proof of intent to dedicate, nor is the word "depot." A wharf is usually subject to private ownership, and the word "wharf" merely signifies the existence or intent to build a wharf. It has been held an intent to dedicate, and it has also been held as no intent to dedicate where the word "beach" has been written in a blank space adjoining water.

Mere permissive use is not an intent to dedicate. Platting a river shows no intent to dedicate the river for public pier or wharf rights.

3-24 Implied Intent to Dedicate

Long continued acquiescence by the owner may eliminate the necessity of direct expression of intent to dedicate, since intent to dedicate can be inferred from the owner's failure to exercise his rights. Mere usage with the consent of the owner does not give the public a right to infer an offer of dedication; the usage must be incompatible with private usage. Setting a fence back to allow a street passageway has been interpreted as an intent to dedicate.

To imply a dedication based on usage, it is usually necessary that the public use the land for a period of time required by the statute of limitations (5 years in California). One individual crossing another's land does not constitute a public right, although it may ripen into a private right. The public is more than one.

3-25 Acceptance of Dedication

In any dedication a competent person must offer or express an intent to offer a dedication, and there must be an acceptance of the offer of dedication. Acceptance, unlike the offer, does not have to be by a particular grantee. The offer of dedication is to the public, and there are many ways the public can accept a dedication. Acceptance can be formal by a governing body; it can be by the action of maintenance crews, or it can be by usage by the public.

A public body that causes a subdivision (land platted into lots, blocks, and streets) and sells land in accordance with the plat makes an offer and acceptance of dedication on recording of the plat.

A dedication by deed is effective on recording the deed, since the act of recording is construed to be the act of delivery.

Acceptance of dedication has been construed from such acts as installing public improvements, maintaining a street by city crews, taking possession, appropriating money for streets, issuance of bonds on a street, resolutions of governing bodies, accepting benefits from the state legislature.

Platting of land into lots, blocks, and streets and recording the same thing does not make it necessary for the city to maintain the streets, since no formal acceptance is made by the city. Liability and maintenance follow acceptance. Formal acceptance must be by those empowered to do so by statute law. The tax collector, police department, and the like, cannot accept a dedication.

3-26 Revoking Offer

An offer to dedicate is good until revoked. Prior to acceptance, an offer may be withdrawn or revoked by the person or persons making the offer, but a completed dedication with offer and acceptance is irrevocable. Death of the person offering dedication automatically revokes his offer. Generally speaking, only the person offering dedication may revoke the offer, but once the dedication is consummated it may become irrevocable.

3-27 Purpose of Dedication

Dedication is a mere gift to the public, and the donor may impose any restrictions as he sees fit, and the land cannot be applied to any other use. Land donated for a street cannot be used for another purpose such as for buildings. Land dedicated for park or commons purposes precludes, in some cases, the erection of buildings inconsistent with park purposes. A state dedicating land for park purposes is not precluded from granting the land's use for other purposes. The purpose of a dedication cannot be inconsistent with the city's right of police power or its rights to supervise improvements.

3-28 Effect of Dedication

At common law, the dedication of streets normally gives only usage of the land for the intended purpose or related purposes. Fee title is reserved in the owner, and the owner may make any usage not inconsistent with the rights of the public in the easement. The fee owner reserves all rights not inconsistent with the purpose of the dedication. But in the matter of dedicating land for churches, schools, and the like, a fee title usually passes. The public can accept a dedication, but it cannot set up the purpose for which it will be used.

Statutory dedication operates by way of a grant, and, if the statute so specifies, a fee title will pass. In the absence of a statute stating that a fee title passes, only an easement in streets and alleys is granted, but in parks a fee title passes.

Common-law dedications operate by way of estoppel, whereas a statutory dedication operates by way of a grant. Dedication can only be made to public or charitable use; dedications cannot be made for an individual. All that a common-law dedication needs is room for estoppel to work. If usage is sufficiently long that the public rights would be affected by an interruption, estoppel will enter.

After a dedication is accepted every member of the public has an interest in the thing dedicated. Dedication is not an individual interest but is an interest in common with the public generally.

In statutory dedication the street width is that which is dedicated and includes any part not used. Gaining a highway by prescriptive means includes only the land used and fenced out.

3-29 Dedication by Plat

A plat made showing lots, blocks, and streets, and recorded, has an intent to dedicate all streets and alleys shown thereon. Each lot sold gives the purchaser a right to use streets and alleys shown. Usage by the public for a period of time equivalent to the statute of limitations gives public acceptance. Revoking of the offer to dedicate can only be done by all owners of lots. Land burdened by private easements and condemned by the public is entitled to only nominal damages.

> It was abundantly proven that the village accepted a number of streets and alleys in both the original village and additions thereto. But it is insisted that an acceptance of some of the streets and alleys of a plat does not constitute an acceptance of the whole. We ... hold the true rule and doctrine to be, that an acceptance by the city or village of some of the streets or alleys appearing on a plat is an acceptance of the entire system of streets and alleys.
>
> —*Village of Lee* v. *Harris*, 206 Ill. 428.

3-30 Prescription

DEFINITION. *Prescription is a term applied to the method of obtaining easement rights from long usage.*

Adverse rights, agreement, and other means of obtaining land result in a fee title, whereas prescriptive usage ordinarily conveys only a right of usage. A person traveling across a parcel of land for the period of time required by the statute of limitations may acquire an easement right to continue. *Dedications* are rights acquired by the public; *prescription* is the means of obtaining dedication or obtaining private easement rights.

Encroachment of structures, eaves, power lines, underground utilities, water lines, and like things may ripen into a permanent right of usage. But once the encroachment is removed the right of usage is sometimes extinguished. Usage rights such as placing a second water line alongside of a first line generally cannot be enlarged from time to time.

The right of continuing usage as an easement by the public or by quasi-public agencies is readily recognized by the courts, especially if the usage extends over the period of time required by the statute of limitations. Easements of necessity—that is, rights necessary for the public or, in some instances, an individual—are more readily obtained by usage.

3-31 Adverse Possession

DEFINITION. *Adverse possession is a method of acquisition of title by possession for a statutory period under certain conditions.*

Sometimes the distinction between lands gained by adverse rights and lands acquired by the process of recognition and acquiescence may not be readily discernible, since each might include parts of the other. The one feature that distinguishes adverse rights from other unwritten methods of obtaining ownership is that the possessor must maintain that his possession is his ownership line irrespective of his written title; adverse rights have the element of hostility; they do not rest on oral claims; they are adverse to the rightful owner.

The doctrine of adverse possession merely maintains a status quo. It is certainly not for the merit of the adverse occupant, since the taking of another's land has no merit. It is probably a penalty against the true owner, since he has failed to act on his rights within the limitations imposed by law.

3-32 Historical Concepts of Adverse Possession

The doctrine of adverse possession is a concept that evolved both through the Roman Common Law and the English Common Law. Under Roman law individuals who went into possession of land believed a part of the spirit of that person was transmitted into the parcel of land itself. By virtue of this phenomenon the person became part of the land itself and by law became owner of the occupied land. On the other hand, the English Common Law gave title to land superior control over occupancy. A merging of these two historical concepts gives the doctrine that allows an individual to occupy a parcel of land under certain adverse conditions, and to obtain a new written title in accordance with the statutes of a given jurisdiction.

As mentioned, in the United States, where Torrens titles exist, adverse rights cannot be obtained. In Australia, where the system was invented, the law has been changed to permit possession rights under some conditions.

3-33 Statutory Character of Adverse Actions

The means to acquire lands by adverse rights is usually defined by statutes, although on occasion common law comes into action. If a statute exists, it must be strictly complied with.

Often two means of court actions to acquire title by adverse rights are provided; one is a long possession duration (about 20 years), and the other is a short possession duration (about 7 to 10 years). For the shorter time requirement the added elements, color of title and payment of taxes, are usually imposed.

In some states, especially where title registration is in effect, land cannot be acquired by adverse rights. In those states where part of the titles are registered by a Torrens system and part are not, adverse rights against the Torrens titles, but not other titles, are usually barred.

3-34 Burden of Proof

PRINCIPLE. *The adverse claimant is charged with the burden of proof; if he fails to prove compliance with all adverse rights conditions, the rightful owner is presumed to have never been ousted.*

In adverse rights actions to acquire paper title, the burden of proof rests entirely on the person trying to prove perfection of title. Presumptions, except in the matter of good faith, are in favor of the written title owner. Failing to prove conclusive compliance with statutes or other requirements of the law will not remove title from the rightful owner.

In those states where good faith and color of title are required, it is presumed that a person buying paper title, although a defective one, entered in good faith. Hence, where good faith is required, the burden of proof for this item is reversed; the rightful owner must prove bad faith.

3-35 Character of Title Acquired

PRINCIPLE. *Land lawfully gained by the process of adverse possession extinguishes the rightful owner's written title and vests it in the former adverse possessor.*

Once a title is perfected by adverse rights, such title is an absolute perfect title in fee simple dating back to the time of its inception (original possession). It cannot be lost by survey, subsequent litigation, acknowledgment of title in another, or by mortgage or liens of the paper title owner. It is a legal title, although not a record title, that is good against all other claims. Title, once acquired by adverse rights, may only be removed by conveyance, another adverse possession ripening into a fee, unwritten agreement, or other lawful means.

> Where the evidence showed that the cross-complainants and their grantors had held possession, apparently as the owners of the land since 1840, and that one of such grantors between 1874 and 1880 said to a witness that he did not claim such real estate, such statement could not operate to defeat such person's title, since the prescriptive period had already run prior to the making of such statement and the title thereto had vested.
>
> —*Rennert* v. *Shirk*, 163 Ind. 542

Theoretically, an adverse title is good against all, provided that it can be proved at law. An adverse title to be marketable must be adjudicated at law through the courts. Title companies will not insure unwritten rights (it is what is known as a *nonmarketable title*), and, in general, loaning agencies will not loan money on unwritten titles, mainly because of the possibility of costly litigation to clear title. Furthermore, a knowledgeable buyer would be unwilling to pay full value for land that is claimed merely on the basis of possession; buyers do not want to purchase a possible litigation. Although it can be stated that an unwritten title which has ripened into a fee right is good against all, it cannot be said that land whose ownership is based on occupancy is of equal value to land whose ownership is based on both writings and possession.

3-36 Prescriptive Title

Generally, title by prescription is construed to be an easement right as opposed to a fee right by adverse possession. A building erected on another's land by mistake, with no intention of taking title to anything other than the true line, may not confer a fee title, but the right of prescriptive use (easement) may be perfected. Water dripping from eaves and water flooding an upland owner (caused by a dam) have been held to be prescriptive usage rights. Failure to prove adverse rights does not preclude the possibility of an easement right.

Prescriptive rights usually arise from the statute of limitations; if a rightful owner cannot show that he has had possession within a given period of time (3 to 20 years, depending on the state statute), he is barred from bringing action for recovery. Although the possessor may not have fulfilled sufficient statutory requirements to bring suit to acquire written title, he may rely on the statute of limitations to keep from being ousted.

3-37 Effect of Survey on Adverse Rights

PRINCIPLE. *A survey, although it may reestablish original property lines, does not create or revive rights lost by adverse possession.*

Although a survey may relocate the original lines of the original surveyor or it may properly and correctly monument the property lines as described in the written conveyance, it can never revive the rights to land lost be adverse possession.

> A survey establishing a line between adjacent owners will not revive the right of an original owner against an established boundary (by adverse rights) since all that the survey does is to establish the line and not the title.
> —*Grell* v. *Ganser*, 255 Wis. 384.

> Adverse possession may change the title to real property, but it cannot change the location of the quarter section line.
> —*Scott* v. *Williams*, 83 Kan. 448.

3-38 Against Whom Adverse Rights Do Not run

PRINCIPLE. *Unless a statute provides otherwise, adverse possession does not run against the United States, the state, county, or city and usually not against quasi-public agencies, infants, and remaindermen.*

The statute of limitations for adverse possession does not operate against the United States; and, unless there is a state statute to the contrary, the same is true for the states. Even in states where statutory provisions allow adverse rights to run against the states, such adverse rights seldom run against lands held in trust for the people. It is usually only lands held in a private manner, such as a road maintenance yard, that may be lost by adverse occupancy.

> By the right of eminent domain, the United States is the absolute and exclusive owner of all public lands which it has not alienated or appropriated. No adverse possession is created by an entry upon its lands.
> —*Cook* v. *Corbin*, 7 Ill. 652.

> The common law maxim is "once a highway, always a highway." Highways belong to the public, and are under the control of the sovereign, either immediately or through the local governmental instrumentalities. The right of the public to the use of the highways is not barred by the statute of limitations. No one can acquire a right to the adverse use of a legally established highway by user, no matter how long such use continues. There can be no such thing as a permanent, rightful, private possession of a public street.
> —*Wolfe* v. *Sullivan*, 133 Ind. 331.

> The better opinion would seem to be that municipal corporations, in all matters involving mere private rights, are subject to limitation laws to the same extent as private individuals.
> Testing the case before us, the county had a perfect right to sell or otherwise dispose of the land at pleasure; the limitation law may run against it.
> —*County of Piatt* v. *Goodell*, 97 Ill. 84.

> But where the road has been established and continually used, the mere fact that the fences bordering it are not on the true line and the portion beyond has been occupied by the land owner up to the fence and not been made use of by the public will not work an estoppel against the public, but the entire width of the highway may be appropriated by the public whenever required for the purposes of travel. The doctrine of acquiescence can have no application to the fixing of a boundary between the abutting owners and the highway, for no one representing the public is authorized to enter into an agreement upon, or to acquiesce in any particular location. The fee to street is in the town or city, but always in trust for the public. The municipality can neither sell nor convey nor authorize their use for private uses.
> —*Quinn* v. *Baage*, 138 Iowa 426.

In Massachusetts an old statute law declared that possession of a road for 40 years or more would give an adverse right. A later statute changed this because of the apparent difficulties the public had in maintaining proper road widths.

> The maintenance for 40 years of a fence along a highway justifies, under general statutes Ch. 46, Par. 1, its continuance; the public officers who wrongfully undertake to remove it and use enclosed land as part of the highway, may be restrained by injunction.
> —*Winslow* v. *Nayson*, 113 Mass. 411, 1873.

Massachusetts Statutes, Chapter 86, Section 2 now provides:

> If the boundaries of a public way are known or can be made certain by records or monuments, no length of possession, or occupancy of land within the limits thereof, by the owner or occupant of adjoining land shall give him any title thereto, unless it has been acquired prior to May 26, 1917.

Adverse rights may in some states operate against land held by a railroad; in others, common laws or statutes prevent such. Often an infant on reaching majority may oust an adverse claimant, since he was not competent to sue until he did reach majority. The remainderman usually may not sue as long as the person owning a life estate is alive; hence, after the death of the owner of the life estate, the remainderman may have the right to oust adverse possessors. There is difference of opinion about whether adverse rights may operate against school lands; often adverse rights may not.

> No person shall acquire by prescription a right to any part of a town house, schoolhouse or church lot, or of any public ground, by fencing or otherwise enclosing the same or in any way occupying it adversely for any length of time.
> —New Hampshire Statute, 477:34.

Torrens registration acts provide by statute that registered land may not be attacked by adverse possession. An adverse title perfected prior to registry would be effective, but the Torrens proceedings are essentially quiet title actions, and failure to assert rights during the proceedings could cause loss of adversely held land, especially if title location was part of registration (see Chapter 9).

In recent years many governmental bodies are reexamining the concept of adverse possession. It has been an accepted principle that land which was once in private ownership and subsequently became under government ownership, if adverse possession had matured against the government's predecessor, no title or rights were obtained by the government, for they had already become lost. Recently, Congress enacted Public Law 92-562 that permits a private party to bring an action at law over property boundaries and surveys but excluding any claim of adverse possession.

3-39 Elements of Adverse Rights

An adverse claim, where permitted, will ripen into a fee when the following acts are complied with continuously and simultaneously for the period of years defined by law. But it is not to be assumed that all 10 are required for all states; there is much variation in the laws.

1. Actual possession.
2. Open and notorious possession.
3. Claim of title.
4. Continuous possession.
5. Hostile or adverse possession.
6. Exclusive possession.
7. Possession as long as required by statute.
8. In some states under color of title.
9. In some states all taxes must be paid.
10. In some states under good faith.

Since there is no merit in awarding the adverse claimant land filched by legal means, the courts usually insist on strict adherence to the requirements of a given state. Although the statutes of most states spell out the length of time necessary for occupancy to ripen into a fee, none completely stipulates the details of the listed requirements. The statutes are part of the common law of the state.

> The real test as to whether or not possession of real estate beyond the true boundary line will be held adverse is the intentions with which the parties take and hold the posession. It is not merely the existence of a mistake, but the presence or absence of the requisite intentions to claim title; this fixes the character of the entry and determines whether the possession is adverse. . . . Where a fence is believed to be the true boundary and the claim of ownership is up to the fence as located, if the intent to claim title exists only on the condition that the fence is on the true line, the intention is not absolute, but conditional, and the possession is not adverse. If however, in such a case, there is a clear intention to claim the land up to the fence, whether it be the correct boundary or not, the possession will be held adverse.
> —*Edwards* v. *Fleming*, 83 Kan. 653.

3-40 Actual Possession

To acquire land by adverse occupancy, actual possession by some act such as fencing, cultivating, farming, improving, or the like, is a fundamental necessity. Actual possession is not an argument or contest over possession. The possession cannot be such that the true owner would not be aware of it on occasional visits. The facts necessary to prove actual possession may vary, depending on the circumstances.

An enclosure, excluding others from entrance, is more conclusive evidence of actual possession; and, in some jurisdictions, an enclosure is required by statute. For possession to be effective as evidence, it needs to be so complete, open, and notorious as to charge the owner with constructive notice. The enclosure can be a natural or man-made barrier that is definite and certain of location. Fences in general must be substantial and erected for the purpose of setting off the boundaries of the claimant and in some jurisdictions must meet legal requirements for sufficiency and legality.

> A fence may be sufficient as constituting visible evidence of possession even though not in good condition.
> —*Shedd* v. *Alexander*, 270 Ill. 116.

Ordinarily, cutting grass or other naturally growing crops (logging trees) may not be sufficient to show adverse possession; however, if the grass or growth of trees is the best-suited purpose of the land, these facts may have weight as evidence.

Cultivation, if intermittent, is not actual possession; but continuous cultivation up to a definite line, especially a barrier, has been accepted as one of the requisites of adverse rights.

Proof of actual possession may be entry denial to the rightful owner or entry denial to the owner's agent or vendee.

Surveying land and the establishment of monuments does not constitute possession. Erecting a fence, planting a hedge, or cultivating to the line of a survey are considered acts of possession. A survey serves to fix bounds and, along with acts of possession, may fix the limits of a claim.

> The survey of unenclosed land, and the placing of stones at the boundary corners, is not such a taking of possession as is contemplated by section 7 of the limitations act.
> —*White* v. *Harris*, 206 Ill. 584.

> A mere intention to occupy land, however openly proclaimed, is not possession. The intention must be carried into actual execution by such open, unequivocal and notorious acts of dominion as plainly indicates to the public that the person who performs them has appropriated the land and claims exclusive dominion over it. Anything short of this is not what the law denominates possession.
> —*Brumagim* v. *Bradshaw*, 39 Calif. 24.

> By actual possession is meant a subjection to the will and dominion of the claimant, and is usually evidenced by occupation—by a substantial enclosure—by cultivation, or by appropriate use, according to the particular locality and quality of the property.
> —*Coryell* v. *Cain*, 16 Calif. 567.

3-41 Open and Notorious Possession

Possession must be open and notorious such that the true owner by occasional visits can observe the acts of possession. Possession must be adverse; it cannot be secret. The occupancy must be notorious in such a way that the immediate public is aware of it and that it must be presumed that the true owner is aware of its adverse nature.

The law does not require a person to act to protect his holdings until he is aware or ought to be aware he needs to act; hence possession must be open and notorious by such acts as fencing, erecting buildings, or farming.

Evidence used to prove open and notorious possession may be acts of possession, construction of buildings, visible use and occupation, mailbox with name on it, payment of taxes, fencing, erecting "no trespassing" signs, and cultivation.

> Surveying and marking of boundaries, payment of taxes, and occasional entries for purpose of cutting timber are not sufficient to constitute adverse possession.
> —*Griffith Lumber Co.* v. *Kirk*, 228 Ky. 310.

> Unloading lumber from cars and piling it upon an unenclosed portion of a railroad right of way without objection by the railroad company, cutting grass, tying horses out to graze, and similar acts, do not constitute adverse possession of the premise, particularly where no notice was ever given to the company that any claim of right was being asserted.
> —*Illinois Central Railroad Co.* v. *Hasenwinkle*, 232 Ill. 224.

> The property was in such an obscure place that the city's manifestations of taking possession were not apt to be observed by the owner of the land or by the public.
> —*Carrere* v. *New Orleans*, 162 La. 979.

3-42 Claim of Title

Title to land cannot be acquired by a squatter. The initial right to land must be in writing, but after the title is acquired land may be added to or subtracted from by adverse rights. The claim of title may be defective and have mere color of title, but, nonetheless, there must be some claim of title.

Like so many of the other elements of adverse possession, what is necessary for a claim of title varies from state to state. In Maine (*Preble* v. *Maine Central R.R.*, 85 Me.260) the intention of the possessor to claim adversely is necessary. If a party, by mistake occupies up to a line that is believed to be the true line, adverse occupancy does not occur. In Connecticut (the majority rule) the mistake is of no importance; the fact that the occupancy occurred *in a mistaken belief* about the true location of the boundary line is immaterial. In Maine it is necessary to occupy land with the intention of stealing it; in Connecticut the state of mind of the occupant is not of importance. (See *Lincoln* v. *Edgecomb*, 31 Me. 345). The unfortunate result of the Maine decision is that it benefits the wrongdoer more than the honest landowner.

3-43 Continuous Possession

Possession must be continuous for the period required by statute, for any interruption causes the law to operate to restore possession to the true owner. It is presumed by law that possession resides in the true owner; the moment possession is not in the hands of another, it is in the hands of the true owner.

Interruption occurs when there is a break in the continuity of possession. No matter how short the period of interruption, if there is an interruption, the limitation period required by statute must start over. Any admittance on the part of the adverse possessor that he does not have title to the land he is occupying, ceases hostilities, causes an interruption, and stops the statutory period. Abandonment of the premises for a short period causes interruption. Occupancy temporarily stopped by flood or fire may be renewed after a reasonable period of time. Short periods of vacancy with intent to return does not cause interruption. Renting or leasing adjoining land ceases hostilities between the adjoiners and causes an interruption in possession.

> The fact that adverse possession of land is interrupted after it is begun is not material if it is continued for twenty successive years after the interruption.
> —*Shedd* v. *Alexander*, 270 Ill. 116.

> Title by adverse possession can only be acquired by the actual holding and enjoyment of land under a claim of right which is opposed to and inconsistent with any other claim. No possession of land is sufficient to ripen into title unless the holding has been such as to furnish the plaintiff a cause of action for the recovery of the lands every day during the 15 year period. Possession must not only be actual but it must be open, notorious, continuous, adverse and peaceful for every hour of every day of the whole 15 years. It is, therefore, incumbent upon one asserting title by adverse possession to show affirmatively continuity of possession, and failing in this his cause must fall. In this case there are several periods of a year or less in

which Sackett and his predecessors did not have actual possession of the land in controversy, or any part thereof, during the period of 15 years upon which he relies for the perfection of his title.

—*Sackett* v. *Jeffries*, 182 Ky. 696.

"Tacking" on of possession by one adverse possessor to another is permitted if there is a privity (father and son, grandfather and grandson, etc.) between the parties. Passing title by inheritance does not cause an interruption, unless it is followed by abandonment. In Wisconsin tacking was limited in the case of *Ablord* v. *Fitzgerald* where the deed read as follows:

> ... beginning 40 rods north of the east $\frac{1}{4}$ corner stake; running thence north 40 rods; thence west 80 rods; thence south 40 rods; thence east 80 rods.
>
> —*Ablord* v. *Fitzgerald*, 87 Wis. 517.

A survey revealed about one acre existing between the north line of this description and a 30-year-old fence. The fence marked the line of actual, visible, and exclusive possession. The possessor of the one acre claimed adverse rights by tacking on to the possession of his predecessor, but the court ruled that there was an interruption. At the time of conveying title the predecessor conveyed in accordance with the above deed without mentioning the one acre he had occupied adversely. A person may sell part of his land, and in this case the court ruled that only the part described was sold. Since the one acre was not sold, adverse occupancy had to start as of the date of the deed. Because this was less than 20 years, title did not pass. This quotation is from the case:

> It seems that as to land not covered by a deed a grantee can claim no right founded upon the adverse possession of his grantor.
>
> —*Ablord* v. *Fitzgerald*, 87 Wis. 517.

The state (Wyoming) also claims that it acquired title by 10 years of adverse possession in accordance with the rule laid down in the *City of Rock Spring* v. *Sturn*, 29 Wyo. 494, 273 P. 908. This contention cannot be sustained. The adverse claim, if it was ever made, did not continue for the requisite period of 10 years. It ceased when the state, through its agent, discontinued to cultivate its land beyond the true boundary, and acknowledged the line established by the U.S. Government survey as the true line. Even though the adverse claim commenced to run in 1919, it was interrupted by the state's acts and acknowledgement, and could not, during the interrruption, ripen into a title. While the state can acquire title to adverse possession, no good reason has been pointed out why it should not be subject to the same rules incident thereto which apply to the case of private individuals.

—*State* v. *Vanderkoppel*, 19 P 2d 955.

Some states disagree with the rule in Wisconsin just referred to, and in other states the Mother Hubbard clause has been used extensively. A deed with a Mother Hubbard clause reads like this: Lot 1 and *all of the adjoining lands possessed by the seller.* In Alabama it is unnecessary to use the Mother

Hubbard clause, since the courts have ruled that tacking on can occur even though the area is not specifically described as in this case:

> For the purpose of effecting title by adverse possession, where all the traditional elements are present, tacking of periods of possession by successive possessors is permitted against the co-terminous owner seeking to defeat such title, unless there is a finding, supported by the evidence, that the claimant's predecessor in title did not intend to convey the disputed strip. *We hold that this rule should apply even though the conveying instrument contains no legal description of the property in question, and irrespective of the period for which the property was possessed by the present claimant's predecessor in title.*
> —*Watson* v. *Price*, 356 S. 2d 625, 1978.

3-44 Hostile Possession

Hostile possession must, in most states, be against the interests of the fee owner without admittance by the adverse claimant that the land is not his. Hostile possession does not imply ill will but does present a claim of land to the exclusion of all others. Permissive entry is not hostile entry. As long as the occupant is possessing with the permission of the owner, adverse possession is not operative. The relation of landlord and tenant is not hostile; renting adjoining land stops hostilities and brings about an interruption.

Although hostile possession is almost uniformly required, differences of opinion exist about what is meant by "hostile." Occasionally, mere recognition and acquiescence for the statutory period is defined by statute as sufficient. An "intent to take title" and an "intent to deliberately take land (theft)" are some of the variations in interpretations.

By most authorities, hostility does not need to be present at the inception of possession, but it most be present for a statutory period that may commence either at or after inception.

> Where one of two adjacent landowners extends his fence so as to embrace within his enclosure lands belonging to the other, but in ignorance of the true boundary line between them, and with no intentions of claiming such extended area, such possession of the land so enclosed is not adverse or hostile to the true owner; but if the fence so extended is believed to be the true boundary line, and he claims ownership to the fence, even though the established division is erroneous, such possession will be adverse and hostile to the owner.
> —*Hess* v. *Rudder*, 117 Ala. 535.

> The rule is that there must not only be 20 years continuous uninterrupted possession, but such possession must be hostile, in its inception and so continued. It must be visible, exclusive and notorious, and be acquired and retained under claim of title inconsistent with that of the true owner. All of these elements must concur. (See section 5-12, 270, Ill. 116).
> —*Clarke* v. *Clarke*, 349 Ill. 642.

> The essence of adverse possession is that the holder claim the right to his possession, not under, but in opposition to the title to which his possession is alleged to be adverse.
> —*Mills* v. *Laing*, 38 Calif. App. 776.

3–45 Exclusive Possession

Possession cannot be shared with the rightful owner, since the rightful owner would never be deprived of possession. A common entryway to two stores does not deprive an owner of possession.

Exclusive possession can be proved by exclusion of the legal owner, by threat of force or legal action, by declarations of the possessor to hold land exclusively, or by refusal to permit the legal owners or his agents to enter the land.

> The rule is, where the entry is made with the consent of the owner, and subservient to his claim of title, the law will presume that the continual possession is subordinate to the title of the true owner. The element of hostility to the title of the true owner is an indispensable ingredient of adverse possession.
> —*Timmons v. Kidwell*, 138 Ill. 17.

3–46 Statute of Limitations of Possession

Possession must extend the length of time required by statute. A person may oust another 1 day prior to fulfilling the statute of limitations. Originally, the common law of England required that possession extend beyond memory of man. Over a period of time this was gradually shortened to mean 20 years and sometimes less. Practically all states now have a statute of limitations, with the time period varying from 7 to 20 years. Often within the same state the period of time will vary, depending on the circumstances. In California the period is either 10 or 20 years, dependent on who pays taxes. Those with color of title sometimes need lesser periods of adverse occupancy. An entry based on a judgment or tax lien may require a shorter period of occupancy.

In a few states, under special circumstances, adverse rights can run against the state, and, in such cases, the statute of limitations is usually much longer than 20 years.

In the state of Illinois the period of time for possession is either 7 years by Chapter 83, Section 6, or 20 years by Chapter 83, Section 1.

> To establish title by limitation under Section 6 of the law of 1839 three things are necessary; first, color of title obtained in good faith; second, payment of taxes for the full period of seven years by the holder of such color of title or by some one acting for him; third, continuous and uninterrupted possession for the full period of seven years. These three conditions must exist concurrently, without interruption, and must continue throughout the same seven years. An abstract of title is not admissible in evidence unless proper preliminary proof is made that the original deed or conveyance is lost or destroyed or beyond the power of the applicant to produce it.
> —*Glos v. Wheeler*, 229 Ill. 272.

> The county, being the owner of land which it may sell and convey without a breach of duty, holds the same as an individual, and its title "may be defeated by possession in payment of taxes under color of title made in good faith, for a period of seven years, in the same manner as if the land belonged to an individual.
> —*Hammond v. Shepard*, 186 Ill. 242.

His title is not based upon the seven-years Statute of Limitations which requires successive years of payment of taxes and color of title, but is based upon the twenty-years Statute of Limitations, which may ripen into title without the payment of taxes at all. No deed is required to the inception, the continuance or the completion of the bar [statute of limitations].

—*Scales* v. *Mitchell*, 406 Ill. 130.

3-47   Color of Title

Color of title is said to exist when by appearance a written conveyance seems to be good, but in actuality it is not good. Any written instrument, regular on its face and purporting to convey title to described land, is usually sufficient to establish color of title. Some states, by statute, require that color of title is necessary to obtain an adverse title.

The record of a survey of lands, does not, in itself, constitute color of title to such lands; but it may be evidence tending to show claim of title.

—*Atkinson* v. *Patterson*, 46 Vt. 750.

In those states requiring color of title, the conveyance purporting to pass title is regulated by the same laws as for written deeds. That is, the color of title, though not a valid writing, must describe a locatable particular area of land.

... the description is as follows: "Part and parcel of the east half of the northeast quarter of section No. 17, township 40 north, range 14 east of the third principal meridian, being 2½ acres off the north end of the ten acres conveyed to...." In neither of these deeds [other description was not included] can the particular two and one-half acres of land intended to be conveyed be ascertained or located from the face of the deed.... Those instruments are, because of uncertainty, inefficient as color of title.

—*Allmendinger* v. *McHie*, 189 Ill. 308.

3-48   Taxes

In some states, but not in the majority of states, taxes must be paid on the land adversely held. This condition greatly reduces the possibility of gaining land by adverse means. Sometimes payment of taxes shortens the statutory time required for possession.

Payment of taxes is not proof of possession; it is evidence of a claim of right. Failure to pay taxes is not a necessary element of adverse possession unless required by law. People do not pay taxes willingly; therefore, payment of taxes is evidence of a claim.

3-49   Good Faith

In some jurisdictions adverse possession must be taken and maintained in good faith; but, by the majority opinion, good faith only need be with respect to constructive possession. Good faith, if required, means that a person believes that he has acquired a good title and takes title in accordance with his belief. A person deliberately taking land, knowing that the title is defective, is not acting in good faith, and, where good faith is required, such action is of no avail.

3-50 Difficulty of Determining Unwritten Rights

As can be deduced from the foregoing, gathering sufficient evidence to prove whether a specific possession is a fee right is extremely difficult and complex. After reading many of the land surveyors' examinations given in the various states, it is obvious that surveyors are not questioned in detail on the subject of acquiring ownership by the unwritten processes so far discussed. The logical conclusion is that surveyors are not expected to be experts on the subject; however, it is certainly necessary for them to know that unwritten rights can occur.

The most difficult of all unwritten rights cases to prove is that of adverse rights. It is much easier to prove recognition and acquiescence, practical location, or other means of unwritten transfers.

A surveyor can assume liability problems for failure to recognize a lawful right obtained by prolonged possession, and he can also create problems for failing to monument the deed description when he knows that an unwritten right has ripened into a fee right. A few examples clarify these remarks.

A deed, one of several written about 80 years ago in the same locality, reads as follows:

> Commencing at the Southwest corner of theoretical section 17 . . . ; thence North along the Westerly line of said section 17 a distance of 10 chains to the true point of beginning; thence continuing along the section line 5 chains; thence at right angles easterly 5 chains; thence southerly parallel with the westerly section line 5 chains; thence 5 chains to the true point of beginning.

A survey (see Fig. 3-1) revealed that an angle of 92° existed at the theoretical (Rancho land-grant area) section corner and that all fences were constructed parallel with the south line of the section. The deed call of "at right angles" deviated from a 30-year-old fence by 2°. After weighing all evidence, the surveyor decided that the fence would be controlling, and he monumented the property with a 92° angle at the corner. The owner, who was not notified of the conditions, erected a house adjoining the south line. On discovery of the facts, the title-insuring agency refused to insure a house loan. The house was several feet over the line when a 90° angle was used, and the surveyor was in trouble. As discussed under evidence, "The written words are conclusive." The deed did not say "parallel with the south line of the section"; it said "at right angles." If the surveyor had disclosed the true facts to the client, in writing, he would have been without fault; but he, on his own, located the land as though the deed read "parallel with the south section line." To be sure, he could use occupancy rights as a defense, but the fact remains that, win or lose, the client was damaged because of inability to get a loan. In California, where this particular situation occurred, all house loans are insured by a title company, and the title company will not insure on possession rights; the surveyor should have foreseen this and warned the client of the possibilities.

The following Florida case (*Madison* v. *Hayes*, 264 So. 2d 852) illustrates the difficulty of determining what a court will do in an adverse possession case. Plaintiffs owned Lot 30 and defendants, Lot 7; the back line of Lot 30 butted against Lot 7. The record depth of each lot was 100 feet, with Lot 30 facing Booker Street, and Lot 7 facing Lucky Street. The record distance from Booker Street to Lucky Street was 200 feet; however, the actual measured distance was only 140 feet. In other words, there was a shortage of 60 feet when measuring from one street to the next. Lot 30 had a fence enclosing 95 feet that only left a depth of 45 feet for Lot 7. In this case the owners of Lot 30 claimed adverse possession with color of title (the original map said they owned 100 feet, and they were occupying only 95 feet of the 100 feet). In Florida the adverse claimant must have color of title. The lower court found for adverse rights and gave Lot 30 the 95 feet. The appeal resulted in a reversal as based on the following reasoning.

> To state the consequences clearly, plaintiffs would visit the error entirely upon the defendants, so that plaintiffs would have a 100' lot (more precisely, they are only claiming 95') and defendants would have a 40' lot. Plaintiffs were successful in this endeavor and, via adverse possession proceedings, were able to obtain the appealed judgment, which we reverse. On the other hand, defendants, appellants, ask only that the deficiency be apportioned to the end that each would have a 70' lot.
>
> The well established rule to be applied in such cases was first mentioned in Florida in City of Jacksonville v. Broward, Fla.1935, 120 Fla. 841, 844, 163 So. 229, 230, in quoting from 4 R.C.L. 115, as follows:

> "When division lines are run splitting up into parts larger tracts it is occasionally discovered that the original tract contained either more or less than the area assigned to it in a plan or prior deed. Questions then arise as to the proper apportionment of the surplus or deficiency. In such cases the rule is that no grantee is entitled to any preference over the others, and the excess should be divided among, or the deficiency borne by, all of the smaller tracts or lots in proportion to their areas. The causes contributing to the error or mistakes are presumed to have operated equally on all parts of the original plat or survey, and for this reason every lot or parcel must bear its proportionate part of the burden or receive its share of the benefit of a corrected resurvey. This rule for allotting the deficiency or excess among all the tracts within the limits of the survey may be applied where the original surveys have been found to have been erroneous, or where the original corners and lines have become obliterated or lost."

> And there the Court further recorded:

> "If the lines of a survey are 'found to be either shorter or longer than stated in the original plat or field notes, the causes contributing to such mistakes will be presumed to have operated equally in all parts of the original plat or survey, and

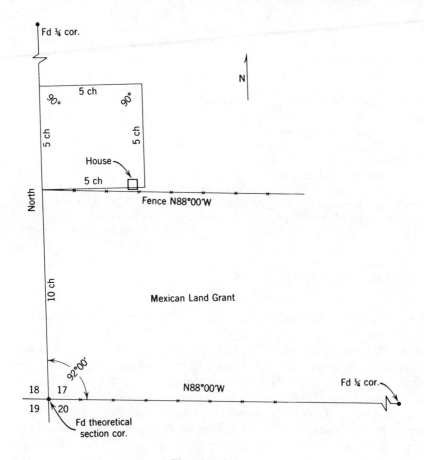

Figure 3–1

hence every lot or parcel must bear the burden or receive the benefit of a corrected resurvey, in the proportion which its frontage as stated in the original plat or field notes bears to the whole frontage as there set forth.' Pereles et al. v. Magoon et al., 78 Wis. 27, 46 N.W. 1047, 23 Am. St. Rep. 389, 392, note."

The rule was further mentioned with approval in City of Pompano Beach v. Beatty, Fla.App.1965, 177 So.2d 261.

In other words, where there is such a deficiency caused by plat error, the respective grantees are treated equally and any surplus or deficiency is prorated or apportioned between them. See 4A Fla.Jur., Boundaries, § 6; 12 Am.Jur.2d, Boundaries, § 63; 11 C.J.S. Boundaries § 124; 1 R. Boyer, Florida Real Estate Transactions, § 13.08 (1971); and 6 G. Thompson, Commentaries on the Modern Law of Real Property, § 3053 (perm. ed. repl. 1962).

The record does not disclose any basis for an exception to this in that its application would not be impractical and there are no facts and circumstances which would be otherwise controlling. City of Jacksonville v. Broward, supra, and City of Pompano Beach v. Beatty, supra. Neither has there been any establishment of a boundary by agreement or acts of the parties. See 12 Am.Jur.2d, Boundaries, § 77 et seq. We are satisfied that the testimony as to the existence of a fence at the 95' mark was not of such weight and dignity as to make the application of the apportionment rule inappropriate.

We think this rule of law is clearly applicable and determinative of the issue. Under it the deficiency must be apportioned and borne equally by the parties.

How about adverse possession? Was it available to plaintiffs under the circumstances of this case? We think not. Under the operation of the aforementioned apportionment rule, the plaintiffs will have color of title to only Lot 30 and the quantity of land as apportioned which would be a lot with dimensions of 50' X about 70'. Thus, they would have no color of title to the defendants' Lot 7 with its apportioned dimensions of 50' X 70'. See F.S. 95.16 and 95.17, F.S.A., Laws of 1969. Also, since plaintiffs' title to Lot 30, as apportioned, is not disputed, there is no basis for an adverse possession suit.

—*Madson* v. *Haynes*, 264 So. 2d 852, Fla., 1972.

The results of the foregoing Florida case could not have been predicted by a surveyor; even the lower court was reversed. In the following case that occurred in a state wherein the surveyor is liable to third parties, the surveyor was held responsible for damages. Surveyors in sectionalized land areas know that when the western tier of sections in a township are divided into parts any deficiency or excess is placed in the last quarter mile. The southwest quarter of one of these sections was short 13 acres. By government survey rules the east half of the southwest quarter contained 80 acres and the west half contained 67 acres. One party patented the entire southwest quarter and then sold off the west half of the southwest quarter containing 80 acres! Later the east half of the southwest quarter was sold. About 40 years later a surveyor was asked to survey the east half of the southwest quarter, and he did so exactly in accordance with the *Manual of Instructions for the Survey of the Public Domain*, and this was undoubtedly correct for that description. The surveyor noted the encroachment of a fence on 13 acres, showed it on his plat, and credited the client with 80 acres. The client sold the land on a per acre basis; a title company insured the land, and in the ensuing litigation the 13 acres were declared to belong to the adjoiner (the west half). The title company paid out over $100,000 for the land and sued the surveyor and won. For correctly locating his client's land in accordance with the client's legal description, the surveyor got in trouble. In other words, the adjoiner's rights should have been looked into; the adverse occupancy should have been the tip-off. It does not pay to certify acreage when someone else is occupying part of the area and may have a right of ownership. The second thing to note about this case is that the surveyor was held liable for damages caused to a third party (those who have a right to rely on the results of a survey) as discussed in Chapter 4.

Here is one other problem involving occupancy rights. A road, called Boundary Street, was offered for dedication and was accepted by the city as part of a subdivision. The road was located along a Rancho Land Grant boundary with control points several miles apart. When constructing improvements in the road, the city discovered that the subdivision had been monumented 5 feet into the adjoiner's land. The city built the road along the true Land Grant boundary, with the result that 5 feet were taken from the lots on one side of the road and 5 feet were added to the lots on the other side of the road (in a different subdivision). After 40 years of uncontested use of the road, where does the surveyor set the monuments for the lots adjoining the road? Would the surveyor dare to set monuments 5 feet in the road as improved and take a chance on being held liable for a client constructing an improvement in a street others have obtained by usage? It would be foolish to do so.

3-51 Surveyors' Duties with Respect to Unwritten Title Transfers

Within the United States the courts' opinions as expressed in the different states have been diverse, and it is not possible to give a standard statement of what surveyor's duties are with respect to unwritten conveyances. Certain observations can be made that most surveyors will agree on.

PRINCIPLE. *If possession is not in agreement with the client's deed, a search of the adjoiner's title must be made to determine the status of senior rights and to determine if changes have occurred in the title's wording at times of transfer.*

The surveyor does not always do the title search; abstract reports can be relied on. As previously stated, senior rights control, and it is necessary to know who has senior rights in the event of conflict of deed calls. In the East deeds commonly can be found that call for one another; the northern deed calls for the southern deed, and the southern deed calls for the northern deed. An essential part of the survey is to discover which deed has prior rights. An example is this: surveyor one has a deed that calls for the adjoiner, and the adjoiner's deed also calls for the adjoiner. The surveyor monuments the fence line between the two properties. Surveyor two does a title search and discovers that sometime ago both deeds called for a creek as the boundary line and the fence was built on one side of the creek for mere convenience. The creek controlled.

PRINCIPLE. *If possession is not in agreement with measurements and if an original survey were made, the surveyor must determine whether the possession represents where the footsteps of the original surveyor were. If the possession duplicates the footsteps of the original surveyor, then the possession does not represent an unwritten conveyance right; it is where the written rights belong.*

Determining whether a given possession represents where an original surveyor ran his lines is very difficult and relies on the interpretation of evidence. In some of the eastern states, if there are indicators disclosing the probability of an early original survey and the lines of possession have the reputation of being correct survey lines, the courts are apt to accept them as such. In contrast, in the more recently surveyed states proof of where original lines were run needs to be more substantial. For example, in Washington, Oregon, and California the records of what monuments were set at the time of the original survey and what monuments have replaced the original monuments are rather complete; it is usually unnecessary to guess about what was done; it is known what was done. As was pointed out in the chapter on evidence, land is located by the best-available evidence that, as a last resort, may be reputation. In the eastern states, where there is widespread loss of records and evidence of what was originally surveyed, the reliance on reputation becomes much more important. It is probable that the major difference between eastern and western states' required survey procedures lies in the amount and type of evidence needed to prove a property location rather than a difference in the principles of law.

Although it may be difficult to determine when a line of possession represents an original survey line or when it represents an occupancy right, it should be the obligation of the surveyor to differentiate between the two and to inform the client of the difference.

PRINCIPLE. *If adjoiners agree that the line or lines of possession between them are or should be the boundary line and if possession is not in agreement with the writings, the surveyor should attempt to transform the agreement to writings.*

A written agreement will eliminate future problems. Unfortunately, when a surveyor is called on, it is often because there is a disagreement between neighbors, and getting an agreement deed signed is often difficult.

After a surveyor has completely eliminated all possibility of a line of possession from being related to a written deed right and he has concluded that occupancy might be an unwritten title right, what are his obligations?

Any time land is claimed strictly on the basis of an unwritten title, the following occurs: (1) loaning agencies will not loan money when title is based on an occupancy right. (2) Title companies will not insure the title either in terms of loss or gain of land by occupancy. (3) Occupancy rights can only be perfected by either an agreement between adjoiners or by a court decree. In other words, lands claimed by occupancy alone have low value. The client is entitled to know on what basis his claim of ownership rests. It is misleading to allow a client to believe that his occupancy is in agreement with his writings when in fact they are not in agreement. It is the obligation of the surveyor to inform the client about the status of his occupancy or encroachments on other land, and in those states where the surveyor is liable to third parties, the information must be presented so that third parties would also be aware (see Chapter 4). This leads to the following.

PRINCIPLE. *The surveyor should differentiate between lands in agreement with a written title and lands claimed by some form of unwritten title.*

This does not say whether a surveyor should or should not set monuments along lines claimed by possession alone. A search of statute laws indicates the following.

PRINCIPLE. *No land surveyor's act specifically states that the surveyor may or may not monument boundaries in accordance with an unwritten conveyance of land.*

In a search of the statutes of the various states and the Federal Government, no statute was found that authorized or barred the surveyor from monumenting boundaries in accordance with unwritten rights. Since no fee right can be obtained by occupancy of public lands, the question only pertains to surveying private lands adjoining private lands. The restraint for or against monumenting occupancy lines apparently is a question of liability and of the opinion of fellow surveyors. The concerns of liability to the client and of all third parties having a right to rely on the results of the survey are discussed in the next chapter. From the point of view of liability, it is a full liability risk for a surveyor to monument occupancy lines beyond the client's written title lines. The surveyor then assumes the responsibility of proving his contentions in court, and, if he fails to prove his contentions, he can be held liable for any damages caused the client.

It should be pointed out that at the present time most private surveyors and surveyor organizations criticize "fence surveyors," but monuments set along an occupancy line believed to be where the original surveyor's footsteps were are approved.

Further discussion of the duties of the surveyor is presented in Chapter 4 on liability.

Besides discussing establishment by statutory proceedings, Chapter 7 also presents a discussion of the processioners, commissioners, or others who are charged with the responsibility of establishing property lines or boundaries. In some states, these professionals are specifically instructed to consider evidence of possession for the duration of the statute of limitations. In Georgia the processioners may mark occupancy boundaries that can be proved to be of 7 years' duration. In Colorado the surveyor appointed by the court is to investigate the possibility of 20 years' possession and report the findings to the court. In Colorado the court makes the final decision about the nature of prolonged possession; the surveyor reports the facts. In Georgia the processioners make a determination of boundaries by either writings or 7 years' occupancy, and these professionals' findings are final, unless appealed to the courts.

PRINCIPLE. *If there is possession that is not in agreement with the client's written deed, the surveyor should inform the client of its possible significance.*

Although the surveyor does not necessarily come to a conclusion about who owns occupied lands that are not in agreement with writings, he should certainly inform the client of possible transfer of title by occupancy rights and do so in such a manner as to avoid liability, the subject of the next chapter.

CHAPTER 4

Professional Liability

4-1 Introduction

One of the marks of educated persons, especially those of the professions, is that they understand the effect of today's action on tomorrow's liability. The courts have imposed on land surveyors the certainty that they may be held liable for costs resulting from errors, omissions, or almost any action that will result in future damage to a client. The concept of personal liability, although a disadvantage to the individual surveyor, is in reality a blessing to the profession as a whole. If it were not for the threat of future liability, those persons found in any profession who are willing to do a poor job for less money would soon lower the standing of all professionals. Those surveyors who carelessly fail to protect or to inform the client in all survey matters may find themselves burdened with costs of negligence. This is as it should be.

A client employs a land surveyor to bring to him a certain level of expertise, and for the surveyor there is only one good rule: bring to the client the level of expertise expected. Those surveyors who foresee mistakes that a client can make and take the time to warn of potential dangers are the ones who have minimum liability and are likely to continue in business.

Surveyor's liability is imposed by the statute of the state, and, in the event of ambiguity, the meaning will be interpreted by the courts. Unfortunately, the statutes enacted by legislatures and the resulting court interpretation found within the various states are diverse; no consistent statement can be made on surveyor's liability. In many states the courts have recently reversed their previous decisions, possibly because of a change in the court's makeup. The surveyor must know and understand the statute of limitations in his state and must also understand the current court opinion. *Liability rules and their application are in a state of flux.*

When a statute is changed, it cannot be applied ex post facto or prior to the date of its enactment or adoption. When a court's opinion is rendered or handed down, it says in effect: this is the law as it is commencing now. As an example, the statute of limitations in Colorado is 6 years, and the rule was that the statute began to run on completion of the work. Surveyors were thus assured that after 6 years any errors or omissions were forgiven. In 1976 (*Doyle* v. *Linn*, 574 P 2d 257) the court decided that the time the statute of limitations begins to run or toll is 6 years *after the discovery of an error*. At this trial the jeopardy of Colorado surveyors immediately changed from 6 years to the *discovery rule*. Exposure to all the old skeletons in the closet was resurrected by court opinion, but those skeletons that had been tried by the courts could not be reintroduced. At this date not all states have adopted the discovery rule, although it is believed that most will eventually do so.

4-2 Discovery Rule

DEFINITION. *The discovery rule as applied to land surveyors means the statute of limitations for liability commences to run from the discovery of an error or after a person should have known or become aware of an error.*

The development of this discovery rule has generally been by court interpretation, not legislative enactment. Some of the first applications were for doctors in negligent actions such as leaving a sponge in a patient, which promulgated the practice to include a metallic thread in all sponges so that the thread itself can be detected in x-rays. Probably the first application to land surveyors was in Illinois in 1969 (*Rozny* v. *Marnul*, 43 Ill. 2d 54). Soon the same rule spread to California, Colorado, Florida, New Jersey, and Washington.

In some recent cases the state courts have ruled against the discovery rule, and in New York it was the opinion of the court that, if the discovery rule should apply, it should be by legislative law, not court interpretation. Courts now applying the discovery rule in recent or somewhat recent times are Delaware (1973), New York (1977), and Tennessee (1962). If in a given state the supreme court or a court of appeals has not recently handed down a decision on the discovery rule, the surveyor is uncertain of his position; however, some guidance might be obtained from the courts' attitude toward other professions.

The case of *Kundahl* v. *Barnett* in the state of Washington illustrates the court's thinking in applying the discovery rule. Portions of the court report were as provided here:

> The facts are these: In 1962 appellant Frank H. Elrod orally agreed to survey and stake boundary lines for several lots which respondents owned on Mercer Island in King County. The parties were aware of surveying problems in the area, and when the work was completed it physically appeared that the lines laid down did not correspond to the general pattern of other properties in the vicinity. Respondents Fischer built a house upon one of the lots, sold it, and then bought it back to settle the suit for encroachment as above noted.
>
> Two questions are presented: First, whether there was sufficient evidence to support the findings of the trial court as to liability; and, second, whether the court erred in deciding that the 3-year statutes of limitation (RCW 4.16.010 and RCW 4.16.080 (2)) did not bar respondents' claim because their action against appellants did not accrue until they discovered or had reasonable grounds to discover the error in the survey.
>
> Expert witnesses were called by each side to give testimony concerning the accuracy of the survey and the methods employed by appellant husband in conducting it. There is substantial evidence in the record that appellant husband was negligent in his conduct of the survey and that this negligence was a proximate cause of the loss.

The controlling question is whether the action, which was based upon a negligent breach of duty, was begun within the time limited by law.

Until recently, the judicial resolution of this question has been that the action "accrues" when the breach of duty occurs, and that the start of the running of the statute of limitations is not postponed by the fact that actual or substantial damages do not occur until a later date.

Until *Ruth v. Dight*, 75 Wn.2d 660, 453 P.2d 631 (1969), the same rule and reasoning have been uniformly applied in lawyer and medical malpractice cases....

In *Ruth*, a specific exception was made in a situation in which a surgeon left a sponge in a patient and the discovery of it was made more than 20 years later. Previous medical malpractice decisions were overruled; and the "discovery rule" was adopted, whereby the statute of limitations commenced to run when the patient discovered, or in the exercise of reasonable care should have discovered, injury. The "discovery rule" has been discussed and applied in subsequent medical malpractice cases. *Fraser v. Weeks*, 76 Wn.2d 819, 456 P.2d 351 (1969); ...

The question of when a cause of action accrues in the case of the negligence of a land surveyor has not been decided in this state. The legislature in 1967 limited the accrual of a cause of action against a surveyor, among others, under certain circumstances, to 6 years from the date of substantial completion or termination of improvements to real estate, but left with the courts the determination of when during the 6-year period the action accrues. RCW 4.16.300 *et seq.*

Although *Ruth v. Dight, supra,* and subsequent decisions applying the discovery rule involve medical malpractice, the reasons given for the rule are as persuasive in this case where a landowner intending to build, and knowing that there were boundary location problems, obtained, for pay, the services of a professional land surveyor to locate his property.

We see no distinction between the medical and other professions insofar as application of the discovery rule is concerned. Although the damages resulting from medical malpractice are more personal in character, the pecuniary loss caused by malpractice in the other professions can be as great or greater. In this case, it is illogical to charge respondents with the obligation of retaining the services of another surveyor to prove out the stakes which appellant Elrod placed. That would be the only way that respondents could discover the error; unless, of course, the second surveyor was in error, calling for a third survey, and so on ad infinitum.

The injustice caused by the strict definition of "accrual" can be avoided by the application of the discovery rule. Of course the legislature has the power to provide special statutory periods for malpractice cases.

—*Kundahl v. Barnett*, 468 P 2d 1164, July 1971.

Opinions rendered in one state jurisdiction are not binding on another state. If the case itself is one of first impression, courts of other states will look to the decision for advice. The only legal decisions binding on all state court jurisdictions are those rendered by the U.S. Supreme Court on a subject granted to the Federal Government by the constitution. On the other hand, opinions rendered by a state court of appeals or state supreme court are primary law for courts of all levels within that particular state. Thus any decision rendered, either of joining a previous decision or completely reversing "old law," will affect future decisions rendered relative to liabilities of professions.

In Colorado the discovery rule was expanded a bit more; the statute of limitations commences running from the time damages are awarded, not from the time the person should have been aware that there were damages. In *Doyle v. Linn* part of the court report was as follows:

> James and Florene Doyle sought recovery of damages resulting from an allegedly negligent survey made by defendant, Kurt O. Linn. Following a trial to the court, it found that the defendant was negligent and that plaintiffs had been substantially damaged as a result of such negligence. However, the court dismissed the action on the ground that the action was barred by the operation of two statutes of limitations, §§ 13-80-110 and 13-80-127, C.R.S. 1973, and plaintiffs appeal that dismissal. The sole issue before this court is whether the action was barred by § 13-80-110, C.R.S. 1973. We hold that it was not and reverse the judgment.
>
> In 1960 the Doyles engaged Linn to prepare a boundary survey of a parcel of mountain property that they were considering buying. Linn knew that the Doyles had selected a site for a home they proposed to build and would buy the property if the site was actually within the plot's boundaries. In September 1960 Linn surveyed and staked the boundaries, and delivered a certificate plat to the Doyles. According to this survey, the desired homesite was within the boundaries of the plot. In reliance on the survey, the Doyles bought the land and began building their house in November of 1960. The house was not completed until 1970.
>
> The land adjacent to the Doyle plot on the north is owned by the United States, being part of the Pike National Forest. In 1964 government agents began negotiating with the Doyles for permission to widen an access road to the National Forest, which road crossed the Doyle land. During the discussions, the agents advised the Doyles that there was a possibility that their house was on government land. In November 1964 the Bureau of Land Management (B.L.M.) began a "dependent survey" of the boundary line. The survey was completed in June 1965, and was officially accepted and approved by the B.L.M. on February 1, 1968. According to this survey the Doyle house was on government land.
>
> In October 1968 the United States sued the Doyles in the U.S. District Court for trespass, and obtained a judgment which gave the Doyles 90 days to remove the house. The judgment was affirmed, on appeal, on November 6, 1972. *United States v. Doyle*, 468 F.2d 633 (10th Cir.). Following this affirmance, the Doyles moved the house onto their own land. The cost of this removal constituted the bulk of the damages sought in the present action, which was commenced on January 21, 1973. All of the above facts are undisputed.
>
> The trial court found that, "plaintiffs suffered $14,721.46 loss and damages, and this loss and damages were proximately caused by defendant's negligence." On disputed evidence the court further found that on completion of the survey in 1965 the government advised the Doyles that their house was, in fact, on government land. The court concluded that the six year statute of limitations, § 13-80-110, C.R.S. 1973, began to run in 1965, the date of the notice to the Doyles of the government's claim.
>
> Here plaintiffs suffered no damages until the adverse claim of the government was determined to be valid and plaintiffs were found to be trespassers, with the consequent damages resulting from their being required to remove their house from government land. The statute of limitations did not begin to run until the judgment became final in 1972.

The judgment of the trial court is reversed and the cause is remanded with direction to enter judgment in favor of plaintiffs for the amount of loss and damages as set forth in the decrees heretofore entered by the trial court.
—*Doyle* v. *Linn*, 547 P 2d 257, 1976.

In Oregon in *Kashmir* v. *Barnes* the statute of limitations commenced at the time it was known that there was an error as follows:

Since plaintiff had full knowledge of the 10-foot shortage in Lot No. 40 before completion of the purchase of that lot in May, 1973, it is difficult for us to perceive the underlying basis for plaintiff's damage claim. In any event, however, the statute clearly states that any action must be commenced within two years of the date of injury. Assuming that plaintiff bargained and paid for 330 feet and received only 320 feet, the injury occurred no later than May, 1973, when plaintiff purchased the property. The fact that plaintiff did not then consider it prudent to spend an additional $1,000 to find the source of the shortage is a strategic decision and does not affect the running of the statute of limitations. Had plaintiff felt that it was sufficiently important to do so, we have no doubt that it could have arranged for a survey of the surrounding properties and found the source of the shortage well within the statutory period. Therefore, since this action was not commenced until June, 1975, plaintiff's cause of action was barred by ORS 12.135.
—*Kashmir* v. *Barnes*, 564 P 2d 695, Ore., 1957.

4-3   Privity of Contract

DEFINITION.   *In an action for damages privity of contract means only those who were parties to the contract may bring suit.*

Along with the development of the concept of the discovery rule has been the erosion of the concept of privity of contract. The old rule, approved by most states, was: only those with privity of contract could sue in the event of damages. On the sale of a property, unless the new buyer was part of the contract of survey, he was barred from claiming damages from the surveyor. The newer court opinion in several states is this: those in privity of contract and those to whom the surveyor can reasonably foresee as having a right to rely on the surveyor's work can sue for damages. This opened Pandora's box. Now title companies or lawyers who relied on the survey in issuing insurance or opinion, the contractor who built in accordance with the monuments, the next buyer who relied on the monuments, and any others who had a right to rely on the monuments, if damaged, can sue.

The impact of the discovery rule, coupled with the erosion of privity, has had an astronomical effect on the cost of errors and omission insurance. The fee of one California firm jumped from $100 per year in 1947 to $7500 in 1965, and in addition the deductible changed from $0 to $3,000 per occurrence.

As of this date California, Pennsylvania, and Illinois have definitely adopted expansion of the concept of enlarged privity, and it is probable that many more states will do the same. In Tennessee (*Howell* v. *Betts*, 362 SW 2d 924) the rule of necessity of privity of contract was apparently approved and the discovery rule not adopted. Parts of the court report were as follows:

> The facts averred in the declaration were that defendants made the survey of this parcel of land "for a former owner" in the year 1934; that the record of the survey described the parcel by metes and bounds, and showed it contained 2.3 acres, more or less; that 24 years later, or on November 12, 1958, plaintiffs purchased this parcel, relying on the accuracy of the survey and description.
>
> It was further averred that in May 1960, plaintiffs learned of the errors in the survey, which were that their west line is 19.6' shorter, their north line 0.8' shorter, and their east line 13.1' shorter, than as shown in the 1934 survey; and that, therefore, the area of plaintiffs' lot is less than that shown by the survey, and they are entitled to damages for the deficiency.
>
> So, the question is whether defendants, in surveying the lot for the then owner in 1934, owed a duty of care to any remote purchaser, not in privity, who might purchase it 24 years later, as plaintiffs did.
>
> It is true the old rule was that there was no duty of care upon a defendant to a plaintiff not in privity. *Burkett v. Studebaker Bros. Mfg. Co.,* 126 Tenn. 467, 150 S.W. 421. But it can hardly be said that such a general rule any longer exists. . . .
>
> On principle and authority, we think the rule of liability cannot be extended to a case like that before us. If these surveyors could be held liable to such an unforeseeable and remote purchaser 24 years after the survey, they might, with equal reason, be held liable to any and all purchasers to the end of time. We think no duty so broad and no liability so limitless should be imposed. . . .
> —*Howell* v. *Betts,* 362 SW 2d 924, Tenn.

Within California the elimination of the need for privity of contract probably began with *Biakanja* v. *Irving* (49 Calif. 2d 647, 1958) and was then later applied to Dames and Moore, soils engineers, in a case claiming damages for failure to disclose unstable material in a construction area.

> Plaintiff brought this action against defendants Central Costa Sanitary District; Brown & Caldwell, a partnership; and Dames & Moore.
>
> Defendant Dames & Moore filed no answer to plaintiff's complaint, but moved for summary judgment in accordance with Code of Civil Procedure, section 437c, alleging that plaintiff's action against it had no merit. In support of its motion for summary judgment, Dames & Moore filed affidavits stating that it had never been a party to any agreement with plaintiff
>
> As to the first asserted ground of defense, admittedly there is no privity of contract between the parties. However, as has most recently been decided by our Supreme Court, the fact that there is no privity of contract between parties does not necessarily result in no liability on the part of a negligent party to one injured thereby.
> —*Miller* v. *Central Contra Costa Sanitary District,* 18 Calif. Rpt. 13.

The last California case directly involved a surveyor, and it was found that privity of contract was unnecessary.

Defendant, by way of answer, denied the charging allegations of the complaint and asserted that plaintiffs had failed to state a cause of action because it was not alleged that there was any privity of contract between plaintiffs and defendant.

The reasoning of the *Rozny* and *Miller* cases would appear equally applicable to the factual situation here before us. When defendant prepared his survey, he could reasonably anticipate that it would be used and relied upon by persons such as plaintiffs, who purchased the property surveyed by defendant.

—*Kent v. Bartlett*, App., 122 Calif. Rpt. 615.

4-4 Standard of Care

PRINCIPLE. *Surveyors are expected to perform with the same degree of care and skill exercised by others in the profession in the same general area.*

In a court trial claiming damages against a surveyor, it is proper to introduce evidence and testimony from other surveyors about what an ordinary skilled surveyor prudently does. In a recent Louisiana case the court decided that the plaintiff did not have to introduce such evidence, since it was obvious that the conduct of the surveyor was unprofessional and below reasonable standards. The Lawyers Title Insurance Company sued a surveyor to recover damages that it had to pay out because of reliance on a surveyor's work. Part of the court report of 1978 was as follows:

> Lawyers Title Insurance Company (Appellant) subrogee of its insured, Lowe's Companies, Inc. (Lowe's), appeals dismissal of its subrogated demand against defendants Carey Hodges & Associates, Inc. (Surveyor) and Surveyor's employee Paul J. Morel, for reimbursement of a claim paid Lowe's pursuant to a policy of title insurance issued by Appellant covering a parcel of land purchased by Lowe's on the strength of a survey by Surveyor. The survey erroneously located drainage facilities on Lowe's land which in reality were on neighboring land. Relying on the survey, Lowe's planned its drainage to utilize the facilities indicated as being on its property. The error was discovered after Lowe's installed its drainage. The work installed by Lowe's had to be removed and new facilities constructed. The trial court found that Surveyor made a mistake but that a mistake does not necessarily constitute actionable negligence and rejected Appellant's demand on the ground that Appellant failed to produce expert testimony to establish Surveyor's negligence by showing that Surveyor failed to exercise the degree of skill and care exercised by other surveyors in the vicinity. We reverse and render judgment for Appellant.
>
> We reaffirm our holding in *Charles Carter & Company, Inc.* that surveyors are expected to perform with the same degree of care and skill exercised by others in the profession in the same general area. Ordinarily, proof of lack of such skill and care or proof of failure to exercise such skill and care in a given instance rests upon plaintiff. We deem it reasonable, however, to except from the general rule those instances when the conduct of a surveyor may be so unprofessional, so clearly improper, and so manifestly below reasonable standards dictated by ordinary intelligence, as to constitute a prima facie case of either a lack of the degree of skill and care exercised by others in the same general vicinity or failure to reasonably

exercise such skill and care. We find the omission of a visible drainage structure from a surveyor's plat to fall within the exception. No profession may, by adopting its own standards of performance, method of operation, or paragons of care, insulate itself from liability for conduct which ordinary reason and logic characterize as faulty or negligent. See *Favalora v. Aetna Casualty and Surety Company*, 144 So.2d 544 (La.App. 1st Cir. 1962). Having failed to refute the prima facie showing made by Appellant, Surveyor must respond in damages.

It is ordered, adjudged and decreed that the judgment of the trial court be and the same is hereby reversed and set aside and judgment rendered herein in favor of plaintiff Lawyers Title Insurance Company and against defendants Carey Hodges & Associates, Inc., and Paul J. Morel, in solido in the sum of $8,215.00, with legal interest thereon from date of judicial demand, until paid, and all costs of these proceedings.

Reversed and rendered.

—*Lawyers Title Insurance* v. *Hodges*, 358 So. 2d 964.

The older rule, used by some courts today in cases involving damages against a land surveyor, is best expressed in *Ferrie* v. *Sperry* (85 Conn. 337 1912):

> The gist of the plaintiff's cause of action stated in the first count was the negligence of the defendant in his employment as civil engineer. Having accepted that service from the plaintiff, the defendant, as the jury were properly instructed, was bound to exercise that degree of care which a skilled civil engineer of ordinary prudence would have exercised under similar circumstances. This being the rule, it was important that the jury should know what such an ordinarily prudent engineer would do under the circumstances of this case. Might he simply examine the land records and muniments of title, and observe the fixed monuments and evidences of present occupancy and ownership, or was he bound to scour the neighborhood to learn whether there had been adverse occupancy and claims of ownership of any part of the premises covered by his client's deed by others than the latter's predecessors in title appearing of record? Would good engineering practice require that he examine a city engineer's map, called the "1870 map," the failure to examine which was claimed to constitute negligence on the part of the defendant? The jury would not know, unless informed by evidence, and such evidence was admissible.

The tendency in courts today is to lean more toward the opinion in *Lawyers Title Insurance Co.* v. *Hodges* rather than the rule just cited. In other words, surveyors cannot always escape liability merely because they can introduce evidence showing standards of practice below what would be expected of the professions. In some courts some things are patently negligent without proof in the form of testimony.

Certainly, the surveyor can expect to be held to the standards adopted by the state association in existence at the time, and this applies whether the surveyor is a member of the association or not. If a state has not adopted a standard of care, the court may inquire about what the standard of care is in other state associations and adopt them for criteria.

The fact that a surveyor can be held liable for negligence and lack of skill is well established; the only question is what constitutes negligence and lack of skill? The elastic definition—the surveyor must exercise that degree of care which a surveyor of ordinary skill and prudence would exercise under similar circumstances—will undoubtedly continue as the court's rule. However, it must be remembered that with improved equipment, better dissemination of knowledge of boundary laws, and improved scientific methods there is a constant force requiring yearly upgrading of the standard of care. What was accepted as minimum measurement accuracies 50 years ago would not be tolerated today. It can be expected that the standard of care required 50 years from now will be quite different from that expected today.

4-5 Negligence versus Breach of Contract

PRINCIPLE. *If a surveyor can be proved negligent in the performance of his duties, punitive damages may be awarded in addition to contract damages.*

In all states a surveyor is required to do what he contracts, and he may be held liable for failure to fulfill his contract correctly. In some states, double or triple damages can be claimed in addition to contractual damages if the work is done in a negligent manner. The extra damage is called *punitive damages*. In states where punitive damages are allowed, the attorney usually adds the charge of negligence in addition to the real or compensatory damage.

According to *Black's Law Dictionary* (4th ed.), *negligence* is "omission to do something which a reasonable man, guided by those ordinary considerations which ordinarily regulate human affairs, would do, or the doing of something which a reasonable and prudent man would not do." To collect punitive damages in some states, as in New Jersey, malice must be shown. It seems that the present tendency of courts is to become more liberal in what they consider to be negligence.

> In this action for trespass to plaintiffs' real property in Rahway, New Jersey, resulting from the wrongful erection of a fence thereon, the jury in the Union County Court awarded plaintiffs $900 as compensatory damages against all defendants. In addition, the jury allowed punitive damages totalling $4,300. The sum of $300 was assessed against the defendants Smith, plaintiffs' neighbors, who ordered the fence erected; $1,000 against the defendant Cohill, the contractor who erected the fence; and $3,000 against the defendant Lawrence, a local surveyor, whose erroneous staking of the boundary line between the Smith and La Bruno properties caused the Smiths' fence to be erected on the La Bruno property.
>
> The cases would thus seem to indicate that one or the other of two factors must be found before punitive damages can be awarded in a suit for trespass to real property, *viz*.: (1) actual malice, which is nothing more or less than intentional wrongdoing—an evil-minded act; or (2) an act accompanied by a wanton and wilful disregard of the rights of another. Clearly, each case must be governed by its own peculiar facts. Accordingly, we must examine the facts herein, as to each of the defendants, in order to decide the validity of the award of punitive damages in this case.

We consider first Lawrence, the surveyor, whose initial mistake of wrongfully placing the stakes on the La Bruno property started the chain of causation which resulted in the damage and litigation. Lawrence committed a trespass and his mistake, honest belief, or professional neglect, is no defense to a trespass. Restatement of Torts, § 164; Prosser on Torts, § 17, p. 115 (1941). A trespass by mistake or through carelessness will ordinarily not justify an award of punitive damages, because the element of mistake or neglect negates any intentional wrongdoing. Such a mistake of one foot in the boundary line does not here, *per se*, bespeak wilful and wanton disregard of the rights of another.

Lawrence, however, aggravated his original trespass by mistake, by his stubborn refusal to examine his own prior survey and to correct his alleged error, when the matter was called to his attention by La Bruno originally, and later by the Smiths, when they received a letter from La Bruno's attorney. Even when La Bruno put it to him that he had to be wrong one time or the other, he said, "I'll cross that bridge when I come to it." Lawrence was fully aware from his personal inspection, when he placed the stakes, that the erection of the fence by the Smiths along that boundary line would damage the La Bruno walk, patio and flower bed. As a surveyor he knew or should have known that the Smiths and their contractor would probably rely upon his professional skill and judgment in erecting the fence. He demonstrated a wilful and wanton disregard of the property rights of the plaintiffs, that was reasonably calculated to aggravate his original mistaken trespass by the consequent trespasses of the Smiths and Cohill. The jury could justifiably find, as they did, in Lawrence's overall conduct and attitude, a proper basis for the award of punitive damages against him. He probably gave the impression to the jury of a man who would rather see property destroyed by his error than admit that he had made the error.

Turning next to Cohill, the contractor, we find no justification for the award of punitive damages against him. There is no evidence of any actual or express malice on his part.

Accordingly, the judgment of the Union County Court is affirmed as to the compensatory damages, against all the defendants; and, as to the punitive damages, the judgment is affirmed as to the defendant Lawrence, and reversed as to the defendants Cohill and the Smiths.

—*La Bruno v. Lawrence*, 166 A 2d 822, 1960.

4-6 Expressed and Implied Guarantees

PRINCIPLE. *The surveyor is liable for damages arising from expressed guarantees and from damages arising from implied guarantees expected of the profession.*

Merely advertising yourself as a land surveyor carries with it implied guarantees that will be enforced by the court, and these guarantees will be imposed by the court as based on fairly high standards set by the profession. Such questions as these can be inquired into. What are the standards set by the state surveyor's society, ACSM, and ASCE? What accuracy does the governing agency require for original (subdivision) surveys? Although at one time the rule was that surveyors could testify about the degree of skill required in their particular area and thus establish minimum standards, it appears that this relaxed rule is on its way out.

At times surveyors foolishly make positive guarantees such as "This plat of survey carries our absolute guarantee of accuracy" as found in *Rozny* v. *Marnul* (43 Ill. 2d 54). Such stipulation invites liability beyond what is required by law and should be avoided.

Surveyors are, of course, liable for adopting the work of others without checking it. This includes, but is not limited to, the acceptance of plats, corners, and surveys of others. In Louisiana a surveyor copied the acreage figure from a previous survey and was held liable for damages from its inaccuracy as follows:

> On January 27, 1971, defendant-appellant, J. J. Krebs & Sons, Inc. (hereafter, Krebs) was employed by plaintiff-appellee, Thomas Jenkins (hereafter, Jenkins), through his attorney, to "confirm" an earlier survey of a tract of land which he had agreed to purchase but had not yet acquired by act of sale. Jenkins did not specifically request an acreage computation in connection therewith but the earlier survey which was subsequently confirmed by Krebs included a written notation that the tract contained 13.16 acres. As a matter of fact, Krebs did not actually confirm the acreage figure but relied, for this information, upon the earlier survey of the tract which had been forwarded to him to confirm. Krebs (or his employee) simply assumed that the acreage figure on the earlier survey was correct since all the measurements of the boundary lines on the earlier survey were confirmed for accuracy when located and measured by Krebs. However, the original *mathematical computation* of the acreage figure had been incorrectly made.
>
> The parties agree that, after the sale, a correct computation showed that the tract contained only 11.26 acres, not 13.16 acres as shown on the surveys—although the lines were all correctly measured in both surveys. Jenkins now contends that he has relied, to his detriment, on the acreage information confirmed by the Krebs' survey since the sale by which he acquired the tract was based upon a dollar per acre price. He claims injury in the amount of $6,840, the alleged "overpayment" computed at $3,600 per acre.
>
> We do not find error in the trial court's determination that Krebs is liable. The inclusion of the incorrect acreage figure may not, at its inception, have been anything more than a mistake of judgment on the part of Krebs' employee. However it was ultimately transmitted, duly certified as correct, by J. J. Krebs & Sons, Inc. to Jenkins who, by reasonable reliance thereupon, incurred a loss. That loss can fairly be said to have directly resulted from a faulty and defective exercise of care and skill on the part of the professional surveyor.
>
> —*Jenkins* v. *Krebs*, 322 So. 2d 426, 1975.

A similar result was found in Maryland in 1972 (*Downs* v. *Reighard*, 265 Md. 344) wherein a surveyor reported 22.075 acres when in fact only 19.58 acres existed. Judgment for $3,339.93, plus costs, as entered against the surveyor on the basis of $1,250 per acre for land not acquired.

Any time a surveyor adopts a monument proclaimed by someone else as being correct or adopts measurement data or computations as his own, he assumes full liability for the correctness of such. This can be likened to endorsing a check or cosigning a note; everyone whose signature is on the check is liable for any damages. The only answer is do not copy data or assume something to be true unless you can prove such to be correct, for if you do, you assume any and all liability for any errors.

4-7 Trespass Damages

PRINCIPLE. *Unless there is a law permitting the surveyor to trespass when necessary to do a survey, the surveyor can be held liable for damages of trespass, either compensatory or punitive.*

Most surveyors assume the right to go on others' land; unless there is a law giving the surveyors the right of entry, such right does not exist. Even if a law does give a right of entry, the surveyor at no time can damage the vegetation existing on the land. Brush cutting or line of sight clearing may result in a damage claim.

From the standpoint of dollars awarded, the *Indiana & Michigan Electric Co. v. Stevenson* (363 NE 2d 1254, Ind., 1977) ranked first. A total of $420 compensatory damages and $100,000 punitive damages was awarded to the landowner as a result of a power company's survey crew illegally removing corn to clear a line of sight. This reward was made even though the power company was in the process of establishing what it wanted to take under eminent domain.

In Florida a surveyor was held liable for damages caused when he cleared a line for survey purposes (259 So. 2d 757, 1972), and in New Jersey a surveyor was held liable for damages during surveying (*La Bruno v. Lawrence*, 166 A 2d 822, 1960).

4-8 Avoiding Liability

With present-day thinking on liability to third parties and the application of the discovery rule, the surveyor must take precautions to avoid paying avoidable damages.

Liability can result from improperly locating the written document, from failure to inform the client of the possibility of unwritten title rights, and from errors in computations. The remainder of the book is devoted to the subject of how to locate written deeds on the ground. The major cause of damages in computations is in computation of acreage. Every client can claim the purchase price is based on a price per acre; the courts are full of damage suits based on an error in acreage computations.

The most disturbing feature of present liability laws pertain to third persons. Title companies have a right under the third-party rule, when applicable, to rely on the results of the survey. When they insure title as based on a survey and damages result they have a right to seek restitution from the surveyor. In effect, the surveyor assumes some of the risk in title insurance business. This makes it necessary that the surveyor present the results of his survey in such a fashion so that all third parties are appraised of possible title problems such as an unwritten rights as a result of prolonged occupancy.

If the surveyor sets his monuments to include land occupied by an adjoiner, there is always the possibility of the adjoiner having an unwritten title right. Unless this is made clear to the client, the client may be prone to do something that will lead to damages, and the possibility exists that the surveyor will be held responsible. In the event of an encroachment on the client's written deed, the surveyor should be certain that the client understands the significance of the encroachment and that the information is presented in a manner that third parties are appraised of it. This means that the surveyor should present a plat clearly showing all encroachments and noting the possibility of occupancy rights in relation to deed or title lines. If an unwritten title right or the probability of such exists, it is foolish to include the occupied area in the acreage of the client as shown by his writings.

The most important third parties that the surveyor has to be aware of are future buyers, title companies, and contractors. The surveyor who omits damaging information from his plat merely because a client does not want it disclosed is exposing himself to possible damage suits from third parties. The day of privity of contract between the surveyor and client is over; the surveyor is obligated to disclose, for the possible benefit of third parties, all information that may lead to damages.

CHAPTER 5

Professional Stature

5-1 Contents

Surveyors believe they are professionals, but seldom do we see in print what constitutes professional conduct. This discussion in no way expresses what professional behavior ought to be. The facts stated were gained from discussions with many, from literature available, and reference to related professions. Today, surveyors are caught in the dilemma of attempting to portray themselves as a rugged, outdoors-oriented people seeking an acceptance as professionals.

5-2 Definition of Profession

An exact definition of the word *profession* is elusive; each authority gives it new variations of meaning. In a narrow sense, the professions are limited to include doctors, clergymen, and attorneys; in a broad sense professions include professional boxers, baseball players, football players, and many others. The following three court cases will serve to define the term.

> Very generally, the term "profession" is employed as referring to a calling in which one professes to have acquired some special knowledge, used by way of instructing, guiding, or advising others or of serving them in some art. Formerly theology, law, and medicine were specifically known as "the professions," but, as the applications of science and learning are extended to other departments of affairs, other vocations also received the name. The word "profession" is a practice dealing with affairs as distinguished from mere study or investigation; and an application of such knowledge for others as a vocation, as distinguished from the pursuits for its own purposes.
> —*State* v. *Cohn*, 184 La. 53.

> The word "profession" in its larger meaning, means occupation, that is, if not industrial, mechanical, agricultural, or the like, to whatever one devotes oneself; the business which one professes to understand and follow. In a restricted sense it only applies to the learned professions.
> —*Geise* v. *Pennsylvania Fire Ins. Co.*, 107 SW 555.

> A "profession" is not a money-getting business. It has no element of commercialism in it. True, the professional man seeks to live by what he earns, but his main purpose and desire is to be of service to those who seek his aid and to the community of which he is a necessary part. In some instances, where the recipient is able to respond, seemingly large fees may be paid, but to others unable to pay adequately, or at all, the professional service is usually cheerfully rendered.
> —*Stiner* v. *Yelle*, 174 Wash. 402.

Research in science is certainly a learned occupation requiring special knowledge, but it is not a profession, since it does not deal with the affairs of others.

5-3 Attaining Professional Stature

PRINCIPLE. *Professional stature cannot be attained by self-proclamation; it must be earned, and others must bestow the title on the profession.*

The lazy say, "Give me the prize without the training, the wages without work, a profession's prestige without a profession's skill." Fortunately, professional stature is something that must be earned, not merely claimed.

Many do attempt to filch good names. Industry, public agencies, and others have found that by bestowing a fancy title on the lowly laborer, they can keep him better satisfied at a lower rate of pay. The boy who cuts brush and carries the surveyor's stake bag is certain of his importance if he is called an engineering aide rather than an axman. Mere claiming of a good name is not proof of right to the name. A person's actions, behavior, and conduct are far more potent proof.

Some occupations inherently carry a connotation of profession. The professions that everyone recognizes and acclaim as professions, that is, the doctors, clergy, and attorneys, need not use the title professional doctor, professional attorney, or professional clergy; all consider them professional. The title was bestowed because of the ethics, behavior, and standing of the members of these groups in the community. Legislating a title "professional surveyor" or "professional engineer" does not in itself prove that the activity the bearer of title engages in is a profession.

If the surveyors want to acquire and maintain a professional reputation, they can only earn that right by the average standing of all surveyors.

Within property-surveying practice there are many grades of workers. One of the failures, and many engineers' failure, is to properly distinguish between technical level and professional level. On highway work the engineer in charge has surveyors who make measurements to determine the shape of the ground. The surveyors are merely measuring the ground as it exists and recording the facts as they are. To be sure, the surveyor must have superior ability in knowing how to use instruments and how to make measurements, but this is purely technical. They do not design the road, nor do they utilize their measurements. Again, the engineer may tell surveyors to grade stake a road in accordance with a given plan. Since no design or judgment is involved, it is a purely technical matter. To the average engineer, surveyors are technicians who carry out his orders. And often surveyors are just that. But the engineer frequently overlooks the fact that there is a professional property surveyor level.

If a person is successful at his craft, he is forced to leave it. The chainman who is good soon advances to an instrumentman. With study, the instrumentman becomes a chief of party. With further study he is licensed and tries to emerge as a professional man still lacking the basic requirement of education that is typical of many of the other learned professions. As he advances with experience and promotion, he becomes a manipulator of people. If a man succeeds as a professional man, he soon discovers that he must leave the craft routines to others and desert, to some degree at least, his craft companions. He must work less with things and more with people. This is the progress to professional standing.

5-4 Attributes of a Profession

If the surveyor or engineer wants to be considered in a learned profession, he must approach the attributes of the learned professions, some of which are these:

1. Unique and superior education in a specific field of knowledge.
2. Service to the public in ability to persuade.
3. Position of trust.
4. Ethics.
5. The possibility of gaining highest eminence with financial return of secondary importance.
6. Independent judgment and liability.
7. Providing services to those unable to pay and in need.
8. If fees are charged those able to pay, fees are dependent on services rendered rather than labor or product.

5-5 Superior Education in a Field of Knowledge

PRINCIPLE. *Superior education in a field of knowledge is one essential feature of a profession.*

Formal education is not necessarily the only means of acquiring knowledge, since knowledge can be attained by experience and self-effort. Many attorneys have been admitted to the bar without the advantage of a college degree, but today that has been changed by most state legislatures. Before an attorney is admitted to the bar he must demonstrate that he does have superior knowledge.

The learned professions depend heavily on formal education and curricula leading toward a professional degree. A property-surveying degree is now offered in a few university-level colleges and in a number of 2-year associate programs. Fresno State University of California offers a 4-year program; Purdue University offers a master's degree in surveying; Ohio State specializes in geodesy. Surveying in most colleges is associated with engineering or forestry, as at the University of Florida. As of this writing, land surveyors as a whole do not have the formal educational standing possessed by the higher professions. If education is the development of the thinking faculties of the individual and not training, land surveyors may deserve higher standing than formal education indicates, but as of yet this concept is not recognized by the public.

Education and knowledge can never be purchased; each individual must acquire it by his own effort. The major deterrents to the development of surveying as a learned profession are the present low requirements for the right to practice. Until such time as the knowledge requirements for registration are raised, surveyors in the minds of others, will be relegated to an inferior professional position.

5-6 Service to the Public

Knowledge in itself does not make a person a professional; one must use the knowledge to aid, assist, teach, and benefit others. The professional has a call of duty beyond the fee and beyond other selfish interests. Application of knowledge to assist and aid in the affairs of others is an essential part of the definition of a profession. A professional has clients or students who directly benefit from the application of this knowledge.

5-7 Position of Trust

Professionals are often delegated an exclusive franchise for the purpose of protecting the public from the unqualified. In exchange for this exclusive franchise the professional does have moral obligations to the public; this individual is in a position of trust. The surveyor in monumenting the exterior lines for a client must also determine the adjoiners' boundaries. In most surveys the adjoiner accepts the results without question. If an error is made someone usually suffers a loss. As an obligation to the public the surveyor should not in any way assist a client in acquiring land not rightfully his. He must protect everyone, including those not paying a fee.

5-8 Gaining Eminence

Professional eminence is earned because of superior ability to apply knowledge for the benefit of others. It is not derived from the ability to earn money or from the notoriety of undesirable publicity.

To be a successful professional surveyor, one must have more than a narrow technical education. Technical education has to do with things. Employees at the bottom of a professional scale deal with things; professionals deal with people and situations. The fundamental concept in human relationships is that it is not sufficient to be right, a person must also persuade. Things cannot be persuaded; humans can be persuaded. All the technical knowledge in the world is of little aid unless a person can also convey this knowledge to others in a convincing manner.

If surveyors are going to participate in activities outside their technical fields and furnish leadership in the broader affairs of the community, they must possess the ability to persuade and the knowledge to determine the right direction in which to lead followers. The manipulation of things, the manipulation of the transit, and the manipulation of calculating machines do not constitute professional standing; professionals persuade and manipulate people in a correct direction.

5-9 Independent Judgment and Liability

The layman seeks help to aid in solving problems beyond his scope of knowledge. The knowledge and experience of a professional, combined with his ability to reason and apply sound judgment, is why his services are sought. The professional, after consultation and determination of the facts, uses independent judgment in arriving at a solution to a problem. Because professionals usually charge a fee for their advice, and as they profess to have superior knowledge, they assume a liability for failure to exercise proper care.

One of the greatest deterrents to substandard professional practice is the fear of liability. Although liability to an individual may be considered a disadvantage, it is an advantage to a profession as a whole. Without liability those willing to accept professional prestige, yet willing to do substandard work for less money, would soon ruin professional standing. Pecuniary punishment of the careless and indifferent soon improves lagging professional attitudes.

5-10 Services to Those Unable to Pay

Any learned professional has a moral duty to all regardless of the ability to pay. Doctors are obligated to serve the sick; the clergy serves those in trouble; and the attorneys defend the criminal. However, fees are charged, and those able to pay must pay. This is as it ought to be. Seldom do property surveyors display this type of obligation. Property surveying is not an urgent necessity; if it is not done today, it can be done tomorrow. If a person cannot pay today, his survey can wait until tomorrow. Attorneys adopt more or less this attitude for services in connection with business matters.

5-11 Fees

Money in itself does not enter into the definition of a profession. Although it is admitted that professional people must somehow gain a livelihood, services are not denied to those in dire need.

The most suitable means of obtaining a livelihood seems to be based on a fee charged commensurate with the services rendered. True professional fees are not dependent on the physical labor or force applied. Personal knowledge gained through experience and education creates the demand for the service rather than the size of the muscle in the arm. Businesses compete on the basis of lowest prices; professions do not.

Although money does not enter into the definition of a profession, it does have a profound influence on what others think of a profession. A group that shows by its outward appearances that it is not successful in handling its own financial affairs and lacks sound business judgment can hardly instill confidence in the public.

5-12 Ethics

DEFINITION. *That branch of a moral science which treats of the duties a member of a profession owes to the public, to his professional associates, and to his client is ethics.*

Ethics is not susceptible to an exact definition. The foregoing definition, adapted from *Bouvier's Law Dictionary*, rather clearly expresses the intent and purpose of ethics.

> It is a fact with which everyone is familiar that an individual may strictly observe the laws of the land and yet be an undesirable citizen and a poor neighbor. The idea that each individual can and should establish for himself rules of conduct for such relations as are not covered by law and without reference to the experience or opinions of others seems equally as absurd as would a similar attempt to establish principles of law. Laws must be established by the majority action of a legislative body, and rules of professional conduct must be based on the concurrent opinions of the members of a profession.[1]

Rules of ethics, as adopted by any profession, are not intended to particularize; they are general guides of conduct and behavior.

Advocating observances to ethics is not sufficient; the surveyor's personal example is far more important. It is not sufficient that the surveyor alone feels that he has honesty and integrity; the public, clients, and fellow practitioners must also believe so. The proof of observance of ethics lies in the opinions of others.

If the professions are to maintain a respected position in the community, they must look beyond the club of the law to ethical standards that prohibit the doing of what the law does not forbid.

5-13 Surveyor's Obligations to the Public

PRINCIPLE. *The surveyor has the following obligations to the public:* (1) *to see that the client's boundaries are properly monumented without subtracting from the rights of the adjoiner;* (2) *not to initiate boundary disputes;* (3) *not to aid in unauthorized surveying practice;* (4) *to see that those licensed to survey are properly qualified by character, ability, and training;* (5) *see that those who prove unworthy of their privileges have those privileges deprived; and* (6) *agree not to attempt to practice in any professional field in which one is not proficient.*

---

[1] Daniels W. Mead, *Standards of Professional Relations and Conduct*, American Society of Civil Engineers.

5-14 Obligations in Monumenting Boundaries

Every boundary survey for a client identifies the boundary of an adjoiner. A reason for giving surveyors the exclusive privilege of marking boundaries is to prevent the unskilled from monumenting lines that encroach on the bona fide rights of an adjoiner. As an obligation to the public the surveyor should not in any way assist a client in acquiring rights to land that are not his to enjoy legally.

5-15 Stirring up Litigation

Provoking litigation, according to common law, is a crime known as *maintenance*. If the offender in a land-boundary case is a surveyor, he is doubly at fault. A surveyor may act as an arbitrator and try to smooth over a difficult boundary situation, but he should not stir up litigation as a solution to the problem, especially where he would collect an expert fee as part of the litigation.

Because the surveyor is morally bound to protect the bona fide rights of the adjoiner, he should not hesitate to point out what the rights of the adjoiner may be. If there is a long-continued possession and title has probably passed by acts of possession, and if knowledge of such facts would tend to prevent the client from entering in litigation, the surveyor should not hesitate to disclose such facts. However, the surveyor should also suggest that attorneys are the proper parties to render a legal opinion on such matters.

5-16 Aiding Unauthorized Surveying Practice

PRINCIPLE. *No surveyor should permit his name to be used in aid of, or to make possible, the unauthorized practice of surveying by any agency, personal or corporate.*

What is unauthorized practice of surveying must ultimately be resolved by the courts, and ethics committees should be bound by their findings.

The selling of signatures for a fee—that is, a surveyor signing a surveyor's certificate and certifying to the correctness of a survey where the work was not performed by him or his employees or his direct subordinates—is the most flagrant violation of the intent of a registration law.

A layman may be properly hired by a surveyor, provided that his services do not constitute the practice of surveying and that his compensation is not a proportion of the fee. Having a layman partner in charge of a corporation practicing surveying is a violation of ethics.

5-17 Qualifications of Surveyors

Applicants for registration are required to furnish a list of professionals as references. It is the duty of those replying to the Board of Registration, in response to questions of a person's qualifications, to disclose all unfavorable as well as favorable qualifications of the applicant. Friendship, family relationships, sympathy, or any other reason should not influence any statement.

Likewise, surveyors and surveyor organizations must take the initiative to comment on the quality of the questions being asked on surveyors' examinations and to call attention to infractions of ethical considerations.

5-18 Unworthy Surveyors

Occasionally, those who are licensed prove themselves by their conduct unworthy of licensing and should have this privilege removed. Surveyors are better able than laymen to appraise the qualifications of other surveyors. If any surveyor is frequently negligent or exhibits negligence in his duties, that fact will be noticed by other surveyors, who, as a group, can prefer charges.

5-19 Surveyor's Obligations to His Client

The surveyor, when performing a given service for a client, assumes certain ethical obligations in addition to liabilities. But these obligations to the client may not supersede or interfere with the surveyor's obligations to his associates or to the public. He should serve his client faithfully, but he should refuse to do that which is illegal and unethical and that which violates a duty of responsibility to others. The surveyor advises his clients about what is right and proper, and if the client insists otherwise, the surveyor must withdraw.

Regardless of the fee charged, the surveyor is obligated to perform a correct survey within specified accuracies. Although there are times that the property owner will agree that an inaccurate or approximate property-line survey will suffice for his purpose, the surveyor ought not to accept such a commission. Future owners, not knowing the circumstances under which the monuments were established, will be misled. Approximately located monuments could be the basis for fraud or deceit on the part of the property owner or the client. Most people assume that surveyors' monuments are located correctly; hence the mere finding of any property corner may be the cause of a misconception and lead to costly litigation.

Communications between the surveyor and client are confidential and should remain so. But the surveyor may not be a party to an illegal act or fraud, and communications concerning illegal acts or frauds are not confidential. At law the surveyor is not given the right to withhold privileged communications. But unless required by law to disclose the business of a client, communications are confidential.

5-20 Surveyor's Obligations to Other Surveyors and the Profession

A profession is distinguished by the fairness and courteousness of one practitioner to another and the unwillingness to encroach on the clients of another. Businesses aggressively compete for customers; professions do not. Members of a profession value the esteem of his associates and the prestige of their calling, especially so for those of mature age. But those who steal another's clients do not induce cordial reception or pleasant relationship, as it ought to exist, among surveyors. Surveyors not only have obligations to one another but also to the profession as a whole.

In general surveyors should not criticize another's work in public; they do not take advantage of a salaried position to compete unfairly with other surveyors; they do not become associated with other surveyors who do not conform to ethical practices; they do cooperate with other surveyors concerning information of mutual interest of benefit, and they do support their professional organizations.

5-21 Professional Reputation

PRINCIPLE. *The surveyor does not attempt to injure falsely or maliciously, directly or indirectly, the professional reputation, prospects, or business of another surveyor.*

Confidence and respect for a profession are gained by praise of one member of the profession for another. Constant sniping between professional people can only degrade the profession.

This ethical rule prohibits the surveyor from falsely or maliciously harming the reputation of another. This does not prohibit the right of any surveyor to give proper advice to those seeking relief from negligent surveyors. Surveyors should expose, with substantial evidence, at the proper time and place, dishonest conduct in their profession and should not hesitate to accept employment that will assist a client who has been wronged. But it is distinctly bad taste and poor manners to accept the word of the client without first checking with one's professional associates. Many times those doing the accusing are biased in the presentation of their side of the story.

5-22 Employment

Under the older rules of ethics it was unprofessional to supplant another land surveyor after definite steps were taken toward his employment, and it was unprofessional to compete with another surveyor for employment by reducing usual charges. Since the time the Federal Government brought antitrust actions against various professional groups and won, it is now illegal to include such statements in rules of ethics. Under present law the surveyor may not take into consideration minimum prices adopted by group actions.

5-23 Discovered Errors

PRINCIPLE. *In the event a property surveyor discovers an error or disagrees with the work of another property surveyor, it is the duty of that surveyor to inform the other surveyor of such fact.*

Surveys are not for the purpose of stirring up arguments and fights between neighbors. If the adjoining property has been surveyed by another surveyor and the two surveys are not in agreement, the matter should be discussed between the surveyors prior to announcing that an error exists. Sometimes evidence found on the first survey may indicate that a different principle should be used in the later survey. Of course, if a surveyor has made a genuine blunder, other surveyors should not honor the mistake, but the first surveyor should be given an opportunity to review and prove the correctness of his survey, if he can.

5-24 Review of Another's Work

PRINCIPLE *It is unprofessional to review the work of another property surveyor or for the same client, except with the knowledge or consent of such property surveyor or unless the connection of such property surveyor with the work has been terminated.*

Consulting with another's client is considered to be, or has the appearance of being, an attempt to supplant the other engineer or surveyor. If a request is made to review the work of another, the person making the request should be informed of the ethics involved, and the other surveyor should be promptly notified of the facts. Even in the event that the work of the other surveyor appears to be fraudulent or neglectful and the other surveyor probably will be charged with misconduct, it is the duty of the surveyor to communicate with the other surveyor and give him the opportunity to reply.

5-25 Discredit to the Profession

PRINCIPLE. *It is unprofessional to act in any manner or engage in any practice that will tend to bring discredit on the honor or dignity of the surveying profession.*

A surveyor should not aid a client in perpetrating a fraud or assist him in an illegal act. He should not, in his personal appearance or manner of conduct before the public or others, bring discredit to himself as a professional.

5-26 Fees

PRINCIPLE. *No division of fees for surveying service is proper, except with another surveyor who by his license is permitted to do property surveying work. In determining the amount of the fee, it is proper to consider (1) the time and labor required, the novelty and difficulty of the questions involved, and the skill requisite to properly conduct the survey; (2) the customary charges for similar service; (3) the amount of liability involved and the benefits resulting to the client from the services; (4) the contingency or the certainty of the compensation; and (5) the character of the employment, whether casual or for an established and constant client. No one of these considerations in itself is controlling. They are mere guides in ascertaining the real value of the service.*

Minimum fee schedules can never be a binding agreement between surveyors. Antitrust laws prohibit such price fixing.

5-27 Advertising

Professions all refrain from advertising in self-laudatory language or in any other manner derogatory to the dignity of the profession. Professionals earn their stature; they do not try to win acclaim by advertising like a used-car dealer. Recent court decisions now bar professional societies from adopting a no advertising policy in their ethics.

5-28 Professional Standing of Surveyors

What the surveyor thinks of himself is not the proof of professional stature. What others proclaim him as being is his standing.

Courts in response to liability litigation have taken a positive stand that all surveyors are liable in the same manner as are other professional people. This is proof of professional standing, but it is not proof of degree of eminence of standing, nor is it *prima facie* evidence that everyone thinks surveyors are professionals.

One thing is certain: professional standing can never be attained by self-proclamation; if a person wants professional standing, he must earn it, and others must bestow that title on him.

REFERENCES

Brown, Curtis M., "The Professional Status of Land Surveyors," *Surveying and Mapping*, Vol. 21, No. 1, 1961.

Drinker, Henry S., *Legal Ethics*, Columbia University Press, New York, 1953.

CHAPTER 6

Historical Development of Property Surveying

6-1 Purpose and Scope

History or knowledge of what has happened in the past is an indispensable tool of the retracement surveyor. How measurements were made, what accuracy could be attained with available instruments, what were the customary materials used for monuments, what laws were in force at a given time, and the knowledge of the surveyors are necessary bits of background information needed to relocate former surveyed lines intelligently. The retracement surveyor's duty, often stated as "follow in the footsteps of the original surveyor," is a feat that cannot be done without some knowledge of past events. A history of surveying and knowledge of the early customs and practices surrounding the use and ownership of land help to clarify complexities of land laws and to explain why surveyors must do as they do.

After Columbus' discovery, the United States was, with minor exceptions, settled from the East to the West. The development of the survey systems in this chapter is treated in much of a geochronological order, beginning with the colonial era, then the French, Spanish, Mexican, and other sovereign influences on this country's boundaries. Following this are treated the divisions of the public domain and the various economic and geographical events that have influenced the shape of this country.

Underlying most of the U.S. systems of surveys are the English measurements and methods. The establishment of the mile[1] and the origin of the chain both play key roles in the definition of metes used to describe real property.

Events of history help explain some of the characteristics of property lines found today. In the Southwest, the army was sent to protect surveyors from the Apache Indians. To keep the soldiers busy, they were instructed to erect large stone mounds for section and quarter corners. This is now one of the best-monumented areas in the United States; the mounds can be seen from more than a half-mile away. In other areas the Indians secretly followed the deputy surveyors and destroyed the newly set monuments. In the extreme northern section of California, the subdivision of townships was being carried on during

---

[1] Probably the earliest known mile was the one used by the Persians, being 1000 fathoms long. The name comes from the Roman "Mille" that means one thousand and was established in Rome by 1000 paces, the pace being equal to 5 Roman feet.

143

the Modic Indian War (1873–1874). The original notes explain the many interruptions because of hostile action from the Indians. Deputy surveyors were forced at times to carry guns; this attracted the points of the magnetic compass.

6-2 Planned and Indiscriminate Land Conveyancing

In some states along the eastern seaboard land was sold in accordance with a plan made prior to a sale, especially in New England. In other eastern seaboard states land was conveyed in an indiscriminate manner; the purchaser was given a warrant for a given quantity of land, and he hunted for and located whatever suited him. After locating a parcel of vacant land, the purchaser arranged for a survey, and the warrent became effective on recording the prepared deed. In many instances the survey and description encroached on previous surveys either inadvertently or deliberately. In 24 counties of Georgia, containing about 8,718,000 acres, 29,090,000 acres of land were granted! In North Carolina, with a land area of about 55,000 square miles, the state granted over 100,000 square miles! Thomas Lincoln, father of Abraham Lincoln, bought a farm in Hardin County, Kentucky, and when he sold it, he lost 38 acres because of lapping boundary claims. In a second purchase he lost the down payment, plus the cost of litigation when a better title was presented. At Knob Creek he lost a third farm in a court case. This was enough; Thomas Lincoln moved to Indiana where Abraham Lincoln, age 8 at the time of the move, was reared.

The adverse results from indiscriminate locations and the cost of litigations that followed were among the reasons why the Continental Congress developed a presurvey system for the public domain.

Principal areas where indiscriminate locations or where settlement prior to survey took place are as follows: eastern New York, eastern Pennsylvania, colonies south of Pennsylvania, Tennessee, Kentucky, southeastern Georgia, West Virginia, parts of Ohio, eastern Texas, and the Mexican or Spanish grants, particularly in California and Texas. Although the Mexican or Spanish grants located within the public domain were indiscriminate in original site selection, each claimant was required to present proof of ownership to the federal court. As part of the court procedure, a survey was required; in effect both location and title were established by the court.

PROPERTY SURVEYS IN EARLY HISTORY

6-3 Civilization and Land Ownership

The earliest recorded accounts of property surveys are Egyptian, but it must be assumed that the Babylonians practiced such an art even as early as 2500 B.C. A Babylonian boundary stone set in 1200 B.C. was inscribed with the translated meaning "Itti-Marduk-balatu, the king's officer, measured that field."

The civilization that developed along the fertile banks of the Nile greatly depended on the fruits of this rich bottomland. The annual spring floods posed a

Courtesy of Dean James Kip Finch

Figure 6–1 The rope stretches.

serious problem when the markers set to define the little tracts of cultivated land were washed away. It is believed by some that the Egyptians devised the science of geometry to restore lost monuments.[2] It is also believed that a system of coordinates, referenced to points on high ground, was used to perpetuate monuments.

Shown on a wall of the tomb of Thebes and carved on a stone coffin in the Cairo, Egypt, Museum are drawings of "rope stretchers" measuring in a field of grain. Witnessing the act are three officials with writing materials (Fig. 6-1). The Great Pyramid of Gizeh (2900 B.C.) has an error of about 8 inches in its 750-foot base, or a relative error of 1 part in 1000 on each side. The Egyptian measurements in land surveying were made by means of a rope, and such linear accuracy is entirely possible. Is it not logical to assume that the same techniques and accuracy could have been used to relocate the lost corners along the Nile? Some of the boundary monuments set as early as 1300 B.C. are in existence today, and distance measurements between them agree accurately with the ancient records.

Scholars of the ancient Egyptian culture believe that these peoples respected land ownership and that land was a measure of wealth. The Egyptians have left evidence of tax registration. In 2000 B.C., in ancient Mesopotamia, clay tablets recorded land surveys wherein areas were divided into rectangles and triangles. The British Museum has preserved a clay cuneiform tablet, dating back to 2400 B.C., that contains the measurements and statistics for 11 fields from the reign of Bur-sin, King of Ur.

---

[2]According to Herodotus, a certain king named Sesostris should be credited with the invention of geometry: "This king divided the land among all Egyptians so as to give each one a quadrangle of equal size and to draw from each his revenues, by imposing a tax to be levied yearly. But everyone from whose part the river tore anything away, had to go to him to notify what had happened; he then sent overseers who had to measure out how much the land had become smaller in order that the owner might pay on what was left, in proportion to the entire tax imposed. In this way, it appears to me, geometry originated, which passed thence to Hellas."

According to Herodotus the science of surveying originated in Egypt. The first extant treatise on surveying is that of Hero of Alexandria (about 130 B.C.).

About the same time as the Egyptian beginnings, the Chinese recognized land ownership and protected rights to land. Under a type of feudal system there was little or no transfer of land ownership; property belonged to the family or clan.

The Greek and Roman cultures contributed greatly to the development of cadastral science. It is recorded that Anthony employed "geometers" to divide land. "De Agrorum Qualitate" by Froninus, written 2000 years ago, is the basis of property law and defined things "movable" and "immovable." The Roman Empire was supported in part by the collection of taxes, mostly on land.

6-4 Biblical References to Land Surveying Problems

Many references can be found in the Bible relating to the ownership of land. Joshua directed the subdivision of the unoccupied lands to the Israelites. The land, according to the Old Testament, Book of Joshua, Chapter 18, was divided among the seven tribes in seven parts.

The land on which Abraham and Lot were living was too poor to support them, so they separated. In Genesis 13:14–18, there is the account of how the Lord spoke to Abraham giving the land to him and his heirs.

One of the earliest recorded transfers of land is found in Genesis, Chapter 23. Abraham wanted a suitable burial ground for Sarah, who had died. He negotiated with Ephron for a field and a cave to bury his dead.

> And Abraham hearkened unto Ephron; and Abraham weighed to Ephron the silver, which he had named in the audience of the sons of Heth, four hundred shekels of silver, current money with the merchant.
> And the field of Ephron which was in Machpelah, which was before Mamre, the field, and the cave which was therein, and all the trees, that were in the field, that were in all the borders round about, were made sure unto Abraham for a possession in the presence of the children of Heth, before all that went in at the gate of his city.
> —Genesis 23:16–18.

The basis of all real property title in the Christian world may be found in Genesis 1:1, "In the beginning God created the heaven and the earth." Kings were given the sovereign right to land by the Pope, and they in turn gave ownership to members of the court. The account of Jacob's purchase of land is in Genesis.

> And Jacob came to Shalem, a city of Shechem, which is in the land of Canaan, when he came from Padanaram; and pitched his tent before the city. And he bought a parcel of a field, where he had spread his tent, at the hand of the children of Hamor, Shechem's father, for a hundred pieces of money. And he erected there an altar and called it Elelohe-Israel.
> —Genesis 33:18–20.

Ahab, king of Samaria, wanted to acquire a vineyard that was near his palace, belonging to Haboth. Ahab offered to purchase the vineyard, but Naboth refused, "The Lord forbid it me that I should give the inheritance of my father unto thee." Jezebel, Ahab's wife, perpetrated a false charge of blasphemy against Naboth, and he was stoned to death. Ahab now possessed the land, but not for long, since the wrath of God overtook him and punished him for his wrongdoing (I Kings, Chapter 21).

Many accounts appear in the Old Testament relating to boundary monuments.

> Thou shalt not remove thy neighbor's landmark, which they of old time have set in thine inheritance, which thou shalt inherit in the land that the Lord thy God giveth thee to possess it.
> —Deuteronomy 19:14.

> Cursed be he that removeth his neighbor's landmark. And the people shall say, amen.
> —Deuteronomy 27:17.

> Some remove the landmarks: they violently take away flocks and feed thereof.
> —Job 24:2.

> Remove not the ancient landmark which thy fathers have set.
> —Proverbs 22:28.

6-5 From Homer's *Iliad*

From the *Iliad*, Book 21, lines 499–503 (Morris translation):

> She only stepped backward a space, and with her powerful hand lifted a stone that lay upon the plain, black, huge, and jagged, which the men of old had placed there for a landmark.

This is the account of Goddess Pallas in conflict with God Mars.

In Book 12, lines 503–508 (Morris translation):

> As two men upon a field with measuring-rods in the hands, disputing stand over the common boundary, in small space, each one contending for the right he claims, so, kept asunder by the breastworks, fought the warriors over it.

6-6 Sovereignty and Ownership

From the beginning of the Christian Era, the Pope vested sovereign rights in rulers. Even in the non-Christian cultures and the pagan societies, the ruler has been recognized as having eminent rights over the land of the nation.

Rulers conveyed parcels to members of their court and expected, in turn, tribute in the form of taxes, rent, services, or royalties. The term *estate* derived from the word "status," represents the interests one has in property, the extent of the rights and privileges one can enjoy therein.

One of the inherent and necessary attributes of sovereignty is the right of eminent domain. In the broadest sense, a nation has the right or power to take private property for public use; it is superior to all private property rights. The common-law principle of eminent domain has been greatly modified and amplified by the constitutions and statutes of the nations and their states. See Chapter 17.

6-7 Feudal Land System

After the battle of Hastings, 1066, William the Conqueror had taken all of Great Britain and immediately claimed all the land as his own. Half of this he gave in parcels to his favorite Norman lords. A survey, the first written basis for taxes, was ordered in 1086 to compile a list of every farm, owner, and implements in the Doomsday Book. The survey was executed by groups of officers called *legati*, who visited each county and conducted a public inquiry. By listing all feudal estates, both lay and ecclesiastical, the Doomsday Book enabled William to further strengthen his authority by exacting oaths of allegiance from all tenants on the land as well as from the nobles and churchmen on whose land the tenants lived.

The feudal landlord conducted his life quite independent of the king and of his neighbors. He offered protection to those who lived within his domain, and, in return, the tenants were required to pay rent, share the crops of their harvests, and fight for their lord when ordered to do so. The lord of the manor often leased land, termed "fiefs" or "feuds," to other noblemen or vassals.

The landlord would enclose his private estate for hunting and other personal pleasures; the area not enclosed was termed "Commons." Peasants were permitted to farm, pasture their livestock, and try to satisfy other basic needs on the Commons.

More and more land was enclosed until many peasants could not find enough Commons land to subsist on. Stealing and other crime increased during the Middle Ages, and many peasants were executed for petty larceny. Some landlords obtained large tracts by conquest of neighbors, and others enlarged their estates by bartering.

At the close of the fifteenth century, England was not very densely populated, but people were starving for want of land. It was nearly impossible for anyone other than those of noble birth to acquire title to property.

6-8 Livery of Seisin

Written conveyances of real property may have been executed at the beginning of civilization. Much land was conveyed by a ritual known as the *livery of seisin* (delivery of possession). The parties to the transfer would meet on or in sight of the land, and through a series of solemn acts, such as the handing over of a twig, a handful of soil, or a signet ring, would memorialize the contract. Other demonstrations, such as throwing stones, driving stakes into the ground, and shouting and uttering such words as "I give," were witnessed to bind the conveyance. If the ritual was performed on the land, it was termed a livery *in deed*, or if within sight of the land, a livery *in law*. The grantor was required to practice *abjuration*, that is, to vacate the land within 6 months. Written contracts may have accompanied the ritual, but these were only evidence of the conveyance, not the conveyance instrument.

Although livery of seisin is replaced by delivery of a written conveyance today, certain parts of the ritual still remain in parts of the country, and the ritual was practiced in England until 1845. An example of the livery of seisin is found in a document dated July 1824, to Robert Millicam from the Mexican Government:

> We put the said Robert Milicam in possession of said tract of land, taking him by the hand and causing him to walk around it and telling him in loud and audible words, that by virtue of the commission and powers vested in name of the Government of the Mexican Nation we put him in possession of said tract of land with all the uses, customs, rights, and services thereof, unto him, his heirs and assigns, and the said Robert Millicam, in evidence of being in real and personal possession of said tract of land, without any contradiction, cried out, pulled twigs, threw stones, drove stakes, and performed the other necessary solemn acts.[3]

6-9 Statute of Frauds

A parliamentary act of 1677 required contracts to be in writing. Section 4 of the Statute of Frauds states:

> No action shall be brought unless the agreement upon which such action shall be brought, or some memorandum or note thereof, shall be in writing, and signed by the party to be charged therewith or some other person thereunto by him lawfully authorized.

Of the five classes of contracts, one was this: "Any contract or sale of lands, tenements, or hereditaments or any interest in or concerning them." Much of the Statute of Frauds has been overruled, but the requirements for land conveyances to be in writing and the need for the grantor to sign such an instrument are the basis of conveyance laws in this country.

The Statute of Enrollments rendered a conveyance void unless recorded within 6 months.

6-10 Early Property Surveys in the New World

The earliest surveys in the New World were to map the country rather than to delineate property boundaries; information for suitable homestead land was quite meager, and surveyors were needed to explore and map rivers and other waterways that were the natural means of travel. Among the surveyors sent to Virginia to locate land allotments was William Claiborne who, in 1621 at age 38, was paid 20 pounds per annum and furnished with a house. In 1629 for defeating the Indians, he was granted a tract of land on the Pemunkey River.

---

[3]Quoted by Thomas R. Newton in "Land Is a Precious Thing," *Surveying and Mapping*, Vol. 13, No. 3, 1953, p. 365.

Claiborne's trading post, founded on Kent Island in 1631, was included in the grant to George Calvert; a dispute continued over the title until 1776.

The men who conducted the surveys came from a variety of backgrounds. Some were astronomers and mathematicians such as Mason and Dixon, and some were frontiersmen like Thomas Hutchins. Then as now there was little formal schooling in surveying, and surveyors with scientific background were rare. The colonial surveyors usually held high social position and were generally able to advance themselves financially. One basic fiber surveyors usually possessed was a solid foundation in mathematics and astronomy that placed them in the class of better-educated citizens. Surveyors were often selected from the teaching or ministry professions, and surveying often was a second or part-time profession.

From 1621 to 1624 the surveyor general for Virginia was appointed by the governor and from 1624 to 1693 by the king. The College of William and Mary on February 8, 1693, was charged with the responsibility of surveyor general appointments as well as the examination and licensing of surveyors. Among the famous men appointed were George Washington (in 1749 at age 17) and Thomas Jefferson. Peter Jefferson, Thomas' father, was a surveyor, and so was Peter Jefferson's grandfather. Thomas Jefferson was appointed surveyor of Albemarle on October 14, 1773, the same county in which his father was assistant surveyor. Joshua Fry and Peter Jefferson were commissioned in 1749 to run a part of the Virginia-Carolina boundary line.[4]

The first registered county surveyor (Culpepper County, Virginia, in 1749) in America was George Washington. His grandfather, Colonel John Washington, came to Virginia in 1657 with wealth and influence. Although he purchased large tracts of land, the wealth of the Washington family was soon dissipated. George was brought up at Mount Vernon, where it is supposed that he learned the "art" of land surveying from his half brother, Lawrence. George's surveying practice, as such, lasted only about 2 years. During this time he used his position to further his land speculation and to locate lands for some of the wealthy Virginia plantation owners, the principal one being Thomas Fairfax in the Shenandoah Valley. In 1751 George and Lawrence joined the British army against the insurgent French. While in the army, George was able to obtain from other officers title to 32,373 acres in West Virginia. With land speculation in full swing, he sent an agent to England to promote buyers for his tracts. Being a very shrewd businessman and having the advantage of being a surveyor, Washington built a large fortune in land; and, by the time he died in 1799, his holdings included 49,000 acres in the rich Kanawha River Valley of West Virginia, as well as vast tracts in Virginia, New York, Ohio, and other places.

Many exploratory surveys were conducted by the army for fortifications and other uses. Thomas Hutchins and Henry Bouquet explored numerous rivers and the frontier country; others conducted surveys for homesteads. The Virginia

---

[4]"The History of Surveying in the United States," *Surveying and Mapping*, Vol. 18, No. 2, 1958.

assembly in 1783 created a board to survey the 150,000 acres given to George Rogers Clark and his men for services rendered.

Methods and instruments used in the colonial surveys were little better than those used in ancient times. The determination of latitude was usually derived from solar observations with a sextant or circumferenter that at best was within a half minute of arc. Mason and Dixon apparently made astrolabe observations on several stars to attain their excellent consistency.

The determination of longitude was not practical in the early days because of the lack of accurate time propagation. Early grant descriptions called for differences in longitude that had to be measured on the ground. Considering the lack of geodetic knowledge, these distances were computed with fair accuracy.

In colonial days direction was usually determined by the magnetic compass. Star observations (Polaris at elongation) were made to check the variation in the needle. The solar compass did not come into use until about 1835. Straight lines were run with pickets by a succession of compass readings or by the circumferenter (Fig. 6-2), a French designed instrument in general use until about 1800.

An English mathematician named Edmund Gunter (1581–1626) contributed to the science of surveying such discoveries as magnetic variation, a portable quadrant for star observations, the introduction of the words cosine and cotangent to trigonometry, and his practical measuring device called, in honor of him, the *Gunter's chain* (Fig. 6-3). The chain, $\frac{1}{80}$ mile or 66 feet long, was and is composed of 100 wire links. The acre, today's measure of area, was defined by him as 10 square chains. The length of a chain was said to be 4 *poles* (later known as rods, poles, or perches) long. Even though the chain was a great improvement over ropes and bars used before, there was still considerable error in its use caused by the lack of standardization, frequent blundering, and surface measurement. Steel ribbon tapes were not in general use until early in the twentieth century.

Figure 6–2 Circumferenter.

Figure 6-3 Gunter chain.

A practical telescope, short enough for ordinary surveying instruments, was not available in colonial times and did not appear on the popular instruments until the early nineteenth century. The circumferenter had sight vanes, a small compass, spirit level, and was mounted on either a staff or a tripod. The sextant was usually hand held and resembled the instrument used by mariners of that day. On land, when making latitude observations, it had to be used with an artificial horizon, with an expected accuracy of location of $\frac{1}{2}$ to 1 mile of the correct latitude. Perhaps the greatest device used by the early surveyor was his own ingenuity.

6-11 Concept of Title

Modern concepts of land ownership have developed from Christianity and feudalism and fee tail ownership evolved from absentee ownerships and feudalism. Fee simple ownership at first was reserved for the favored few. *Title* may be defined as the legal basis or grounds for land ownership, and it may be transferred by descent through heirs or by purchase. According to law, conveyances include not only that acquired by buying but also by device, will, grant, adverse possession, escheat, condemnation, and the numerous other ways land title can be transferred.

Real property is not a creation of humankind; original ownership, therefore, can come only from conquest or discovery. Private ownership of land has many injustices, but perhaps it is the least of several evils. An alternate (although entirely undesirable) plan would be nationalization of land as prevails in some noncapitalistic countries. As civilization progressed, strong-armed men assumed ownership by force. *All titles today run back to and are maintained by force.*

Real property in Europe has been divided so many times that title is composed of a multitude of slivers. In many instances, reapportioning is necessary to establish equity in the chaos.

Today, people accept title and title transfer of real property. With the present high value of property, delineation must be exacting and interpretation made with extreme caution. The property surveyor must have an adequate understanding of the background of land titles to approach property problems intelligently.

Land titles today are held by many people and are protected by more legislation than any other singular right or posession. The property surveyor, in locating a title, is charged with a tremendous responsibility that cannot be discharged lightly.

SURVEY SYSTEMS IN THE EAST AND SOUTH

6-12 The Origin of Title

An ideal situation could have existed if all this nation's land had all been acquired at one time, from one source, and included no prior grants. This nation is indeed fortunate to have so vast an area covered by the Rectangular System that, in spite of its problems, has been a tremendous help to title, resurveys, and other functions necessary to land ownership. Twenty states have not been subdivided under the National Land System (see Fig. 6-4), and parts of all the rest of the states have exceptions and prior grants.

English, Dutch, French, Spanish, Mexican, and even Russian grants have been upheld in America, and the title going back to these sovereign nations reflects the system (or lack of one) imposed by custom or royal decree. It is a dangerous practice to generalize in regard to any one system or area, and even a conservative treatment of the various survey systems in the United States would fill several volumes. It is the authors' intent in the next few pages to draw a few examples of these complex survey systems and to point out that such extremely local problems must be studied thoroughly on the local level. A successful Illinois surveyor on reaching retirement age migrated to Florida so that he could continue his practice in the comfort of a warm climate. In spite of the fact that

154 / HISTORICAL DEVELOPMENT OF PROPERTY SURVEYING

Figure 6–4 States without public domain, rectangular system (Hawaii not shown).

both states were public land states, the differences in the local systems were so many that the new business soon failed, and the surveyor was forced to return to Illinois.

6-13 Surveys in the Early States

Virginia. The original charter for the colony of Virginia was granted in 1606 and reissued again in 1609. The description was quite indefinite, calling for "West and northwest to the South Sea." The description in the 1609 charter reads, in part:

> All the territory two hundred miles northward and two hundred miles southward of Old Point Comfort and all that space and circuit of land, lying from sea coast of the precinct aforesaid, up into the land throughout from sea to sea, west and northwest.

In 1776, Virginia gave up claims to area now in Maryland, Pennsylvania, North Carolina, and South Carolina. In 1792, Kentucky was allowed to become a state, and West Virginia was separated in 1866. At the time of the Revolution, Virginia had land in Ohio, Indiana, and Illinois.

The grant to the "South Sea" was to the Pacific Ocean, and it was later reduced to the Mississippi River by a treaty with France.

The London Company, which held the original charter, planned to have lands surveyed and assigned to shareholders at the rate of 50 acres per share; later, the rate was raised to 100 acres per share. Also introduced by the London Company was the *head right concept*; the head of the family could claim 50 acres per head for all servants and persons with paid transportation to the colony. After the dissolution of the London Company because of a lack of funds, the Virginia colony claimed the charter, but later kings gave grants of the Virginia area to others such as Maryland to the Calverts in 1632, Pennsylvania to William Penn in 1681, and the Carolinas to the Carolina proprietors in 1663 and 1665.

Land speculation gained momentum beginning in 1700. The king often empowered lords coming to Virginia to make grants in areas where the governor had already granted lands. Just before the Declaration of Independence was signed, squatters were given the right to preempt up to 400 acres of land on which they settled at a cost of 3 pounds per 100 acres (soldiers' claims were free). It is very understandable why the legal profession was quite busy at the time.

Early surveys were of poor quality, run with a compass, with distance often measured on the ground (not horizontal), mostly done by unskilled operators. Boundaries were poorly marked, and, frequently, abutting grants encroached on one another. After one understands this original conveyancing method, one finds it easy to understand why laws were enacted to give the occupant a relatively simple method of clearing title by evidence of possession. In some of the southern seaboard states only 7 years are needed to perfect adverse rights. In this area and in many of the other early states (West Virginia, Tennessee, and Kentucky included), possession is a very important factor in determining ownership status. By contrast, possession in many of the western states is of very minor importance. In no way can it be said that the survey problems in the older states are comparable to those of the newer states. Although the basic laws for interpreting the meanings of deed wordings are similar all over the country, the applications of the laws, because of differences in original conveyance methods, vary considerably.

New England. The Plymouth Company was organized in England in 1606 and in 1607 received a charter from the king of England at the same time as the Virginia charter. The operation of the Plymouth Company was not a great success, and in 1620 it was reorganized as the Council of New England. Grievances among the colony inhabitants finally led to the House of Commons declaring the charter forfeited in 1635. New England was not developed as a unit; many different grants were given to different areas. The Massachusetts Bay Company received a charter in 1628, and the charter contained, in part, the following description:

> All that part of new England in America which lies and extends between a great River ther [sic] commonly called Monomack River, alias Merrimack River and a certain other River ther, called Charles River, being in the Bottom of a certain Bay ther, commonly called Massachusetts, alias Mattachusetts, alias Massatusetts Bay; and also all and singular those lands and hereditaments whatsoever, lying within the Space of Three English miles on the South part of the said River, called Charles River, or of any, or every part thereof; and also all and singular the lands and

hereditaments whatsoever, lying and being within the space of three English miles to the southward of the Southern most part of the said Bay, called Massachusetts, alias Mattachusetts, alias Massatusetts Bay; and also all these lands and hereditaments whatsoever, which lie and be within the Space of Three English miles in the Northward of the said River called Monomack, alias Merrimack, or to the Northward of any and every part thereof, and all lands and hereditaments whatsoever, lying with the limits aforesaid, North and South, in latitude and breadth, and in length and longitude, of and within all the Breadth aforesaid, throughout the main lands ther, from the Atlantic and Western Sea and Ocean on the East Part, to the South Sea on the West part.

This charter was surrendered and a new charter was granted by William and Mary in 1691. Connecticut received a charter in 1662; Narragansett Bay (R.I.), in 1663. New Hampshire (then including Vermont) received recognition as a province in 1680.

The present northern boundary of Massachusetts was first surveyed and marked in 1741. The east-west line with Rhode Island was agreed on by commissioners appointed from the two colonies in 1711, but it was not surveyed until 1719. Because the line was not very accurately located by this survey, almost ever since there have been suits in the U.S. Supreme Court contesting the position of the boundary.

The earliest grants of land in New England were of irregular size and generally without a plan. Soon it became apparent that it was advisable to have uniform regulations for the disposal of land. As early as 1634 a committee was formed to locate the boundaries of unsurveyed towns and to settle boundary disputes. One result of the committee's work was a law that required the walking of town boundaries every 3 years. Since this law was never repealed, many areas still practice walking of town boundaries, although it may be every 5 or 7 years. In Fig. 6-5 is a town line marker set by surveyor "W" in 1934 as part of a perambulation.

In New England the word *town* is not to be confused with the word *city*; it is a political area that may be composed of large farms.

When the need arose for additional land for settlement, commoners (proprietors) petitioned the court for permission to establish a town in accordance with a plan presented. The plan usually had some house lots, land areas of variable sizes—often 40, 60, 80, or 100 acres each—and reservations for the school, church, and minister, plus areas to be held in common. Rangeways (land set aside for roads) were allowed for between ranges and tiers of lots. Since the roads were not granted, the roads, in many cases, remain in the town. Since these "future roads" were laid out on paper with no regard for topography, many could not be used. When attempts were made to use them, considerable variation was necessary to avoid steep hills, ledges, ponds, and other obstacles.

After lotting plans were prepared, the custom was to "draw" at random and often in accordance with the number of shares held by the settler or proprietor. The settler could be entitled to more than one lot, and most settlers were entitled

6-13 SURVEYS IN THE EARLY STATES / 157

Figure 6–5 An early New England ordinance required town lines to be perambulated at regular intervals. In the process town line posts were often erected as pictured in this figure. The "T" and "L" stand for town line, the "W" for the surveyor who set the post, and the 1934 is barely visible. "Sapling" is scribed on the left side of the post, and "Misery Gore" is on the right side, each facing the towns of those names.

to a home lot. If the lots drawn were deficient in either quality or quantity, the settler could petition for another parcel. In some places "pitches" were made, meaning that the settler knew how much land he was entitled to and was instructed to enter the area of yet ungranted land and stake out his area in any place that would not encroach on another.

Kingston, New Hampshire had lots of 20, 40, 60, and 200 acres, home lots, meadow grants, and commons for the public. As an illustration of a town plan, Candia, New Hampshire is shown in Fig. 6-6. The lots were generally 80 or 100 acres in size with a few of variable sizes and dated 1719. The following is an example of a Candia deed: in 1764 Caleb Brown purchased the west half of Lot 57 by the following description:

158 / HISTORICAL DEVELOPMENT OF PROPERTY SURVEYING

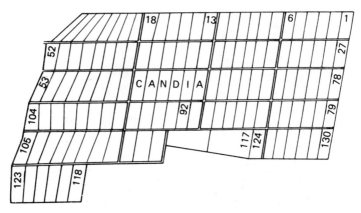

Figure 6-6 Plan for the town of Candia prepared prior to 1719. The lots of 80 to 100 acres, numbered in the sequence indicated, were granted to proprietors according to a drawing held in 1719. None of the original proprietors settled on the lots, but by the time of the Revolutionary War, many of the lots were occupied. In New England the word *town* is not to be confused with *city*; a *town* is a political unit, not a city.

A certain piece or parcel of land lying and being in Candia, containing forty acres be it more or less and is the westerly half part of the 80 acre Lott of land be it more or less which I bought of John Taylor of North Hampton in ye provence aforesaid, which lott was laid out to the right of Joseph Bacheldor and said lott is bounded as follows: First at the Northwest Corner at a Beech tree which is the Northeast bounds of the 56th lott, then East South East 72 rods to a Hemlock No. 57, then South 29° West 180 rods to a beech tree No. 57; then west northwest by the highway 72 rods to a beech tree which is the South East Bound of the 56th Lott, then in straight to the first mentioned bounds.

From the foregoing a survey must have been made; tree monuments were called for. It is possible that 260 years later the trees are gone and possession may be the best indicator of where the original surveyor traveled. Several of the roads shown on the Candia plan (Fig. 6-6) were never built because of difficult terrain, and those that were built certainly were not straight.

In New Hampshire a surveyor of highways and a fence viewer were elected to locate new roads, settle fence arguments, and perambulate town boundaries. Caleb Brown served in both posts and settled on Lot 92 of Fig. 6-6 about the time of the Revolutionary War.

In addition to locating the "lotts" in New England the surveyor usually judged the quality of the land; poor lots usually have an overrun of area. Ten percent overrun is not uncommon. Surveyors and assistants (often called "junior surveyors"), as well as chainmen, were under oath and returned a report about what was done. The plan and description of the addition to Alexandria, New Hampshire, contained this note:

June ye 1:1773. This Plan Sheweth ye Number of Lotts in the addition of Alaxandrea Joyning on the South westerly Sid of said Alaxandr and the Easton side of the Patten Line to Grate Sunipe Pond then by Said Pond to fisherfield then Easterly on fisherfield to Parryton North Line then Down Parrytown Line Easterly to a Beach Tree which is the Corner of Said alaxandra: it is Divided in to 137 Lotts Each Lott Containing 150 acrs numbered as is Set Down in the Plan Lay^d Down by a Scale of one Mild to one Inch and as Convenant as the Land wold alow I Laid Tew Senter Squars of Ten acrs Each for Publick Uses as marked in the Plan the Ponds are as Near as Posable Laid in their Shape and Bignes and the Streems Drawd as thay Run throw Said Land Said Streems Run Easterly from the Patten Line mesured by me.

Jeremiah Page, Syayar of Land.

John Tolford and Robart Mcmrphey, Junr Cheen man June ye 23:1773 Then Jeremiah Page appeared and made Solomn oath that this Plan of the addition of Alaxandrea by him Drafted is Just and Trew acording to the Best of his Skill and Judgment———.

Benj^a Day Just Peac.

In 1727, new towns along the Connecticut and Merrimack rivers were instituted. Two townships were established in Maine and five in Massachusetts. In Vermont, in 1760, Governor John Wentworth planned a subdivision with 6-mile squares; for this Joseph Blanchard surveyed along the Connecticut River, marking township corner trees every 6 miles for about 60 miles. Many later town charters referred to this plan, but purchasers had to locate the bounds at their own expense. Herein lies the problem found in most of New England; early survey records are not available. Whereas it is probable that surveys were made by approximate methods—that is, approximate by today's standards—the task of the surveyor today is to try to prove what was done at early dates. In many recent surveys the lowest forms of evidence must be resorted to, such as reputation. As compared to the Virginia indiscriminate method of disposal of land, the New England town developments by a plan were far superior; only one line existed between lots, and no overlap between lots existed, as frequently occurred in Virginia.

Much of New England was cleared of trees, and the large rocks in the fields were moved to boundary lines. Although not all rock walls are boundary lines, many are (see Fig. 6-7). During the great depression many of the lands were left idle, and second growth of trees filled in where pastures formerly existed.

Connecticut. In 1630 the Plymouth council made a grant to the Earl of Warwick. A later charter, granted by Charles II, was described as

> all that part of our dominions in New England in America, bounded on the east by Narraganset River, commonly called Narraganset Bay, where the said River falleth into the sea; and on the North by the line of the Massachusetts plantation; on the South by the sea; and in longitude as the line of the Massachusetts colony, running from east to west, that is to say, from the said Narraganset Bay on the east, to the South Sea on the West part with the islands thereunto adjoining.

Figure 6-7 In New England rock walls commonly mark property lines. This is a picture of a lane with rocks on each side.

Later Connecticut made a trade of equal parts with New York, and, at the time of the Revolutionary War, still held western lands in Ohio, Michigan, Indiana, and Illinois. Because of the many disputes over the state boundaries, resurveys were conducted as recently as 1925 to confirm positions.

Rhode Island. Settled by Roger Williams and followers in 1636, this small area was granted a charter by King Charles II in 1663. The description calls for natural objects, recites metes and bounds, and was originally included in the Charles I grant to Connecticut. Although some dispute developed over the location of the boundaries, the state never had claim to western lands.

New Hampshire. The territory of New Hampshire was included in the original charter to Virginia in 1606 as well as in two other grants of 1620 and 1622. The area, described by metes and bounds, was granted by the president of New England to John Mason in 1629. Controversies over the borders (including the one with Canada) flourished for many years. New Hampshire had no western lands.

New York. The area of New York, claimed by Henry IV of France as part of the territory of Acadia in 1603, was included in the grant to the Virginia colony by James I of England in 1606. The French and Indians held territory west of the Hudson River and were the cause of many years of dispute. In 1613, the Dutch established trading posts along the Hudson and claimed jurisdiction for the territory between the Connecticut and Delaware Rivers. The United New

Netherlands Company, chartered in 1616, and later the Dutch West India Company (chartered in 1621), were vested with the government of this Dutch settlement.

In 1664 King Charles II of England granted to his brother, the Duke of York, a large territory in America that included, with other lands, the area between the Connecticut and Delaware Rivers. In 1664, with an armed fleet, the Duke of York took New Amsterdam and renamed it "New York." In 1673, the Dutch recaptured New York and held it until 1674 when it was restored to the British by treaty.

The area as described in the Duke of York's charter was reduced by many sales, cessions, and trades. With the exception of the relatively small triangle on Lake Erie, the colony relinquished no western lands.

The border between New York and Canada, intended to be along the forty-fifth parallel, was found by the Collins and Southeir survey of 1773 to vary as much as a half mile from this location. By the treaty of 1842, this survey was confirmed as part of the northern boundary of the United States. The monumented line between New York and New Jersey also varied from the intended straight line, giving New York about 10 square miles more than stated in the grant from the Duke of York to Lord John Berkeley and Carteret. In 1834, each state ratified an agreement accepting the line as it was marked on the ground.

For the most part, the wilderness of New York was subdivided by privately owned land companies, and these companies could divide and sell their purchases as they saw fit. Little uniformity resulted between the various parts of the state. Most of the land companies patterned their names after the U.S. Rectangular System, using the terms "range" and "township." *Ranges* varied in widths from 4 to 8 miles, were numbered from either the east or west sides, and were oriented as much as 45° off north. *Townships*, like ranges, were reasonably rectangular but varied in dimensions and directions, containing 40 to 144 lots.

The lots, so-called and found inside townships, were not usually marked on the ground prior to passing title.

The Military Tract for Revolutionary War veterans was an instance of the state's making a prior survey of a large area. Originally the area was to include 640,000 acres, but it was later expanded to include more than 750,000 acres. Two ranges of six townships apiece had each township divided into approximately 93 lots.

One of the outstanding private land subdivisions was the *Holland Purchase*, located in western New York. The tract was referred to a *base line* and *transit lines* previously surveyed. Pennsylvania's boundary line formed the base line, and lines run due north formed the transit lines. Created under the direction of Joseph Ellicott in 1797, the system contained townships nearly 6 miles on a side, with lot corners every three-quarters of a mile. With 64 lots in each township, numbered 1 to 64, beginning at the southeast corner and progressing north along each tier, a lot contained 360 acres nominally. Occasionally, some townships were divided into large lots, 1½ miles on a side. Townships were numbered toward the north from the base line and westward from the east transit line.

Tracts in the Phelps and Gorham Purchases, located further west in the state, were subdivided by a similar plan.

In 1883, the state legislature ordered New York public lands to be resurveyed because of confused and undeterminable lines. During the years following, the survey resulted in the first accurate location of public lands. Triangulation, using 12-inch and 20-inch theodolites across the Adirondack Mountains, was the primary control. To complete this survey, it was necessary to locate almost all the boundaries of the large land purchases. The original notes of some of the earlier land company surveys were still in existence in 1884 when Verplanck Colvin, superintendent of surveys, conducted the state land surveys. Careful study of these old notes, along with Colvin's notes, would be necessary before a modern surveyor would attempt to resurvey the property lines in much of rural New York state.

Most of the New York state lands were titled under English Common Law principles, but in the New York City area some titles were based on Dutch law, and, in these instances, the title to the streets and highways may rest with the state. A few titles go back to Swedish grants.

As surveyor general of the state of New York, Simeon DeWitt supervised the subdivision of most of the lands in the northern and western parts of the state. Together with James Clinton and Andrew Ellicott, he surveyed the Pennsylvania–New York boundary from March 31, 1785, until the summer of 1786 at a cost of $4,750.27 for 90 miles. Every 20 to 32 miles they took sights on six or more stars with a 5-foot sector and accepted only those observations that agreed with the mean within 4 seconds. It was thought that their position error was within ½ second or 50.5 feet.

Simeon De Witt had a brother named Moses who was a compass man on the Pennsylvania–New York boundary line survey in 1786. In 1826, De Witt made a detailed atlas of the state.

New Jersey. The Virginia grant of 1606 included New Jersey. The first specific grant for this territory, the grant that all New Jersey title chains go back to, was that of 1664 from the Duke of York to Lord John Berkeley and Sir George Carteret. In 1676, the lands of New Jersey were divided into Carteret's "East New Jersey" and William Penn's and other Quakers' "West New Jersey" by a line running from the extreme northwest corner of the state on the Delaware River to the southern tip at Little Egg Harbor. In 1702, the two parts were joined into the "Province of New Jersey" and later the Colony of New Jersey. The division line was surveyed by John Lawrence, George Washington's half-brother, in 1743, and it cut through several of the present-day counties.

New Jersey was the last state in which title to vacant lands resided wholly in proprietors. A court decision took away title to tide lands in favor of the state.

The bounds of the grant were the Delaware Bay or River, the Atlantic Ocean, and a straight line connecting a point on the upper Delaware River with a point in the Hudson River. Considerable dispute has arisen over the jurisdiction of the islands in the Delaware River and over the location of the line between New Jersey and New York. New Jersey held no western lands.

In the early history of New Jersey, feudal tenures were established; later they were abolished by the following laws (still on the books):

> The feudal tenure estates, and the incidents thereof, taken away, discharged and abolished from and after March twelfth, one thousand six hundred and sixty-four, by section two of an act entitled "An act concerning tenures," passed February eighteenth, one thousand seven hundred and ninety-five, shall so continue to be taken away, discharged and abolished; and no such estate, or any incident thereof, shall, at any time, be created in any manner whatsoever.
>
> The tenures of honors, manors, lands, tenements, or hereditaments, or of estates of inheritance at the common law, held either of the king of England, or of any other person or body politic or corporate, at any time before July fourth, one thousand seven hundred and seventy-six, and declared, by section three of an act entitled "An act concerning tenures," passed February eighteenth, one thousand seven hundred and ninety-five, to be turned into holdings by free and common socage from the time of their creation and forever thereafter, shall continue to be held in free and common socage, discharged of all the tenures, charges and incidents enumerated in said section three.

Pennsylvania. As a result of the careless royal grants, many areas were ambiguously described, and William Penn was in dispute with all his neighbors over boundaries. The original grant defined the boundary as being 12 miles distant from New Castle and along the fortieth latitude north. It should be noted here that the early geographers counted the parallels in a different manner, such that this would be the thirty-ninth parallel. Because of inadequate knowledge of geography and poor descriptions, the line between Pennsylvania, Maryland, and Delaware was in dispute for nearly 100 years. In 1760 the line was finally decided on and surveyed (1763–1767) by two English astronomers, Charles Mason and Jeremiah Dixon. Besides the "New Castle Arc," they were to run the south line on a parallel for a distance of 5° west of the Delaware or 244 miles. Mason and Dixon determined the northeast corner of Maryland to be at latitude 39°43'17.6", and later surveys have found the position to be 39° 43'19.91" or about 180 feet in error. Huge limestone posts, cut in England with the respective coats of arms on the faces, served as monuments. The work of the survey was halted by Indian action short of the western border of Pennsylvania. The accuracy of the line is still a source of amazement; recent surveys reveal a variation of only 1 or 2" of latitude along this ancient line. A report by the legislature of Maryland (1909) lists a 2000-entry bibliography of manuscripts and documents relating to the line.

The eastern part of the territory was originally held and occupied by the Swedes (Swedish West India Company) in 1625. In 1655, the land was surrendered to the Dutch who yielded to the British in 1664 during the conquest by the Duke of York.

A dispute resulted over the southern border, the location of which was not completely settled until 1921. The large area known as Westmoreland County was under dispute with Connecticut until it was granted to Pennsylvania in 1782. The Erie triangle, comprising 324 square miles, was originally part of New York and was ceded to the United States in 1781 that, in turn, deeded it to Pennsylvania in 1792 so that the state could have more access to Lake Erie. The conveyance was signed by George Washington and carried a consideration of $151,640.25 or about 75¢ per acre. Pennsylvania held no western lands.

In parts of Pennsylvania the usual procedure for obtaining land was to buy a warrant for a given quantity of acreage in a given county near a developed area. After the selected site was surveyed and recorded at the expense of the purchaser a patent became valid. In places squatter settlements arose, and the settlers formed a mutual compact to settle all disputes over boundaries. Some titles, even today, are based on "squatter's rights" dating back to about 1718. With the diverse origin of titles, there have been many boundary disputes.

Maryland. Although the territory was included in the earlier charters of Virginia, Lord Baltimore, in 1632, received a royal grant of the Province of Maryland. The governor of Virginia, also Lord Baltimore, proclaimed a recognition of the Province of Maryland in 1638. The treaty of 1658, which was reaffirmed in the constitution of 1776, formally gave the lands to Maryland. Controversy over the boundary between Maryland and Virginia continued through the years and was not finally resolved until 1929.

Another lengthy dispute continued between Pennsylvania and Maryland and was settled by the Mason-Dixon line survey. Lord Baltimore's grant was described, for the most part, by natural objects and called for no western lands.

Delaware. Like Pennsylvania, the territory of Delaware was originally settled by Swedes and, in 1655, was surrendered to the Dutch who, in turn, lost it to the Duke of York. William Penn obtained a 10,000-year lease of the area, although the Duke of York had an uncertain title. Later Penn obtained a better title by royal grant. Under a proprietary government of Penn, the people of New Castle, Kent, and Sussex counties petitioned for separate statehood. The state of Delaware, which never had claims on western lands, is still very nearly the original size of the three counties.

Carolina. The charter to the Carolina territory, 1663–1665, was described as running to the "South Sea." In 1729 the territory was separated into North Carolina and South Carolina provinces because of their widely separated settlements. The present state of Tennessee was ceded to the United States in 1789 and became a state in 1796. In 1787 South Carolina ceded to the United States a small strip of land about 12 miles wide lying south of the thirty-fifth parallel. Other western land claims were later ceded to the United States.

Carolina, like so many of the colonies, had poorly defined metes and bounds surveys that did not clarify the relationship to adjoiners. The lack of systematic procedures resulted in controversy and confusion among landowners and resulted in litigation.

Georgia. In 1752 King George II separated the territory of Georgia from Carolina by a description calling for land to the "South Sea." The exchange of cessions between Georgia and the United States netted the state $6,200,000 or about 11¢ per acre. The larger part of Mississippi and Alabama were once in the original Georgia.

The separation of Georgia from the Carolinas began in 1717 when Sir Robert Montgomery obtained a grant from the Carolina proprietors for the land between the Savannah and Altamaha Rivers and called it Azilia, "the most delightful country in the universe!" This land was to have square districts, 20 miles each way, subdivided into various smaller squares, the least being 640 acres. The plan never got under way. Also in South Carolina, by order of the king in 1729, 11 square townships of 20,000 acres were to be created, but only nine were actually laid out. Not all were square, and not all contained the required 20,000 acres.

Georgia, the youngest of the 13 original colonies, was chartered by King George II as a charitable trust. In 1730 when the early Georgia settlers arrived with James Oglethorpe (Georgia's founder), Noble Jones, a surveyor, was second in command. The basic title to all lands emanated from either King George II, King George III, or the state of Georgia. Oglethorpe had authority to grant lands in the new colony, and all grants of record in the Surveyor General's Office in Atlanta were made between the years 1755 and 1909.

The original boundary of Georgia extended to the South Seas but for all practical purposes stopped at the Mississippi River. Under the grants of land the grantee was free to seek out only the most desirable lands. Large Revolutionary War bounties were made in the Creek Cession to the Oconee River. These early grants were based on a specific number of acres for each person or head of a family, including slaves. Additional grants were made for sawmills, grist mills, and other purposes.

Starting in 1805 a series of six land lotteries were surveyed for the purpose of granting the lands to private individuals by lottery. Each eligible person's name was registered in his county of residence, sent to the capital, and placed in a container. The presurveyed lot numbers were placed in another container, and the drawing of one paper from each container matched a person with a lot. After payment of a fee, the person was granted title.

Survey instructions for the lotting were minutely detailed, and each county was divided into numbered land districts, with each district subdivided into square lots. Lot sizes varied from 490 acres, 250 acres, 202½ acres, and 160 acres and 40-acre gold lots. Except for gold lots, lots were predicated on the value of the land or the producing capability of the soil. See Fig. 6–8.

This system was an abrupt departure from the metes and bounds surveys of the southern colonies and provided for surveys prior to disposal. Today, the

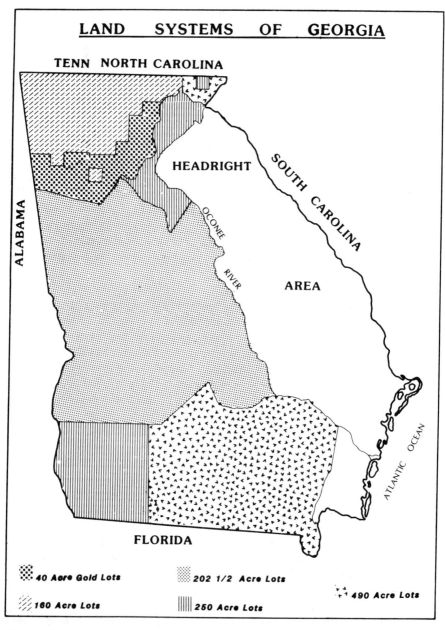

Figure 6–8 Land divisions in Georgia. Lot sizes were supposed to reflect the productivity of the soil. The 40-acre lots occur in the area of the first gold find in the United States.

office of Surveyor General in the Georgia Archives building in Atlanta maintains all grant information, original survey notes, and survey plats, plus all lottery information. Unfortunately, like so many of the early subdivided lands, few if any original corners are now available.

Trans-Appalachian States. Development of trans-Appalachian states was slow; Indians and a mountain barrier discouraged settlement. In addition the French claimed the entire Mississippi Valley. The British sought to restrain settlement beyond the Appalachians in 1763 by establishing a series of forts, and by proclamation in 1763 they refused to issue title to any lands west of the Appalachians. After the end of the French and Indian wars, settlement was rapid. Kentucky and West Virginia were largely controlled by Virginia's liberal granting of warrants for land; settlement was indiscriminate, with numerous overlapping claims.

Tennessee. Rights to land titles transcend English, state, and federal distribution of lands. The first settlement was made in the Watauga River Valley in 1769 by people who took possession without title or authority to settle. In 1772 these settlers negotiated specific leases for lands along the river.

The area of Tennessee was included in the original charter of North Carolina. After the Revolution North Carolina attempted to maintain the sovereignty of the Indians, but the settlers claimed the lands by conquest. The state granted tracts of 640 acres to settlers on a preemptive basis and also granted over 3,000,000 acres as a military reservation. In all over 8,000,000 acres of Tennessee lands were granted prior to 1789 when the remaining lands were ceded to the Federal Government.

The early entries were strictly selected on an indiscriminate basis; the best lands went first.

In the 6 years Tennessee was a territory under the jurisdiction of the Federal Government, no specific intent was made to introduce the Rectangular System of surveys. Also during this period North Carolina continued to grant lands in the area ceded to the United States. A system predicated on townships, ranges, and sections is still in existence in middle Tennessee adjacent to the Georgia state line, with most land descriptions referred to the aliquot parts. Long ago the field books and notes of the original survey disappeared.

6-14 The Subdivision of Ohio Lands

A large part of Ohio was public domain, subdivided by the early U.S. Rectangular System. The Seven Ranges, as described in Section 6-38, was the beginning of a great American experiment, and its evolution can be followed westward across the state. Other systems used in Ohio were a result of large purchases, military reservations, the Western Reserve of Connecticut, French grants, Indian treaties, and other circumstances. Undoubtedly, Ohio has more variety of survey systems than any other public domain state, and at least 20 different and distinct systems can be found. A more comprehensive discussion of this subject is found in C. E. Sherman, *Ohio Land Subdivisions*.

The Virginia Military District. Before ceding all its northwest territories, the new state of Virginia reserved some areas as bounties for returning soldiers and sailors. One such area, located in southern Ohio between the Little Miami and the Scioto Rivers, was indiscriminately located with sites varying from 100 acres to as much as 1500 acres. Land was inexpensive, and surveying was crude. In 1779, Virginia passed a law that provided an allowance of 5 acres in every 100 for variation of the instruments. Descriptions were by metes and bounds, often written from partial surveys and seldom written from prior surveys.

The Ohio Company Purchase. Since the public sales of the Seven Ranges were not a complete success, the Federal Government sold some large tracts of land to private companies. One such sale, to a company made up of New Englanders, included 1,128,168 acres just west of the Seven Ranges. Although it was specified by Congress that the land had to be laid out according to the Act of 1785, the company was left to do the surveying at its own expense. When it became necessary to proportion the land equitably among the shareholders, confusion prevailed. The general plan of the Seven Ranges was followed to some extent, although subdivisions as small as 0.37 acre were provided to the shareholders.

The Donation Tract, adjacent to and considered a part of the Ohio Company Lands, was a federal gift; and, similar to the Homestead Act, it was to encourage early settlers. The tract, containing 100,000 acres was to be granted in 100-acre lots. Such a division could not follow the Rectangular System of 640 acres in a section, and, although rectangular, the pattern was rather indiscriminate. Most 100-acre lots were further divided into numbered lots.

French Grants. To provide relief for some unfortunate French immigrants, several small tracts were granted along the Ohio River. Unlike the French grants along the Mississippi, these were not settled prior to the formation of the union but were made for people who were land-swindle victims. The tracts totaled to a little more than 25,000 acres and were subdivided into Ohio River front splinters. Ohio had surprisingly few grants prior to the Revolutionary War.

Symmes Purchase. Between the Miami and the Little Miami Rivers are several tracts not included in the U.S. Rectangular System. One was the Symmes Purchase that included 248,540 acres north of and including Cincinnati, all purchased at 67¢ per acre. The surveys within Symmes Purchase were quite irregular, although the section numbering was the same as that prescribed under the Act of 1785.

Western Reserve. The colony of Connecticut claimed title to lands between the forty-first and forty-second parallels, and from Pennsylvania 120 miles westward. On the basis of this claim, the state of Connecticut reserved a strip 120 miles long in the Cleveland area, and all titles to this land go back to Connecticut, although the jurisdiction over the area is and was under the state of Ohio. Using the Pennsylvania border as a meridian line, the land was subdivided into townships of 5 miles. Five hundred thousand acres were set aside in the western part of the reserve for relief of those who had lost their property in the Revolutionary War. Some of these 5-mile townships were divided into quarters and numbered clockwise; others were divided into sections of 1 mile square or laid out into tracts containing lots.

The United States Military District. The land set aside to pay the officers of the Revolutionary Army was subdivided in a similar manner as that of the Western Reserve. Many of the townships (5-mile) were subdivided into quarter townships or 100-acre tracts.

Northwestern Ohio. The last part of the state to be subdivided contained many different situations. Indian reservations, military posts, and lands donated for roads and canals under the "swamp-lands acts" created irregularities. In general, when applicable, these lands were subdivided according to the U.S. Rectangular System, but many grants and reservations resulted in exceptions.

6-15 Prior Land Grants in the Louisiana Territory

Prior to the purchase of Louisiana in 1803, French and Spanish grants were made extensively along the rivers tributary to the lower Mississippi. Many of the French grants called for a certain number of arpents (one arpent equals 191.8 feet, approximately) fronting on a river or bayou and extending back 40 to 80 arpents.[5] Although the frontage on the water was usually definite, the direction and depth was often vague and indefinite. Land was plentiful, and the back portion was often swampy and unimportant.

Moses Austin, whose son Stephen Austin is known as the Father of the Texas Republic, and his associates were granted land in Missouri in 1797; but the depression following the Napoleonic Wars caused bankruptcy to Moses. Before his death in 1821 Moses received permission to settle 300 families in Texas. His son, Stephen, completed the grant.

Shortly after 1803 the United States appointed a commission charged with reviewing title claims from either France or Spain. Proof of title right consisted of legal documents, surveys, or witness proof that land had been occupied for more than 10 years. After these claims were certified to, U.S. deputy surveyors segregated the land from the public domain by survey. In each case the deputies were instructed to favor the claimant as much as possible and to make every effort to establish lines between claimants to the satisfaction of each.

HISTORY OF PROPERTY OWNERSHIP IN THE SOUTHWEST

6-16 Lands of Spanish Origin

The Spanish era began with the discovery of America and ended with the Mexican independence in 1821. The Mexican era in the Southwest (encompassing California, Arizona, New Mexico, Texas, Utah and parts of

---

[5]Technically, an arpent is a square measure, and the 191.8 feet is the side of a square arpent.

Colorado, Wyoming, Kansas and Oklahoma) ended with the Treaty of Guadalupe-Hidalgo in 1848 and the Gadsden Purchase in 1853.

After the Mexican rights were extinguished, Texas retained title to its vacant lands. Patents to lands in the Southwest, after the Mexican era, were obtained from either Texas or the United States.

At the time of independence, Mexico agreed to recognize Spanish titles, and at the Treaty of Guadalupe-Hidalgo both Texas and the United States agreed to recognize acceptable Spanish and Mexican titles. In the Southwest there are four sovereign sources of original titles: (1) Spain, (2) Mexico, (3) Texas, and (4) the United States. Most of the patents issued by the United States were based on the Rectangular System, whereas the others were dependent on the system in force at the time and were often of the metes and bounds type.

6-17 Seniority of titles

Land grants and patents from Spain, Mexico and Texas were, in general, but not always, made in sequence with senior rights attaching to the older grant or patent. Because of this, the order of title alienation from the sovereign is important information needed to establish overlapping title rights.

Patents issued by the United States were, in general, surveyed prior to being offered for sale or for homestead. Descriptions of United States' lands were usually of the lot and block type (called "sections" and "townships") simultaneously created without senior rights attaching to older patents. This fundamental difference (senior rights and lack of senior rights) has created a substantial difference in resurvey procedures; lands patented from the United States usually have proportionate rights to surplus or deficiency, whereas lands with senior rights do not share excess or deficiency (some one person or the sovereign has title to excess, and someone takes the loss).

6-18 Rights Included with Spanish and Mexican Grants

The foundation of all land titles is conquest followed by grants from the sovereign. At the time the Mexican rights were extinguished, the new sovereigns, Texas and the United States, could have refused to recognize existing land ownership but chose to recognize all valid titles as a matter of principle. Although Article 10 of the Treaty of Guadalupe-Hidalgo did stipulate recognition of all Spanish or Mexican titles, it was never ratified by the U.S. Senate on two grounds. First, the public lands of Texas belonged to that state, and the United States had no control over them. Second, with respect to lands in other areas, Article 10 was unnecessary, since valid titles to land in such areas were unaffected by the change in sovereignty.

6-19 Minerals

Land grants from either the Spanish crown or Mexico did not pass title to minerals, not because of the language in the grants, but because the general laws in force applying to all grants. The United States and Texas, each acquiring the Mexican and Spanish lands, became the owners of all minerals.

The Republic of Texas, by quitclaim, released all mines and minerals as of the date of the Constitution, 1866, to the owners of the soil.

6-20 Spanish Water Laws[6]

The derivation of Spanish, Mexican, and English laws pertaining to waters and riverbeds stems from Roman civil law. In interpreting Roman law the English adopted the construction that public interest attached only to the water and that the bed belonged to the adjoiner. In England the sovereign owns the bed only insofar as the tide ebbs and flows.

By Spanish law, particularly in the United States, the general principle adopted was that the sovereign not only attached rights to the water but also to the bed. A distinction was generally made between navigable and non-navigable streams.

In the Spanish areas the term "perennial stream" was used to describe the limit of sovereign rights to the bed. All nonperennial stream beds, or as sometimes called "torrential stream beds," belong to adjacent owners, whereas the bed of perennial streams was retained by the sovereign. In general, a *perennial stream* is considered to be one that flows all or most of the year, except in times of drought. A *torrent* or *nonperennial stream* is caused by an abundant rain or abnormal melting of snow and is dry a greater part of the year.

Until 1837 lands within the boundaries of Texas followed these rules. After that date, "the 30-foot rule," by statute law, modified riparian ownership.

> All lands surveyed for individuals, lying on navigable water course shall form one-half of the square on the water course and the line running at right angles with the general course of the stream, if circumstances of the lines previously surveyed under the laws will permit. All streams so far as they retain an average width of thirty feet from the mouth up shall be considered navigable streams within the meaning hereof, and they shall not be crossed by the lines of any survey. All surveys not made upon navigable water courses shall be in a square, so far as lines perviously surveyed will permit.
>
> —Texas Civil Statutes Article 5302.

Prior to 1837, in Texas and the Southwest in general, if the stream bed was perennial and lay wholly within a grant, title to the bed was reserved in the crown. If the grant bordered on a nonperennial stream, title of the bed, up to its center, passed with the grant.

After 1837, in Texas, where the bed of the stream was more than 30 feet, the sovereign retained title to the bed. If the bed were less than 30 feet, the title

---

[6]"Title and Boundary Problems Relating to Riverbeds," Kenneth Roberts, *Seventh Annual Texas Surveyors Short Course*, Texas Surveyors Association, Austin, 1958.

passed with the grant; and if the grant bordered the bed, up to the middle of the stream was granted.

This so-called 30-foot rule complicated ownership, since the law was and is ambiguous. Does "so far as they retain an average width of thirty feet from the mouth up" mean average width in front of the grant or average width of all measurements to such point that the average is 30 feet? A stream could be 100 feet wide at its mouth and narrow to 10 feet and yet have an average width of 30 feet. To date the court has not ruled on this point.

The 30-foot statute also contained a clause that Texas grants "shall be square, so far as lines previously surveyed will permit." Most Texas land holdings are squares or as nearly square as may be.

According to Spanish water laws, water could be appropriated and used. Unlike the remainder of the United States, water rights could be established by usage; and, after the rights were established, they could be sold. California and the Southwest in general recognized this principle. At the time Boulder dam was built on the Colorado River a treaty was made between the lower basin states (California, Arizona, and New Mexico) and the upper basin states apportioning water between them. California became the biggest user of lower basin water and claimed it on the basis of use. The U.S. Supreme Court decided that the treaty at the time of Boulder Dam's construction was binding; usage was not the deciding issue.

6-21 Road Beds

Because roads were reserved for the crown of Mexico in fee, abandoned roads in Mexican and Spanish grants, existing prior to 1848, revert to the state of Texas or the United States.

6-22 Early Settlements

Despite the rule that no settlement or town could be legally established without the approval of the king of Spain, many were. The Crown soon realized that towns or settlements could not exist without means of livelihood; hence certain amounts of land were set aside and granted to the people to form *pueblos* or villages. Land was given to those who made conquests, and other land was sold; the remaining land could be used by the king's subjects to reap the benefit of natural resources. Although this land could be used, it could not be sold, tilled, or enclosed without permission. These common lands were known as *Valdios* or *Realengos*.

6-23 Ordenanza de Intendentes

Although there were rights in land granted prior to 1754, the *Ordenanza de Intendentes* of October 15, 1754, was the beginning for most land grants in the United States (and Texas). Its more important provisions were (1) the appointment of officials who could legally approve titles in the name of the king and (2) the granting of the right of perfecting a title within a specified time to those holding public lands (*Valdios* or *Realengos*) since 1700.

As a result of the Act of 1754 a commission granted some 400 porciones of land (1000 to 1500 varas by approximately 25,000 varas deep or about 4500 to 6500 acres) along the Rio Grande River, and about 140 of these are in the state of Texas.

6-24 Mexican Land Grants

Although independence of Mexico was not effected until 1821, many events started in 1808 led to the final declaration of independence. Following Mexican independence, land grants and colonizations were regulated by the central government. Later colonizations were delegated to the states, and in what is now Texas, three different states (Coahuila and Texas, Chihuahua, and Tamaulipas) passed their own liberal land laws. As a result of Santa Ana's declaring a dictatorship on March 2, 1836, Texas colonizers proclaimed a republic independent of Mexico; and, from that date on, no further Mexican land grants were made in Texas. Texas joined the United States in 1845. In 1848 the Treaty of Guadalupe-Hidalgo ceded the Southwest to the United States.

6-25 Empressario System

Within the state of Coahuila and Texas the empressario system of colonization was established. The empressario was given land in exchange for bringing in families, and each empressario was given a specific area in which to operate.

Until 1820 Congress in the United States sold land at $2 per acre, whereas in Texas it was much less.[7] Texas grew rapidly, and about one-seventh of Texas was alienated by Spain and Mexico up until the time of independence.

6-26 Suits Against the Sovereign

The king or sovereign cannot be brought to court without his consent; that is, the king can do no wrong. To straighten out many of the Texas title problems, the Texas legislature passed enabling acts allowing the citizens to bring title suit against the state. At various times the legislature of Texas passed laws voiding titles not recorded as of a specific date, but such acts were declared a violation of the Treaty of Guadalupe-Hidalgo.

6-27 Instructions to Surveyors

While Texas was under the government of Mexico, the following instructions were sent to the colonial surveyors by J. A. Nixon, commissioner for the eastern part of Texas:

---

[7] Also, there was up to 10 years of exemption from taxes.

Sir:

In consequence of your known honor, integrity, and ability, you are hereby appointed surveyor for the colonies granted to the following Empressarios by the Supreme Governor of the State, to wit, to Lorenzo de Zavala on the 12th of March 1829, to David G. Burnett on the 22nd December 1826, and to Joseph Vehlein and company on the 21st December 1826, and also another grant to Joseph Vehlein and company dated on the 11th of October 1828. And in the exercise of the duties of your office, you will be governed by the following instructions and such other as from time to time may be forwarded to you.

Article 1st Each and every surveyor shall provide himself with a compass after Rittenhouse construction.

Article 2nd All the lines shall be run in conformity with the TRUE MERIDIAN, and the greatest care shall be taken to have the horizontal measurement obtained by the chain carriers.

Article 3rd In surveying, chains of iron or brass 10 varas long shall be used and the length of each link shall be 6⅔ inches and the pins used for surveying shall not exceed 12 inches.

Article 4th Field books must be provided to keep the notes.

Article 5th The initials of the Grantees name must be cut on the bearing trees at each corner with a marking iron, and a mound raised three feet high, and three feet in diameter at the base around the stake and the timber shall be so blazed near the line that it may be followed with ease. Line trees shall be blazed, and a notch cut above and below the blaze.

Article 6th Rivers, large streams, and lakes must be considered natural boundaries, and no survey shall cross them, but their courses must be correctly taken and the contents of all surveys must be calculated by latitude and departure.

Article 7th All surveys that do not close by 50 varas must be corrected and make each league contain 25,000,000 square varas as near as practicable.

Article 8th On all natural boundaries, one half league front shall be allowed to each league of land so on in proportion to the whole quantity that may be surveyed.

Article 9th The field notes must be forwarded to this office so soon as the surveys are completed.

Article 10th No surveys will be acknowledged unless expressly ordered in writing by the Commissioner.

Article 11th Special care shall be taken that no vacant land be left between the possessions.

Article 12th A report must be made qualifying the lands and giving as near as practicable the quantity of arable and grazing land contained in each survey.

Article 13th All chain carriers shall be sworn by the Surveyor before commencing the survey, to perform their duties truly and faithfully according to the best of their ability and no person akin to the parties interested nearer than the fourth degree, shall be appointed to that survey.[8]

---

[8] From Ralph J. McMahon, "Perpetuating Land Corners in Texas," *First Texas Surveyors' Short Course,* December 16–17, 1940.

John P. Borden, who was the first commissioner of the General Land Office under the Republic of Texas, issued these instructions:

> The measure to be used will be varas and tenths of varas, or Spanish yards; each surveyor will therefore regulate his chain to the length of 10 varas, or, what is the same, 27 feet 9⅓ inches, the vara being 33⅓ inches; 5,099.01 varas on each side of a square will constitute a league and labor of land, or 4,605.53 acres; a labor of land will be equal to a square of 1,000 varas on each side, or 177.136 acres; and one acre, 5,645.75 square varas.

6-28 Survey of Spanish and Mexican Grants

A summary of the procedure for obtaining a Mexican land grant and the methods of surveying was ably written by Mario Lozano, public land surveyor, and is quoted from the report of the *Eighth Annual Texas Surveyors' Association Short Course* as follows:

> After complying with all necessary Judicial Proceedings, every one of which was a properly notarized instrument, the entire retinue composed of all the parties joining in the proceedings, would go to the PLACE OF BEGINNING where the first official act in connection with the actual survey was performed. The steps followed from this point on through the actual survey were as follows:
>
> The Judge of Measurements would command the surveyor to properly wax a rope, usually a hemp rope, and afterwards to measure 50 varas over said rope while held taut. The Vara measure used in this ceremony was supposed to be a standard and properly certified vara. After this was done in the presence of the entire party, it was made part of the record after attestation.
>
> The next step, of course, was to start the actual survey. The instruments dealing with this phase of the proceedings are the actual Field Notes. In these instruments they recorded Direction or Bearing, passing calls for ravines, hills or any other physical characteristics of the terrain over which they measured and they often indicated the watershed. At the end of each day the work done was attested and made part of the Expediente.
>
> In connection with these proceedings it is interesting to note that what we often deduce as corners in an Expediente often were what they called "Parajes." Now a Paraje is a Place. To be exact, a stopping place.
>
> Often times they would get to a corner or at least to what in reading the Expediente we would expect to be a corner and this was called a Paraje also. This, even though it was a corner of the grant, was not necessarily an exact point in the sense we think of the corners we set today. A paraje can be said to be a site of small proportions and often this particular site constituted a corner. Furthermore, almost all Parajes designated as corners were given a name. The name usually being connected to some physical characteristic of the site, with some incident or with the Saint of the particular day when it was reached.
>
> Not all Expedientes record the actual measurement of angles. But in instances when angles were measured, mention is made of the Graphometer; this is the case on some Mexican Grants. In other instances mention is made of the "Agujon"

which was the name given the Compass. In all surveys the Judge of Measurements would cause the Cordel[9] to be checked every fifty times it was used.

After the survey was completed the Judge would command the Surveyor to compute the areas and make his map. After this was done and made part of the Expediente, the Appraisers were appointed. Their acceptance of appointment was made part of the record and also their findings.

The Judge would then draw an instrument of transmittal and send the Expediente to the Treasury official or Department, which in turn would add the necessary instruments to show that the appraisal value had been paid. After this, the Expediente was sent to the proper authorities for approval and confirmation. When this was done, the Expediente was returned to the Judge or Notary that acted through all these proceedings with instructions to place the Grantee in possession of his land. At this point the Judge would take the Grantee to his newly acquired land, admonish him to comply with the law by properly marking his corners with Monuments, and place him in actual possession by taking him by the hand, walking him over the land where he would dig, touch a piece of timber, pull some grass, take some water to irrigate the land (by sprinkling), and go through various other acts in keeping with the ritual or ceremony connected with giving him actual possession.

After the act of putting the man in possession of his land the execution of the proper instrument, the Expediente remained in the custody of the Judge or Notary that officiated and became part of the Protocolo of said Notary. When a copy of the entire proceedings was requested and issued, it, of course, was a certified copy or "Testimonio."

The procedure used in acquiring a grant when Mexico was New Spain and later when Mexico became independent are quite similar since the procedure required to be followed for Mexican Grants was in substance the same as that used in acquiring Spanish Grants.

6-29 Gradient Boundary

In the Red River case of *Oklahoma* v. *Texas* in 1923 (260 U.S. 606, 261 U.S. 340, 265 U.S. 493), the Supreme Court of the United States declared by unprecedented action that the boundary between Oklahoma and Texas was to be determined by the "medial line" between gradient lines on each bank. Prior authority for such rule did not exist.

> One of the questions involved in the riparian claims relates to what was intended by the terms "middle of the main channel" and "mid-channel," as used in defining the southerly boundary of the treaty reservation and of the Big Pasture. When applied to navigable streams, such terms usually refer to the thread of the navigable current, and, if there be several, to the thread of the one best suited and ordinarily used for navigation. But this section of Red River obviously is not navigable. It is without a continuous or dependable flow, has a relatively level bed of loose sand over which the water is well distributed when there is a substantial volume, and has no channel of any permanence other than that of which this sand bed is the bottom. The mere ribbons of shallow water which in relatively dry seasons find their way over the sand bed, readily and frequently shifting from one side to the other, cannot

---

[9] A *Cordel* is 50 varas (about 138.9 feet) and, as used, is a rope 50 varas long.

be regarded as channels in the sense intended. Evidently something less transient and better suited to mark a boundary was in mind. We think it was the channel extending from one cut-bank to the other, which carries the water in times of a substantial flow. That was the only real channel and therefore the main channel. So its medial line must be what was designated as the Indian boundary.
—*Oklahoma* v. *Texas*, 258 U.S. 574

If the river had been navigable, the rule of "thread of the navigable current" would have applied. Determination of the "gradient boundary," a line on the bank, is difficult and requires great judgment. The gradient boundary cannot be determined without surveying long stretches of a river; hence it is costly to establish. After establishment, changing river conditions will change its location. Colonel Stiles discusses in detail the gradient boundary in *Texas Law Review* (volume 30, 1952, p. 305), and surveyors attempting to establish a gradient boundary should certainly read this report.

Since the Red River case Texas courts have made it clear that the gradient boundary principle is applicable to all boundaries between the state and private owners even for grants made prior to 1837.

6-30 The Effect of Indians

Accuracy of early surveys was sometimes in inverse proportion to the density of Indian population. Indians were well aware that settlers soon followed the surveyor; by exterminating the surveyor the Indians believed they could prevent the stealing of their land.

Standard survey equipment included chains, axes, compasses, rifles, and pistols. Constant vigilance to avoid surprise attacks caused anxiety and suspense. No wonder occasional chain lengths were forgotten, and corners falling in forbidding thickets were not set.

Captain Barlett Sims' survey party in 1846 was surprised by Indians, and three of the four were killed. A party of 10 surveying along the Guadalupe River, Texas, was attacked and all killed. Captain John Ervey's survey party of 10 had three killed in 1839. In 1838 Henderson's survey party of about 25 was attacked by Kickapoo Indians, which in Texas history is called the "Surveyors' Fight." Eighteen were probably killed, and seven escaped.

In Indian country, bobtail surveys were not uncommon; the surveyor set one or two corners and calculated the remainder.[10]

6-31 Early California History

The coast of California was the object of a search for a shortcut to the Orient. Balboa, in 1513, claimed the Pacific shores in the name of the King of Spain. Cabrillo stopped at San Diego Bay in 1542. The Spanish government discontinued exploration of the California area after Viacaino's voyage in 1603.

---

[10] Forrest Daniell, "The Thing That Steals the Land," *Surveying and Mapping*, Vol. 15, No. 4, 1955, pp. 461–467.

By 1768 the Spanish government realized the need for colonization if the threat of English and Russians occupying the territory were to be prevented. The Jesuits were ousted from control of the missions in Lower California, and in 1767 the Franciscans assumed control. Father Serra established the first mission in Upper California (San Diego) in 1769. Numerous missions followed, and by 1823 they numbered 21.

The soldiers and Fathers soon clashed over ownership of land. The Franciscans maintained, and at first successfully so, that all land was held in trust for the Indians until such time as they became civilized. The soldiers contended that the land belonged to the crown; and, in accordance with the Spanish laws of 1773 and 1786, land could be granted to both the Indians and soldiers. By the time Mexico attained independence in 1821 only 20 private ranches existed in California. Soon Mexican law called for expulsion of all missionaries, and a stringent secularization act in 1833 and 1834 ended the hold of the missions in California.

Under Mexican rule land grants became frequent to those held in favor, and, during the early era, no records other than brief notes were kept. It was simply a matter of public knowledge about who owned what.

During the Mexican era many U.S. citizens moved into California—especially seagoing men. By merely claiming Mexican citizenship and proclaiming themselves Catholics, some obtained land grants. After war with Mexico started, the presence of Americans in California eased the problem of conquest by the United States.

By the end of the Mexican era (26 years), in 1847, the number of land grants in California had increased from 20 to 800. And many of these grants were made by Governor Pio Pico in the last days of his reign.

The size of the grants varied from 28.39 acres (Rancho de la Canada de los Coches, the glen of the hogs), to one-fourth the size of Rhode Island (Rancho Santa Margarita y Las Flores in San Diego, Orange, and Riverside Counties contained, at one time, 226,000 acres or 354 square miles).[11]

6-32 Resurvey of Land Grants of the Public Domain

After the Treaty of Guadalupe-Hidalgo all Mexican and Spanish land grants in the public domain had to be processed through the federal courts. As a part of the procedure, a survey by a U.S. deputy surveyor had to be made, and all grants were made in conformance with the final approved survey. All indiscriminate parcels were described relative to the rectangular survey system. Overlaps, except for one grant on another, were avoided.

PRINCIPLE. *In the resurvey of a Spanish or Mexican land grant, the surveyor cannot go behind the court decree that established the grant; the field notes and survey are a part of the grant, and the grant is senior to all sectionalized lands.*

---

[11] *A History of San Diego County Ranches*, Union Title Insurance Company, San Diego, 1960.

6-33 Summary

From the foregoing it is apparent that the conveyance systems and methods of surveys within nonsectionalized states varied significantly from state to state. In older areas where anyone could perform a survey and anyone could prepare a map, many land location problems exist to this day. Today's surveyors should not be blamed for the unregulated systems existing in the past; the lack of regulations and the resultant chaos point to the necessity of good licensing laws qualifying who may practice land surveying. The system of allowing any person by self-proclamation to declare himself a surveyor has to go with other obsolete ideas.

THE U.S. RECTANGULAR SYSTEM

6-34 The Public Domain

Throughout this nation, about three-quarters of the area of real property may be traced back to the public domain. There are at present about 720,004,000 acres or about 30 percent of the land area still in the public domain.

The public domain is the vacant land held in trust by the government for the people. This area does not include federal reservations or other government lands set aside or purchased for specific purposes. It was acquired by purchases and cessions at a total cost of $85,179,222 or less than 5¢ per acre. The largest item of cost has been Indian claims, some of which are as high as $2.50 per acre, plus interest. Table 6–1 shows the contributions to the public domain, and Fig. 6–9 shows the land's location.

6-35 The Ordinance of 1785

A task committee, headed by Thomas Jefferson, was to draft "an ordinance establishing a land office for the United States." As a law graduate from William and Mary College, as an attorney, and as a man experienced in surveying (his father was a surveyor), Jefferson was well qualified to suggest new and better methods. Although much credit is due Jefferson for introducing the Rectangular System to America, it was not entirely original with him. Egyptians were known to use a rectangular system in the Nile Valley surveys. Land divisions by meridians and parallels had been of religious and mystical origin. The words "decumanus" (east and west) and "cardo" (north and south—the cardinal direction) were in use with an instrument called the *groma*. The one-hundredth of a nautical mile was the ancient division of the shire in England. This was adopted in America by Delaware, Maryland, and Virginia; but most of the land in the colonies was being sold with indiscriminate location, particularly in the South. North Carolina had a law requiring orientation with true north, although there were many exceptions permitted.

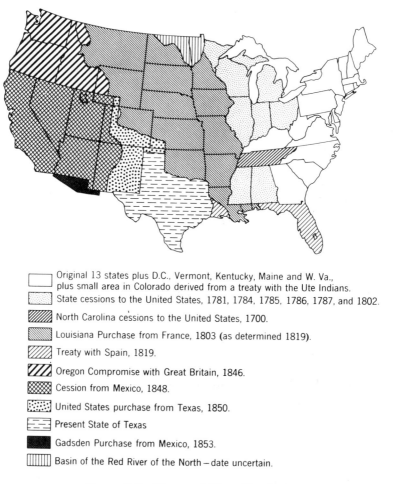

Figure 6–9 The acquisition of territories.

After much debate, a quadrangle with seven English miles on a side, containing 49 parts or square mile lots, was called for. Seven soon gave way to six, but the total of 49 was overlooked. Later, the law provided for a numbering system: "numbered from 1 to 36, always beginning the succeeding range of the lots with the number next to that with which the preceding one concluded." The final plan began numbering in the southeast corner, progressed north, then progressed south on the next tier, and so on. About half of the plats prepared for the Seven Ranges carried this scheme on the face, but it is felt that the surveyors did not follow such a pattern in the field.

6-36 Prior Surveys

The survey of the Seven Ranges introduced the system of *prior surveys*, that is, surveys prior to selling lands.

Table 6-1 Principal Sources of Land to the Public Domain

Date	Acquisition	Area (acres)	Total Cost (dollars)	Cost per Acre (cents)	States Included
1781–1787	Cessions from the original states	236,825,600	6,200,000[a]	11	Wisconsin, Michigan, Ohio, Illinois, Indiana, Alabama, Mississippi
1803	Louisiana Purchase	529,911,680	23,213,568	4	Louisiana, Mississippi, Arkansas, Missouri, Oklahoma, Nebraska, Kansas, Iowa, Minnesota, North Dakota, South Dakota, Wyoming, Idaho, Montana, Colorado
1819	Spanish cession of Florida	46,144,640	6,674,057	14	Florida, Louisiana
1825	Oregon compromise	183,386,240			Washington, Oregon, Idaho, Montana, Wyoming
1848	Mexican cession	338,680,960	16,295,149	5	California, Nevada, Utah, Wyoming, Colorado, Arizona, New Mexico
1850	Texas purchase	78,926,720	15,496,448	20	New Mexico, Colorado, Wyoming
1853	Gadsden Purchase	18,988,800	10,000,000	53	Arizona, New Mexico
1967	Alaska purchase from Russia	375,296,000	7,200,000	2	Alaska
Total		1,808,160,640	85,179,222	(5¢ average)	

[a] Part of the cession from Georgia.

Townships and sections were a compromise between the North and the South; the North wanted smaller tracts that would be available to the common man, whereas the Southern plantation owners thought more in terms of thousands of acres. The 640-acre unit was thought to be within the reach of homesteaders and small farmers.

The new law specified lines to be run in the cardinal direction, after correcting for the variation of the needle. Certification was required of the surveyor; the surveyor was to mark trees, and he was to erect corner monuments. The plats and notes were to include information on the topographical and cultural features on the land to "enable purchasers to judge of the value of the lands."

6-37 The Geographer of United States

The organizational structure of the Rectangular System called for the Geographer of the United States to direct surveys and transmit plats of completed surveys to the Board of Treasury. Simeon De Witt, until 1784, occupied this post; and, on resignation, he was succeeded by Thomas Hutchins.

6-38 Beginning the Survey of the Seven Ranges

The nearest land available, that adjoining Pennsylvania, was selected for the start of public land surveys, but before the land could be divided, the western line of Pennsylvania had to be established. In 1784 a survey was necessary to establish a beginning point. In 1784 a group of commissioners from Pennsylvania and Virginia began extending the Mason-Dixon line the final twenty-two miles needed to reach Pennsylvania's southwest corner. More recent surveys have shown that the mean error in this line was less than one second of arc. From the southwest corner of Pennsylvania, four commissioners, among whom were David Rittenhouse representing Pennsylvania and Andrew Ellicott from Virginia, ran a line northward for 63 miles until the Ohio River was reached. It is believed that the commissioners used a transit instrument equipped with a telescope and that points were set on the tops of hills at ½ to 2-mile intervals. Polaris as well as other stars was observed to maintain the north direction. At the end of this line an error of about 50 feet exists, an accuracy that was not equaled for many years to come. In recent years the starting point on the north bank of the Ohio River was perpetuated by a large granite monument erected in memory of the momentous occasion.

David Rittenhouse, born in Germantown, Pennsylvania, in 1732, became an astronomer, calculator, clock builder, and instrument maker. He had run the circular portion of the boundary between Pennsylvania, Delaware, and Maryland (a circle of a 12-mile radius with Newcastle as its center). He also established the point from which the dividing line between New York and Pennsylvania was to be traced westward (42° of latitude).[12]

---

[12] Joan Marie Paulikas, "The Scientific Work of David Rittenhouse," Master's Thesis, University of Illinois.

The western border of Pennsylvania has become known as "Ellicott's Line." Ellicott later designed the streets in Washington, D.C., and surveyed the boundary between the United States and Florida.

Thomas Hutchins began running the "Geographer's Line" west from the point on the north bank of the Ohio River in 1785. Hutchins' sextant determination of latitude of the beginning point was in error by about half a mile. Because of an Indian attack in October, only 4 miles of this base line were run in the first year. With verbosity that has never been equaled, his report to Congress, December 27, 1785, written closely spaced on eight foolscap pages, reads in part:

> For the distance of Forty six Chains and Eighty six links West . . . , the Land is remarkably rich, with a deep black Mould, free from Stone, excepting a rising piece of ground on which there is an improvement of about 3½ Acres, where there are a few Grey and Sand Stones thinly scattered. The whole of the above distance is shaded with black and white Walnut trees, also with Black, Red and an abundance of white Oaks, some Cherry Tree, Elm Hoop-Ash, and great quantities of Hickory, Sassarfrax, Dogwood, and innumerable and uncommonly large Grape Vines producing well tasted Grapes of which Wine may be made. All the Hills in this part of the Country seem to be properly disposed for the growth of the Vine. Near the termination of the above mentioned measurement is a thicket of Shoemack, Hazel and Spice bushes, through which a passage was cut for the Chain-carriers. The first of these Bushes produces an Acid berry well answering the purposes of sowering for Punch, the Hazel yield an abundance of Nuts, and the Spice bushes bear a berry, red when ripe of an aromatic smell, as is also the Shrub on which it growes; the berry is about the size of a large Pea, of an Oval shape possessing some Medicinal virtues, and has often been used as a substitute for Tea by sick and indisposed persons. The Dogwood, the bark is used by the inhabitants and is said to be little inferior to Jesuits bark in the cure of Agues; the Tree produces a berry about the size of a large Cranberry when ripe, but something longer and smaller toward the Ends, excellent for bitters; and decoctions made of the budds or blossoms have proved very salutary in several disorders, particularly in Bilious complaints.
>
> The whole of the above described Land is too rich to produce Wheat, the aforementioned rising ground excepted, but it is well adapted for Indian Corn, Tobacco, Hemp, Flax, Oats &c, and every species of Garden Vegetables, it abounds with great quantities of Pea Vine, Grass, and nutritious Weeds of which Cattle are very fond, and on which they soon grow fat.[13]

The greatest influence Hutchins had on the beginning of the Rectangular System was the precedent established by him for meticulous notes and descriptive plats. The crude locations can be explained in part by the threat of Indians, untrained help, and the lack of suitable instruments.

At the completion of each 6 miles of base line established by Hutchins one group of surveyors was to take off to the south. The selection of whom was to be first was determined by lot, but, as it turned out, Hutchins did not make 6 miles the first year. The "secant method" was employed in running the parallel of

---

[13] Hutchins to President of Congress, New York, December 27, 1785, Papers Cont. Cong., 60, 225.

latitude, and the direction was determined by observing a star or stars with a sextant. The beginning of the base line had an error of 25.2 seconds of latitude and the western end, when completed, was 1500 feet south (15 seconds) of its intended latitude. It is interesting to compare this accuracy for 42 miles of line with that of the Mason-Dixon line, where 244 miles of line were established with only 2.3 seconds of error.

6-39 The Survey of the Seven Ranges

Each state was to send a surveyor, but only eight surveyors showed up for duty, and none of these performed work the first year. When actually surveying, none turned in good results with their circumferenters (see Fig. 6-2). On May 12, 1786, the requirement of running lines "by the true meridian" was repealed, and township lines were projected by magnetic compass.

The length of Gunter chains used for distances was checked by Hutchins, and horizontal measure was advocated. Closures on townships were quite poor and nonuniform in character. The numbering of townships progressed south to north, and ranges were numbered from east to west.

Surveyors failed to show in their notes what had transpired and contributed little, except for good marking of trees, fair record of points set, and recorded distances measured.

6-40 Ordinance of 1796

The act of May 18, 1796, amending and superceding the Act of 1785, was a victory for advocates of prior surveys and the Rectangular System. Small buyers were provided for in alternate townships wherein section lines were marked at 2-mile intervals. Although all interior section corners were not marked, the new plan would provide the establishment of three corners for each mile square. Large buyers were sold land in other alternate townships wherein only township exterior mile posts were preset.

For some unknown reason, the numbering system of sections in townships was changed to the following:

> The sections shall be numbered, respectively, beginning with the number one in the northeast section and proceeding west and east alternately, through the township, with progressive numbers till the thirty-sixth be completed.

The previous system as described in Section 6-35 was more logical, since it followed more closely the procedure used to divide townships, but the numbering system adopted in 1796 has been and still is in force. The pay of surveyors was raised from $2 per mile of line to a maximum of $3 for all expenses except occasional military escort along Indian boundaries.

More specifications appeared in the Act of 1796. The marking of trees, the notation of cultural and topographic features, the standard length of the chain, and the office of surveyor general in Marietta, Ohio, were now written into the law. Provision was made for special surveys and deviations from the plan when the lands were adjacent to prior grants, bodies of water, and other irregular boundaries. Townships of 5 miles square, marked at 2½-mile intervals, were called for in the military district and the missionary tracts. In all, the Act of 1796 was clearer and more explicit than the Act of 1785.

6-41 Act of May 10, 1800

The act of Congress approved May 10, 1800, required the

> townships west of the Muskingum, which . . . are directed to be sold in quarter townships to be subdivided into half sections of three hundred and twenty acres each, as nearly as may be, by running parallel lines through the same from east to west, and from south to north, at the distance of one mile from each other, and marking corners, at the distance of *each half mile*, on the lines running from east to west, and at the distance of *each mile* on those running from south to north . . . And in all cases where the exterior lines of the townships thus to be subdivided into sections or half sections shall exceed, or shall not extend, six miles, the excess or deficiency shall be specially noted, and added to or deducted from the western and northern ranges of sections or half sections in such townships, according as the error may be in running the lines from east to west, or from south to north. (emphasis added)

By the year 1800 the placing of excess or deficiency in the north and west tiers of sections was established, but the subdivision into quarter sections was not recognized.

6-42 Act of February 11, 1805

Subdivision of sections into quarter sections was provided for in the Act of 1805, and corners of half and quarter sections not marked shall be placed, as nearly as possible,

> equidistant from those two corners which stand on the same line. Boundary lines (center lines of Sections) which shall not have been actually run and marked as aforesaid shall be ascertained by running straight lines from the established corners to the opposite corresponding corners; but in those portions of the fractional townships, where no such opposite or corresponding corners have been or can be fixed, the said boundary lines shall be ascertained by running from the established corners due north and south or east and west lines, as the case may be, to the external boundary of such fractional township.

The direction due north and south or east and west has been interpreted by the courts to mean due north and south or east and west in the average direction run by the original surveyor as determined and tested by nearby lines. It is not generally interpreted to mean astronomic north.

6-43 Principal Meridians and Base Lines

No principal meridian was established in the Seven Ranges, although the point of beginning was on the western boundary of Pennsylvania. The Geographer's

Line served as Hutchins' base line and was in keeping with Jefferson's plan to base the surveys on the state's boundary.

What is now known as the First Principal Meridian and is also the boundary between the states of Ohio and Indiana was run north from the Great Miami River. The forty-first parallel of latitude served as the base line. To avoid rough terrain, the Second Principal Meridian was run through the eastern end of the Vincennes Tract in Indiana. The base line, run by Ebenezer Buckingham, Jr., in 1804, was started from a point in Illinois 67.5 miles away and was located in the southern part of the state, since the Indian claim to lands had not been settled in the north.

Beginning at the confluence of the Ohio and Mississippi Rivers, the Third Principal Meridian ran the length of Illinois. The base line for this territory was one extended westward from the Second Principal Meridian. To define the lands contained in a 2,000,000-acre military tract in northwest Illinois, the Fourth Principal Meridian was extended from a point north of Illinois River on the meridian of the mouth of that river. The base line was established westward from this point to the Mississippi. Another base line, coincident with the Wisconsin border (latitude 42°N), serves the area northward in Wisconsin and parts of Minnesota.

Of the 35 principal meridians in the United States (see Fig. 6-10), the first six have numbers, and the remainder are named. Most of the locations were arbitrary or according to some natural features instead of coinciding with the state boundaries as Jefferson had planned.

Very little was stated in the early instructions about the establishment of the principal meridians and base lines. It is presumed that the surveyors followed customs laid down by the men who surveyed in the Seven Ranges. Even though the surveyors were relieved from running according to "true" north, the directions were taken with considerably more care than in the other surveys, but the distances were no better than that observed elsewhere.

6-44 The Surveyor General

As provided in the Act of 1796, the post of surveyor general, at a pay rate of $2000 per year, was to be appointed to a man who would engage skilled deputy surveyors, administer oaths, frame regulations, and issue instructions. General Rufus Putnam, the first surveyor general and Washington's aid during the Revolutionary War, had long experience as a surveyor, both in the colonies and in the states, and was an advocate for the rectangular surveys. During the 7 years that he served (1796 to 1803) he faced many problems in Ohio. He established the First Principal Meridian and base line in accordance with Jefferson's plan; it is suspected that he was the author of the present system of numbering of sections, and he was the first to place excess and deficiency into the north and west tiers.

Jared Mansfield, a man of great learning and scientific background, succeeded Putnam in 1803, becoming the second surveyor general of the United States. He held this office until 1814. Mansfield's most noteworthy contributions to the Rectangular System were his proposal and execution of a general framework for the division of land into townships and his resolution of the conflict between rectangularity and the convergence of principal meridians. Although the idea was not completely new, it was Mansfield who clarified the necessity of base lines and principal meridians. His first application was the laying out of the Second Principal Meridian through the present state of Indiana and the base line from Clark's Grant to the Wabash River in 1804. He later extended this base line to the Mississippi River, and he established the Third Principal Meridian in Illinois. From this framework he ran the township lines, devising a new numbering system, whereby the townships were counted from the principal meridians and base lines rather than from natural boundaries.[14]

Putnam in his instructions did not require true meridian lines and allowed meridians to run by the compass with slight regard for deviations. The lines run under Mansfield's directions were performed with the best celestial methods of the time in an effort to attain true north.[15] The idea of closing the surveyed lines on each other began to take form in the last 3 years of Putnam's office and was furthered by Mansfield.

In direct conflict with the authority at the time, Mansfield wanted principal meridians and base lines located so as to simplify land divisions rather than conform to a political boundary. It is interesting to note that in laying out the Second Principal Meridian through the central portion of Indiana, he did so (supposedly in ignorance of the instruction) against the intent of Secretary of the Treasury Albert Gallatin who wanted the line to conform with the western boundary of Indiana. He afterward succeeded in winning over Gallatin; and, since that time, the boundaries of the Rectangular System and political boundaries have coincided only where it was found convenient to expedite the surveys.[16]

In 1814 Edward Tiffin became the surveyor general in the area north of the Ohio River and served in this post for 15 years. He issued the first set of written instructions and sent numerous letters to deputies. He instituted the scheme of guide meridians and standard parallels in 1824.

6-45 Instructions to Deputies

The importance of studying original instructions issued to deputy surveyors cannot be overemphasized. Before old lines can be retraced, it is necessary to understand how they were required to be established. The first instructions were issued by letter or by word from the Geographer and later from the surveyor general. Early contracts often contained specific instructions, but it was not until July 26, 1815, that written instructions were issued to all deputies.

---

[14] W. D. Pattison, *Beginnings of the American Rectangular Land Survey System—1784–1800*, University of Chicago Press, 1957, p. 210.

[15] *Ibid.*, p. 215.

[16] Ibid., p. 214.

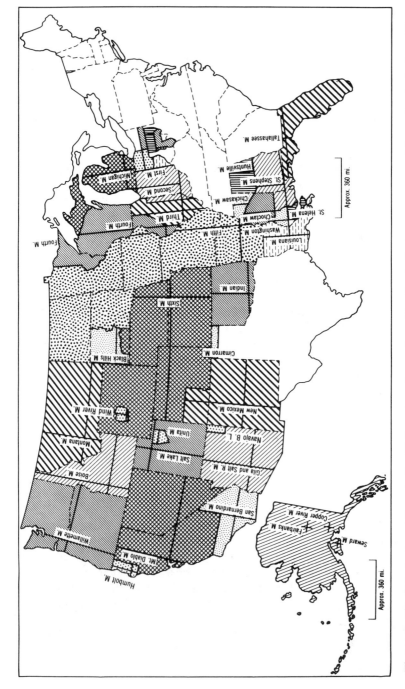

Figure 6-10 Principal meridians in the U.S. rectangular system. Shading indicates area governed by each principal meridian and its base line.

6-46 Tiffin's Instructions

Because Tiffin's instructions were the first issued instructions on a general basis, and because they set many precedents for the instructions that followed, it is important to note some of the many written features that specifically applied to the surveys in parts of Ohio, Indiana, Illinois, Michigan, Arkansas, and Missouri.

According to the act of May 10, 1800, within a township, excess, deficiency, or defects in measurements and convergence were to be thrown into the north and west tiers of sections. Quarter section corners were to be established on a straight line between section corners and, except for closing lines, were to be at half-mile intervals. These instructions are similar to that used today, but in closing lines, quoting from Tiffin's 1815 instructions, there was a significant difference:

> Also in running to the western (township) boundary, unless your sectional lines fall in with the posts established there for the corners of sections in the adjacent townships, *you must set posts and mark bearing trees at the points of intersection of your line with the town boundaries,* and take the distances of your corners from the corners of the sections of the adjacent townships, and note that and the side on which it varies in chains, or links or both.
>
> The sections must be made to close by running a random line from one corner to another *except on the north and west ranges of sections.*

"Double corners" were permitted on the north and west sides of a township; in effect this meant double corners were permitted to exist on any side of townships. The practice of setting double corners was partially abolished in 1843 and almost completely abolished in 1846 (double corners are used on occasions today).

The procedure in subdividing townships was to begin at the southeast corner of the township, where the variation of the needle was checked with the township line, then retracing along the south line of the township, restoring the quarter corner if not in place. At 1 mile (south corner of Sections 35 and 36) a meridian line was run northward, possibly parallel with the adjacent meridian according to Tiffin's suggestion, and a quarter corner was erected at 40 chains. At 80 chains, the corner common to Sections 25, 26, 35, and 36 was erected. From here a random line was run to the east until it intersected the township line. The amount of falling to the north or south of the standard corner was noted as well as the chainage of the line. The true line connecting section corners was marked by offsets, and the quarter corner was set on line and equidistant. Next, the meridian line between Sections 25 and 26 was established, and so on, until the last mile was reached. On the meridian line between Sections 1 and 2, a quarter corner was placed at 40 chains, and the line was extended until it intersected the north line of the township. Here a "closing corner" was erected, and the falling east or west of the standard corner was noted, as well as the total chainage of the line. This procedure was continued for all the tiers of sections,

190 / HISTORICAL DEVELOPMENT OF PROPERTY SURVEYING

progressing from east to west, until the last tier was reached. When the corner common to Sections 29, 30, 31, and 32 was established, and a random line to the east was run, and a true line returned, a line, at right angles to the meridian, was run westward and a quarter corner erected at 40 chains. The line was continued westward; at the intersection with the township line, a closing corner was set, and the falling north or south of the standard corner was noted, as well as the distance between the previously set quarter corner and the township line (see Fig. 6-11).

This procedure was, in many respects, the same as is followed today. The order of running lines and setting corners was and is the same. Of course, the instruments used were not as good or as accurate as those used today; the direction north, as determined by the compass, could not have been very accurate.

Today, the instructions are definite that all section lines shall be run parallel with the township line. Parallel lines running northward cannot have the same bearings; meridians converge toward the pole. There is a difference between running parallel with a township line and running north; lines run parallel will

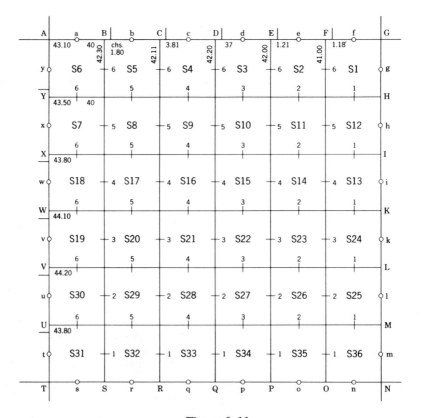

Figure 6–11.

have different bearings for each line run. But because the difference was and is small and the error amounted to less than that introduced by the inaccuracy of compasses, in Tiffin's time this instruction would have been without much significance; instruments used could not detect a change of a few minutes in bearings.

Double corners, permitted on the north and east sides of a township, were the most significant difference from today's instructions. Today, under certain circumstances, double corners will be permitted but not on every north and west township line. If a new area is being surveyed and it is adjoining a defective former survey, the errors of the former survey will not be incorporated into the new survey. This will necessitate double corners on the defective line.

Instructions to place errors in the north and west tiers of sections in the township have remained the same since Tiffin's time. Tiffin warned the deputy surveyors about errors. He stated that if the closure is too far north or south, they should mistrust the chain, and when there was excessive overrun or deficiency, they should mistrust the compass. The instructions point out that the principal source of error is the measurement by chain; the two-pole chain (33 feet) should be used horizontally with tally pins. Tiffin describes the method of breaking chain and refers to a standard of length. When meandering rivers, bearings should be taken of the river course, and the contents of the fraction of the submerged lands should be made by latitudes and departures.

All courses were taken with compass sights and with the needle corrected. The Rittenhouse compass (Fig. 6-12) was specified, and the deputy was to allow no one else to do the work of running the lines. It was not until some time after 1836 that the compass was abandoned.

Posts for the corners were to be driven well into the ground, and one or two sight trees were to be marked on each line. Bearing trees, located in each section, were to witness the section corner. Corner posts that could be a tree or a post at least 3 inches thick were to rise at 3 feet above the ground and were to

Figure 6-13 Section post with notches.

have notches on opposite corners for the number of miles to township lines (Fig. 6-13). Bearing trees were to referenced by distance and bearing, and the sizes and species of the trees were to be noted. Except for the corner post material (iron pipes and brass caps are now used), these instructions are similar to those given today.

Specifications for the notes included the manner of recording distances along the left margin as well as comments on timber, undergrowth, swamps, ponds, stone quarries, the quality of the soil, rivers, streams (navigable or not), and any improvements found. The entries were to be made at the close of each day and "written with a fair hand." (Today, final notes are accepted after being typewritten.)

The plat was to be prepared on durable paper at a scale of 2 inches to a mile or 40 chains to an inch. A diagram was included to show magnetic north. The surveyor was required to certify the work as follows:

> Pursuant to a contract with, and instructions from _____ Surveyor General of the United States, bearing date the _____ day of _____ , I have admeasured, laid out and surveyed the above described township (or fractional part) and do hereby certify that it had such marks and bound, both natural and artificial as are represented on said plat and described in the field notes made thereof, and returned with the plat in the Surveyor General's office. Certified this _____ day of _____.

As a result of these first instructions issued in written form, the surveyors were required to check their chains and compasses and submit uniform and complete notes. The precedent for plats and notes was now established.

6-47 Later Instructions

Numerous issued instructions followed those of Tiffin, although some have become lost. The General Land Office was reorganized in 1836 and published the first Manual of Instructions in 1855.

The General Instructions of 1843, applicable to Arkansas and elsewhere, made it mandatory to run to the existing section corner on the western side of the township and correct back on a true line. But in closing on the north township line, double corners were still permitted.[17] In the General Instructions of 1846, applicable to parts of Wisconsin and Iowa, section lines closing on both the north and west township lines were to close on the existing section corners. Since that date, no instructions, other than to correct defective survey conditions, have called for double corners.

In the General Instructions of 1846, applicable to parts of Wisconsin and Iowa, section lines closing on both the north and west township lines were to close on the existing section corners. Since that date, no instructions, other than to correct defective survey conditions, have called for double corners.

In the instructions of 1846 is this note:

> Base, meridian, correction and township lines are to be run with an instrument that operates independently of the magnetic needle, which is to be employed only to show the true magnetic variation. Section, meander and all other lines interior of a township, may be run either with the same instrument, or with the Plain Compass, provided it is of approved construction and furnished with a vernier or nonius.

The Oregon Manual of 1851 should probably be considered as the first manual of instructions for the survey of the public domain; however, in the haste to put it in the hands of Oregon and California surveyors, several figures were omitted (later put in the 1855 manual). There were minor changes in 1855.

In the instructions of 1855 is noted:

> Where uniformity in the variation of the needle is not found, the public surveys must be made with an instrument operating independently of the magnetic needle. Burt's improved solar compass, or other instrument of equal utility, must be used of necessity in such cases.

In the instruction of 1850 are provisions for a telescope with stadia (two parallel lines in the principal focus) and permission to measure meander lines with stadia.[18]

The *Manual of 1855* was prepared under the direction of the commissioner of the General Land Office by John M. Moore and was approved by an act of Congress, May 30, 1862 (Sec. 2399, R.S.)

> That the printed manual of instructions relating to the public surveys, prepared at the General Land Office, and bearing the date February twenty-second, eighteen hundred and fifty-five, the instructions of the Commissioner of the General Land Office, and the special instructions of the surveyor general, when not in conflict with

---

[17] John S. Dodds, *Original Instructions Governing Public Land Surveys, 1815–1855*, Ames, Iowa, 1944.

[18] Ibid., p. 173.

said printed manual or the instructions of said Commissioner, shall be taken and deemed to be a part of every contract for surveying the public lands of the United States.

Although before this time Congress had not officially approved by law the instructions issued by those in charge, the courts have recognized instructions as being evidence of what was done.

Until 1855 little attention was paid to the problem of convergence of meridians. The *Manual of 1855* provided for standard parallels, and among other interesting notes are the following:

> Standard Parallels (usually called correction lines) are established at stated intervals to provide for or counteract the error that otherwise would result from the convergency of meridians, and also to arrest error arising from inaccuracies in measurements on meridian lines, which, however, must ever be studiously avoided. On the *north* of the principal base line it is proposed to have these standards run at distances of every *four* townships, or twenty-four miles, and on the *south* of the principal base, at distances of every *five* townships, or thirty miles. [Note: this provision of five townships does not appear in later manuals.]
>
> Where uniformity of the variations of the needle is not found, the public surveys must be made with an instrument operating independently of the magnetic needle. Burt's *improved solar compass*, or other instrument of equal utility, must be used of necessity in such cases.
>
> The township lines, and the subdivision lines, will usually be measured by a two-pole chain of thirty-three feet in length, consisting of fifty links, and each link being seven inches and ninety-two hundredths of an inch long. On uniform and level ground, however, the four-pole chain may be used.
>
> You will use eleven tally pins made of steel, not exceeding fourteen inches in length, weighty enough towards the point to make them drop perpendicularly.
>
> In measuring lines with a two-pole chain, every *five* chains are called "a tally," because at that distance the last of the ten tally pins with which the forward chainman set out will have been stuck. He then cries "tally," which cry is repeated by the other chainman, and each registers the distance by slipping a thimble, button, or ring of leather, or something of the kind, on a belt worn for that purpose or by some other convenient method.

These instructions point out that in chaining, errors of count could occur in any multiple of half chains (33 feet) or could occur by dropping a tally of 10 half chains (330 feet). Numerous discrepancies of this nature have been noted by surveyors. It is also interesting to note that where a four-pole chain (66 feet) was permitted, it was generally recognized, as of that time, to be 66.06 feet long. This assured "full" mile measurements.[19]

Burt's "improved solar compass," mentioned in the foregoing quotation, was not invented until 1836; few lines before to that date were run independent of the needle.

---

[19]Ibid., p. 173

The *Manual of 1881* states that the initial point was to be located astronomically. The base lines, to be tested in direction every 12 miles, were to be run by a solar instrument. If a transit were used, the tangent method of base line was to be followed.

The *Manual of 1890* prohibited the use of the magnetic needle in major lines and permitted its use in subdividing and meandering when local attraction did not exist. The *Manual of 1894* required all surveys to be independent of the needle.

The *Manual of 1902* required that the initial points be in conspicuous locations and perpetuated by indestructible monuments. Among the many instructions of that time is the note:

> The surveys of the public lands of the United States, embracing the establishment of base lines, principal meridians, standard parallels, meander lines, and the subdivisions of townships, will be made with instruments provided with the accessories necessary to determine a direction with reference to the true meridian, independently of the magnetic needle.

In the *Manual of 1930*, the error of closure requirement was introduced. In terms of latitudes and departures, it was not to exceed 1/640 in either, with no more than 1/452 as a result. Distance was to close within 12.5 links per mile or 50 links per section, 175 for a tier, and 300 for a township.

Comprehensive discussions on the instructions from the General Land Office will be found in the following two books: L. D. Stewart, *Public Land Surveys*, and J. S. Dodds, *Original Instructions Governing Public Land Surveys 1815–1855*.

6-48 Reorganization of the Land Office

The General Land Office, established in 1812, was reorganized on July 4, 1836. The contract system of subdividing lands began with the Act of 1796 and was abolished in 1910 when a form of civil service was adopted. In 1946, the General Land Office and the Grazing Service were combined to form the Bureau of Land Management (BLM) under the Department of Interior. This agency employs professional cadastral engineers who use the latest methods and equipment to locate and subdivide the public domain. Much of the engineers' present activity is in the state of Alaska, although there is a considerable amount of unsurveyed land in the western states. In 1973, BLM published a manual of instructions that is presently in effect.

6-49 1947 Surveying Instructions

According to the *Manual of Instructions, 1947*, the following points are appropriate to the division of lands in the public domain:

Initial Point. New points will be established in Alaska by the director, and the position (latitude and longitude) will be accurately determined.

Principal Meridian. These shall conform to the true meridian and will be extended north or south from the initial point. Corners are to be set each 40 chains and at intersection with bodies of water. Two independent sets of measurements are to be employed and must agree by 14 links per 80 chains. Also, a test must be made to assure alignment within 0°3'00" of true direction.

Base Line. The base line is extended east and west from the initial point and will have corners set at 40 chains distant and at points of meander. The same accuracy standards are required as given for principal meridians. The line is to be run on a true latitude employing the solar method, or, if not, the tangent or secant methods are to be used.

Standard Parallels. Standard parallels, called "correction lines," are run east and west from the principal meridian at intervals of 24 miles north and south of the base line. They are to be run in the same manner as the base lines, with corners set at each 40 chains and at meander points.

Guide Meridians. At intervals of 24 miles east and west from the principal meridian and extending north from the base line or standard parallels, guide meridians are projected on a true meridian. Excess or deficiency is thrown in the last half mile on the north, and a closing corner is placed at the intersection on the next northerly base line or the standard parallel (Fig. 6-14).

Township Exteriors. The south and east boundaries of the 24-mile tracts are the governing lines for the township exteriors. The random lines are run (ideally) from east to west with temporary points set, then corrected on the true line. The true line is to be blazed through timber, with notes taken on the character of the land. The field notes will omit the random lines except where necessary to explain triangulation and any other unusual procedures. The limit of closure for the township exteriors is set at 1/640.

Subdivision of Townships. The boundaries of the townships previously run are to control. The meridian section lines are to begin on the south boundary line and will be run north parallel to the governing east boundary. Corners are to be set at each half mile, with fractions thrown in the last half mile. Lines are to be corrected back on true course. Meander corners are to be set where the meander line leaves the section line. Traverse angles of meander lines should be adjusted to ¼°. Distances in whole chains or multiples of 10 links are to be used for the meander traverse, with the fraction only in the last course.

A river that is three chains wide or wider is termed as "navigable." Lakes with an area of 25 acres or more will be meandered. If the lake is entirely within the section, a traverse is run from one of the quarter lines, or if entirely within the quarter, an auxiliary meander line is run. Artificial lakes and reservoirs are not to be segregated from the public domain. Islands may be determined by triangulation, and the title, name, or identification is to be noted.

Limits of Closure. A relative error of 1/452 is permitted, provided neither latitude nor departure exceeds 1/640. A distance error not to exceed 12.5 links

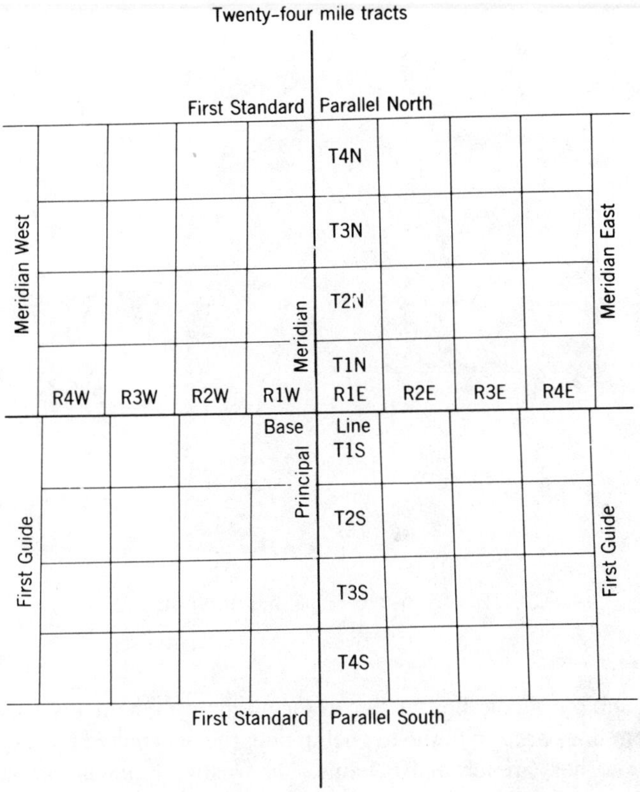

Figure 6–14 Twenty-four mile tract.

per mile in either latitude or departure is required. The latitudes and departures of a normal section will each close within 50 links with a maximum of 175 links for a tier and 300 links for a normal township. Fractional sections, including irregular lines, must close within 1/640.

Marking Lines between Corners. Blazes and hacks would be made so that they will not be mistaken for other marks. Sight trees or line trees are to be marked, as well as others within 50 links of the line.

Corner Monuments. The standard iron post is 30 inches long, 2 inches inside diameter, split on the bottom, and set with a brass tablet riveted to the top (Fig. 6-15). The standard tablet may be set in a boulder, or, if a sound living tree falls in the corner location, it may be used to mark the corner. Cairns are to be used to protect and draw attention to the corner monument.

Running Section Lines. The latitudinal section lines will be run random, parallel to the south line of the township. Temporary quarter corners are set at the half mile, and after the line is corrected back to true line, the permanent

198 / HISTORICAL DEVELOPMENT OF PROPERTY SURVEYING

Figure 6–15 Pipe and brass cap.

quarter corners are set on the true line at the midpoint. On the last section on the west, random lines are run to the township line, then corrected back on true line. The quarter corners are left at 40 chains. The westward line is run parallel with the controlling south line.

The meridinal section line will have precedence in the order of execution. The order of progression is the same as outlined in Tiffin's instructions of 1815.

Subdivision of Sections. The basic unit of subdivision is the quarter-quarter section. Sections are not subdivided in the field by BLM; this is a function of the local surveyor. Subdivisions are made by protraction and are shown by broken straight lines on the plat. Fractional sections are protracted into "government lots," usually numbered in a counterclockwise direction. The fractional sections along the north and west tiers will have Lots 1 to 4, except for section 6 that will have Lots 1 to 7. Some of these lots will be shown on the plat by their dimensions, and some will be stated by area. Fractional lots will also be caused by meanderable rivers, meanderable lakes, mineral claims, defective boundaries, prior title, and so on.

Meandering. All navigable bodies of water and other important rivers and lakes that are to be segregated from the public domain will be traversed to determine the sinuosities of the shore. The chief purpose of this survey will be to determine the quantity of the land remaining. Normally, the mean high water is to be taken. In no case is the meander line intended to be a boundary line.

Corner Accessories. Four bearing trees, one in each section, are to be marked with deep scribing indicating the township, range, and the section in which the tree stands. If trees are not close enough to the corner, mounds of stone are to be erected so that the location of the corner can be restored by intersecting lines. Memorials, such as glassware, stoneware, a stone marked with an "X," a charred stake, pieces of metal, and the like, are to buried at the corner. In desert country pits are dug, and mounds are to be erected at the site of the corner.

Instruments. A solar transit that complies with exacting standards is the usual instrument for the running of section lines. Steel tapes of 2 to 8 chains in length are in common use. The lengths must be reduced to horizontal and to mean sea level. In some circumstances, distances by stadia are permitted. The instruments must have a stadia coefficient of $1/132$ and a focal constant of 1.2 links. Stadia readings, when made on a targeted rod, may have a relative error as small as $1/1000$.

6-50 Manual of 1973

The Manual of Surveying Instructions, 1973, although important to the present government surveyors, is not of very great importance to the retracement surveyor; the manual in force at the time of the original survey should be referred to. It tells what was supposed to be done at the time.

The latest manual permits new measuring devices (electronic, photography, etc.) and requires greater accuracy. The "limit of closure" set for the public land surveys may now be expressed by the fraction $1/905$, provided that the limit of closure in neither latitude nor departure exceeds $1/1280$. When a survey qualifies under the latter limit, the former is bound to be satisfied. Of considerable interest to the surveyor are the sections on apportionment of accretions and lakes (included elsewhere in this book). Also the section on mineral surveys is new.

The retracement surveyor must again be cautioned; in a resurvey the procedure to be followed is that laid down by the courts. For example, in the relocation of a lost quarter corner in a regular section the courts say it is to be located *exactly* halfway between section corners, not halfway with a permissible error of $1/905$ feet. Original instructions are a guide about what was supposed to be done; the field notes say what was done; the courts guide one in the procedure to use in a resurvey, and that procedure may or may not agree with original instructions of the manual.

6-51 Rectangular System Protraction Program

The outstanding feature of the U.S. Rectangular System has been the precedent of prior surveys, yet it may be that it is evolving into a master protraction scheme. In the vast unsurveyed lands of Alaska, expedient action has become necessary to identify lands for oil and gas leases. In the Umiat Principal Meridian, established December 31, 1956, a USC and GS monument was chosen as the initial point. This meridian controls about 180,000 square miles of Alaskan lands where very little, if any, survey data exist.

Protraction diagrams are prepared on a mylar base and show an area of 16 townships per sheet. All township and section lines are indicated, with delineated lease areas of four sections (nine per township). All dimensions of the sections other than a full mile are shown as well as the boundaries of fractional sections adjacent to water bodies. The geographical coordinates of the township corners are stated and used as a basis for description and location (presumably relative to USC and GS monuments) of any of the smaller parts within the whole.

Alaska contains about 20,000 whole or fractional townships, and most of the titles will be located on this protraction system. Once the diagram has been approved and entered into the record, the geographical coordinates of the section corners become controlling, and it is then necessary for BLM to locate the tract according to these values. This system has the advantage of subdividing lands at a much lower cost than the conventional procedure and will provide a more realistic method of location and restoration. It is quite possible that this method of protraction will revolutionize the process and philosophy of cadastral subdivision.

6-52 Summary of Public Land Surveys

Instructions on how to subdivide the public domain have varied from time to time. In resurvey procedures, the instructions issued at the time of the original survey are to be used as the guide for resurveys. The most recent manual, *The Manual of Instructions, 1973*, is not the manual to be used for resurveying most areas within the United States; the manual in effect or the instructions issued as of the date of the original survey should be used as a guide. It would indeed be foolish to try and apply the 1973 manual to surveys in Ohio.

6-53 Timber Culture Act

In 1873, Congress passed an act to encourage the growth of timber on western prairies. This law provided that any person who was the head of a family, or who had arrived at the age of 22, and was a citizen of the United States, or who had filed his declaration of intention to become such, could secure a title, at the end of 10 years, to not more than a quarter section of land, by planting and keeping in a healthy growing condition 40 acres of trees, or a corresponding less amount if less than 160 acres were entered. Not less than 40 acres could be entered. Later this law was amended to allow the settler to prove his claim at the end of 8 years and required the planting of only 10 acres of trees on each quarter section. Only one quarter in each section could be entered for timber culture. This law was repealed in 1891.

6-54 Classification of National Forest Lands

Some of the lands of the United States were set aside and reserved as National Forest land (mostly from 1891 to 1911). Since these lands were never under private ownership, the Federal Government has conclusive jurisdiction and rights of survey or resurvey. Under authority of the Weeks Law, passed by Congress on March 1, 1911, the Federal Government was authorized to purchase forest land privately owned. All of the 20,000,000 acres so acquired by purchase or exchange under the Weeks Law was at one time patented and under state jurisdiction. After purchase by the Federal Government, the lands remain under state jurisdiction. As a result, all legal matters pertaining to the boundaries of such lands are tried by the states in accordance with state laws. Under proper circumstances, long-continued possession of Weeks lands can result in an unwritten title right against the Federal Government. Against lands continuously under U.S. ownership, possession cannot ripen into a fee title against the U.S. government. The private surveyor, as licensed by state laws, has the rights of resurvey for Weeks lands.

6-55 Land Acquired by Homesteading

The idea of developing a new area by granting lands to people who would occupy, grow crops, and otherwise improve the area, originated several centuries ago. Even in the colonies there were "headright" grants that permitted people, without payment of land values, to gain title to lands after a certain period of occupation. A homestead act was passed by the Republic of Texas in 1839, and the first Federal Homestead Act in the United States was passed May 20, 1862. This act permitted the head of a family who was at least 21 years of age and a citizen or who had served in the army or navy to enter and take possession of 160 acres or less. Persons who were already in the territory could take smaller adjacent parcels if the aggregate did not exceed 160 acres.

Under this first act, the entryman was required to reside on the land for a period of 5 years to get an absolute title. Several other homestead acts have been passed since this first one, and in general the steps are as follows:

1. Person must file application stating his intent and qualifications.
2. A Land Office fee must be paid. This is usually a nominal $1.25 per acre and in some cases less.
3. The entryman must reside in or occupy the tract for a given period of time (3 to 5 years) and in some cases must make certain improvements.
4. Taxes are applied as soon as entry is made, but title is not perfected until the residence time and other stipulations are fulfilled.

Most of the United States, excepting Alaska, is no longer open to homesteading. A large part of the public domain was titled by homesteads, and in Alaska the process is continuing. Homesteading differs from the colonial headright grants in that the homesteader had to be of sound financial status.

Under state laws a *homestead right*, wherein a limited amount of land and improvements thereon are exempt from liability for certain debts, is established by filing a declaration with proper officials. This is not to be confused with the process of acquiring land by homesteading.

6-56 Obtaining Small Tracts Within the Public Domain

Under an act of Congress, June 1, 1938, qualified individuals and certain groups are able to obtain small tracts (normally not more than 5 acres) of unreserved public lands for residence, recreation, and other purposes. These tracts may be leased, purchased, or leased with an option to purchase. The sale price will be at least as much as the appraised price, and the sales are conducted at a public auction.

A suitable tract of land may be a regular subdivision of a section such as the N ½ of SW ¼ of NE ¼ of NW ¼, Section 12 or an irregular parcel described by metes and bounds. Under unusual conditions, tracts as large as 7.5 acres may be acquired. Residence and improvements are not required, except as may be imposed by local zoning and planning ordinances. The tract holder has no rights to the timber or minerals on the land, since these are reserved in the public domain.

6-57 Land Ownership in Hawaii

Captain James Cook discovered the islands in 1778 and named them for the earl of Sandwich. These eight inhabited islands were well populated by people who lived under a strict feudal system. The land belonged to the king who doled it out to his chiefs. The "Great Mahele" of 1846 was a homestead reform that gave much of the public lands to the people. Title entries were made in the Mahele book and titles of property today refer to this record.

The United States annexed Hawaii on July 7, 1898. The act stated "to cede and transfer to the United States the absolute fee and ownership of all public, Government or Crown lands." It was further provided that "the existing laws of the United States relative to public lands shall not apply to such lands in the Hawaiian Islands but the Congress of the United States shall enact special laws for their management and disposition."[20] At the present time, the public lands comprise about 40 percent of the land area of the eight inhabited islands.

In 1920, the Hawaii Homes Commission was created to grant 99-year land leases to persons who were not less than one-half Hawaiian by blood.

Two systems of land registration are practiced in Hawaii: a recorder system and the land court. About 40 percent of the island of Oahu (containing Honolulu) and about 8 percent of the remaining lands are registered in the land court. A lot and block system tied to a rectangular grid system is used in the land court; the recorder system has many of its tracts described by metes and bounds.

6-58 Lands in Alaska

Alaska was discovered by Vitus Bering on July 16, 1741, while he was on a Russian expedition in search of fur. A Russian-American company followed with establishment of small coastal settlements. Russia sold the entire claim to

---

[20] U.S. Statutes at Large, Vol. 30, pp. 750–51.

the United States in 1867 when virtually all the territory was vacant public lands.

The federal acts pertaining to the disposition and management of public lands apply to Alaska, and at the present time slightly more than half of the public domain is situated in this state. Under the statehood land grant, Alaska is to acquire 103 million acres in a period of 25 years.

Alaska is an active area of homesteading. Any male citizen who is 21 and head of a household can homestead 160 acres, but one-eighth of this land must be cleared at the cost of the homesteader. Many oil, gas, and mineral leases have been located in Alaska, and title surveying is active.

6-59 Court Reports

One of the more important sources of written historical records of survey procedures is found in case reports in law libraries. What the law was and how it was interpreted at a given date are usually summarized at the beginning (syllabus) of each case. The opinion of the judge, as recorded, gives an insight to the customs and thoughts of the period.

An example illustrates this point. In the year 1844, in Missouri, two neighbors, Gamble and Clark, got into a squabble over the location of a quarter corner. It seems that the section corners were located nearly a mile apart; yet the quarter corner was 20 poles (330 feet) nearer one corner than the other. (Note: This error corresponds to a tally mistake using a 2-pole chain, as was customary at that time.) The fact that the quarter corner found was the original was amply proved. Today, this would not present troubles; but then, it must be remembered, many of the sections surveyed east of the Mississippi did not have quarter corners set; and it was common knowledge, in such cases, that the quarter corner was set halfway. In the summary of the report a historical sequence of public domain land laws pertaining to setting section corners was given. The judge's opinion, in accepting the corner as found, was:

> From this review of the laws of the United States relating to the survey of the Public Domain, it will be seen, that at one time no subdivision of sections was made, it being divided into townships and sections only; consequently, no half or quarter section corners were established on the public survey. The act of 1800 authorized the sale of half sections, and prescribed a mode of ascertaining the contents thereof; in the act of 26 March, 1804, authorized the sale of the public lands in quarter sections, which were required to be surveyed and the contents thereof ascertained at the expense of the purchaser. Then came the act of 1805, above cited, which prescribed the mode of ascertaining the corners of sections and quarter sections at the expense of the United States. No lands were surveyed in this state until long after the act of 1805. The act of March 3, 1811, first provided for the survey and sale of land west of the Mississippi. At that time the quarter section corners were required to be marked on the survey and these corners were, by the act of 1805, to be taken as the true corners; and the provisions of the act, which required the corners of half and quarter sections not marked on the survey to be placed

equidistant from the two corners of the sections, were intended to be applied to surveys made prior to the time when the sale of half sections and quarter sections was authorized. Mr. Gallatin, then at the head of the treasury department, and to whom was entrusted the superintendence of the survey and disposal of the public lands, shortly after the passage of the act of 1805, in a letter of instructions to the Surveyor General, thus speaks of it;—"You will perceive, from the enclosed act, that the principal object which Congress has in view is, that the corners and boundaries of the section and subdivision of sections should be definitely fixed, and that the ascertainment of the precise contents of each is not considered as equally important." Indeed, it is not so material, either for the United States or for individuals, that purchasers should actually hold a few acres more or less than their survey may call for, as they should know with precision, and so as to avoid any litigation, what are the certain boundaries of their tract.

Now, that section may be divided among so many holders, and as valuable improvements are frequently made near lines dividing sections, it would produce great confusion and litigation if the established corners were to be abandoned in subdividing sections, and lines were to be run equi-distance from the corners of sections, in order to subdivide it into quarter sections.

—*Campbell* v. *Clark*, 8 Mo. 398.

The historical sequence and timing of laws, in the foregoing case, is important information for resurvey procedures in the areas mentioned.

Many of the principles outlined in Chapter 2 on evidence are traceable back to the beginnings of the United States. History in itself is often evidence used by the courts. In an early Massachusetts case, dated October 1804, in a dispute between Paul Revere (a mill owner) and Jonathan Lenord (an adjoiner), the question of the grantor's right or absence of right to testify about his intent, was at issue (see Section 2-23).

In the course of the trial, it became important to ascertain the place of the forge-dam mentioned in the deed, the parties not agreeing as to its location. The council for the defendants moved that Mr. Robbins might be sworn as a witness to prove what was understood and intended by the parties to the deed, as to the forge-dam therein mentioned. He was objected to, as not being a competent witness, on the grounds; first that he was interested, and secondly, that it is against the rule of law to permit a grantor to establish his own grant; that the intent is to be collected from the deed itself, except in the case of a latent ambiguity, which may, indeed, be explained by parol evidence, but not by the testimony of the grantor himself.

As to the first ground of objection, the court held, that as Mr. Robbins was not interested in the event of the suit, the objection could not avail. But as to the second, the grantor was never permitted to explain his own grant, even in the case of latent ambiguity; that in such cases, although it was competent and necessary to explain the same by the testimony of witnesses, yet such testimony could not come from the grantor.

Sedgwick, the judge, charged the jury, that it was against every legal principle to go out of the deed itself, by an inquiry into existing facts, to ascertain the meaning of those and such like words, there being in them no ambiguity of any kind; that what their meaning is, was merely a question of law, of which the courts are to judge by the words themselves.

—*Revere* v. *Lenord*, 1 Mass. 91.

Maine has had numerous court trials over land boundaries, no doubt because of poor early conveyancing and surveying methods. Among the many questions that arose was the question of measurement index. If an original surveyor consistently makes long measurements, cannot the resurveyor, when resetting a lost monument, also make allowances and use long measurements?[21] In a dispute between Joseph Otis and Moody Moulton this was the point in issue, and the report and judge's opinion were as follows:

> The plaintiff, at the trial to prove title, read a deed from the Commonwealth of Massachusetts to Jarvis, dated Feb. 16, 1794, conveying "all the unappropriated land lying between Penobscot River and the Lottery Township in the county of Hancock, Nos. 7 and 8, surveyed by John Peters in 1786; also the gore of land lying north of said township No. 8"; also a deed from said Jarvis to the Union Bank dated Dec. 26, 1800, of seven-eighths of the land in No. 8; also a deed from the Union Bank to Sarra Russel dated Nov. 10, 1816, in the same premise, and from Sarra Russel to Otis, dated Nov. 21, 1816, all of which deeds were duly acknowledged and recorded.
>
> ... he introduced the grant from the Province of Massachusetts Bay to James Duncan and others of six townships, each township being conveyed by a separate description, and being numbered from one to six inclusive.
>
> The opinion of the court, by Judge Shepley, was:
>
> "On the report in motion for a new trial two questions are presented for consideration: One arises out of the testimony relating to the boundaries of the town of Bucksport; the other out of and relating to the occupation and the title of the tenant. The bounds of the first six townships, being Bucksport, are described as beginning on the east side of the Penobscot River, 'at a hemlock tree marked and running into the land in a course N 70° E, 5 miles and 184 rods to a stone monument, from thence along a line (which forms a boundary of the first and second of the said townships on the northeast and runs on a course S 76° E, 9 miles and 40 poles in the whole) upon a stone monument set up thereon which marks the east corner of the said first township; and from thence by a line S 53° W, 5 miles and 232 poles to a monument on the northwest side of the east branch of the Penobscot River and down the said branch one mile and 56 poles unto another monument on said branch, and from thence S 56° W one mile and one hundred and thirty-two poles to a monument on the east side of the river Penobscot,' and from thence along the river to the first bounds. The place of starting from the Penobscot River at the first bound is not disputed, and the course of the line, making allowance for the variation, is found to be correct and not disputed. The stone monument named as the second bound is not found; and there is no proof of the original survey of line upon the earth between the two monuments, by which its length can be ascertained. The length

---

[21] See Curtis M. Brown, *Boundary Control and Legal Principles*, 2nd ed. John Wiley & Sons, New York, 1969, p. 172.

should be ascertained by admeasurement upon the earth. It is contended however, that in measuring it, the proprietors should not be limited to the exact measure named, but should be allowed a larger measure, corresponding to the measure found on other parts of the land, compared with that stated on the plan. . . . the rule has been too well established to be now disturbed, that the admeasurement should be made upon the earth, and not by the scale upon the plan. And this case illustrates the propriety and necessity of the rule; for although it has been stated by a witness, that in ten different admeasurements there was a larger measure in each case upon the earth, than that stated on the plan, yet there was no uniformity in the excess.[11]
—*Otis* v. *Moulton*, 20 Me. 205, July 1841.

This early case points out some interesting facts: (1) Townships in Massachusetts existed prior to 1794. (2) The term "gore" is an old one. (3) Measurement index was not tolerated in Maine.

Georgia was one of the older states that granted land patents. Seemingly, from the facts stated in the following case, land had to be surveyed as part of the patent proceedings. In a difference of opinion between Margare Riley and George W. Adams (16 Ga. 141), dated August 1854, a controversy arose over measurements, monuments, and possession. The syllabus quoted the law at that time, and it is interesting to note the similarity to present laws.

A possession which is the result of ignorance, inadvertence, misapprehension or mistake, will not work a disseisin.

Marked trees as actually run, must control the line, which course and distance would indicate.

If nothing exists to control the call for course and distance, the land must be bound by the course and distance of the grant, according to the magnetic meridian; but course and distance must yield to natural objects.

All lands are supposed to be actually surveyed (note: that is in Georgia); and the intention of the grant is to convey the land according to the actual survey.

If marked trees and marked corners are found, distance must be lengthened or shortened, and course carried so as to conform to these objects.

Where the calls of the deed or other instrument, are for natural, as well as known artificial objects, both course and distance, when inconsistent, must be disregarded. And this rule is supposed to prevail in most of the states of the Union.

Whenever a natural boundary is called for in a grant or deed the object is to determine at it; however wide of the course called for, it may be, or however short, or beyond the distance specified.

Whenever it can be proved that there was a line actually run by the surveyor, or was marked, and a corner made, the party claiming under the grant or deed, should hold accordingly, notwithstanding a mistake in description of the land in the grant or deed.

When the lines or courses of an adjoining tract are called for in a deed or grant, the line shall be extended to them, without regard to the distances provided these lines and courses be sufficiently established.

When there are no natural boundaries called for, no marked courses or trees to be found, nor the place where they once stood, ascertained and identified by evidence; or where no lines or courses of an adjacent tract are called for, in all such cases, courts are of necessity confined to the courses and distances described in the grant or deed.

Courses and distances occupy the lowest, instead of the highest grade, in the scale of evidence, as to the identification of land.

Any natural object, and the more prominent and permanent the object, the more controlling as a locator, when distinctly called for and satisfactorily proved, becomes a landmark not to be rejected, because the certainty which it affords, excludes probability of mistake.

Courses and distances depend for their correctness on a great variety of circumstances, are constantly liable to be incorrect; difference in the instrument used, and in the care of surveyors and their assistants, leads to different results.

In ascertaining boundaries, the location (here meaning monuments and lines) of the original surveyor, so far as they can be found, are to be resorted to ; and where they vary from the proprietor's land, the location actually made, will control the plan.

Whenever, in the conveyance, the deed refers to monuments, actually erected as the boundaries of the land, it is well settled that these monuments must prevail, whatever mistakes the deed may contain, as to course and distance.

Course and distance are pointers and guides to help to ascertain the natural objects of boundary.

Where a given line is exceeded (long) in the grant, according to the course and distance, evidence may be given of long occupation under it, to prove the boundaries.

Boundaries and courses may be proved by hearsay, from the actual necessity of the case.

—*Riley* v. *Adams* 16 Ga. 141.

REFERENCES

Allcorn, Bill, *History of Texas Lands*, General Land Office, Texas, Austin, 1959.

Brown, Curtis M., *Boundary Control and Legal Principles*, John Wiley & Sons, New York, 1969.

Bureau of Land Management, *Manual of Instructions for the Survey of the Public Lands of the United States*, Department of the Interior, U.S. Government Printing Office, Washington D.C., 1973, 1947, 1930, 1902, 1855.

Chandler, Alfred N., *Land Title Origins*, Robert Schalkenback Foundation, New York, 1945.

Dodds, John S., *Original Instructions Governing the Public Land Surveys 1815–1855*, J. S. Dodds, Ames, Iowa, 1944.

Hilliard, Sam B. H., "An Introduction to Land Survey Systems in the Southwest," *Studies in the Social Sciences*, Vol. 12, June 1973.

McEntyre, John G., *Land Survey Systems*, John Wiley & Sons, New York, 1978.

McLendon, Samuel G., *History of the Public Domain of Georgia*, Atlanta, 1924.

Marschner, Francis J., *Boundaries and Records in the Territory of Early Settlement from Canada to Florida*, Agricultural Research Service, U.S. Department of Agriculture, Washington, D.C., 1960.

Pattison, William D., *Beginnings of the American Rectangular Land Survey System, 1784–1800*, University of Chicago Press, Chicago, 1957.

Price, Jerry K., *Tennessee—History of Survey and Land Law*, Kingsport Press, Kingsport, Tenn., 1976.

Sherman, C. E., *Ohio Land Subdivisions*, Ohio State University, Columbus, 1925.

Stewart, L. O., *Public Land Surveys: History, Instructions, Methods,* Collegiate Press, Ames, Iowa, 1935.

Uzes, F. D., *Chaining the Land*, Landmark Enterprises, Sacramento, Calif., 1977.

CHAPTER 7

Location Establishment by Statutory Proceedings

7-1 Introduction

The establishment of a lost or disputed boundary line is usually a function of the court, but in some states special statutes have been enacted to enable adjoiners to have their boundaries established by arbitration, commissioners, county surveyors, or other means. As used, establishment includes both the right to locate boundaries and the power to make the findings final, unless appealed, as can be done in any court case. In the Torrens title registration system (Massachusetts Land Court procedures included) usually boundaries are established at the time of registration.

The entire subject of establishment is important in that it corrects deficiencies of poor descriptions; it often maintains the status quo. Conveyances sufficient when made cannot be made insufficient by later loss of evidence. Because of this the law has provided means of establishing boundaries. In some jurisdictions special statutes provide for establishment proceedings by the appointment of processioners, commissioners, surveyors, or clerks of the court to establish or ascertain lost, destroyed, or disputed boundaries or corners; this is the subject of this chapter.

In some states, unless the proceedings are appealed within the time prescribed by statute, the determinations of the officers are conclusive. In a majority of the states, the reports of the officers are presented to a court, but there is diversity of opinion about what weight should be given the report.

The authority for the proceedings is entirely dependent on the statutes; and the proceedings, to be effective, must in general comply with the letter of the law. Although each state is different, a few of the states' laws are discussed to present an understanding of the subject.

In general the final decree establishing a boundary comes through the court but not always. In a few states the county surveyor or other parties are authorized by law to "establish" a boundary: and if the statute is followed exactly and the findings are not appealed from within a definite period of time, the "established" boundary is conclusive. One difference exists between this type of establishment and court establishment. In a court procedure the *facts* found by the court are final; only questions of law can be appealed. If an establishment procedure of a county surveyor or others is appealed to the court, both *facts* and *law* are considered.

In some states the findings of the county surveyor are *prima facie* evidence (assumed true until found false), and this is generally true, even when the law says the county surveyor's findings are conclusive.

Generally, the appointment of citizens to go out and view property disputes was a development in the original states and those states admitted to the Union soon after. In New England a fence viewer was a recognized officer of the town. In Georgia three suitable persons were appointed processioners by the court. In many places the county surveyor was elected, usually without requirements about qualifications. In the West and in the sectionalized land states, the use of laymen to settle boundary disputes seldom occurred. The following selected state laws illustrate some of the procedures of establishment. Since Georgia is one of the states with multiple boundary problems, it will be cited first.

7-2 Georgia Establishment Procedures

All the following statutes are in Chapter 85-16, entitled "processioning."

> 85-1601
> In all cases of disputed lines the following rules shall be respected and followed: Natural landmarks, being less liable to change, and not capable of counterfeit, shall be the most conclusive evidence; ancient or genuine landmarks, such as corner station or marked trees, shall control the course and distances called for by the survey. If the corners are established, and the lines not marked, a straight line, as required by the plat, shall be run, but an established marked line, though crooked, shall not be overruled; courses and distances shall be resorted to in the absence of higher evidence.

This portion of the law is in agreement with the common law of the rest of the United States except for the emphasis on a crooked line. With a 7-year statute of limitations for possession, crooked lines are often declared to be ownership lines.

> 85-1604 Appointment of processioners;
> The judge of the probate court of each county shall, at the second term of his court in every second year, appoint three suitable persons in every militia district in the county, who shall be processioners of land for that district until their successors are appointed: Provided, that in the event the judge of the probate court is unable to find three persons in a militia district to serve as processioners, or in the event a processioner shall disqualify himself or refuse to serve and the judge of the probate court is unable to find a person to serve in his place in such militia district, the judge of the probate court may appoint a processioner or processioners, as the case may be, from a different militia district to serve in such militia district. Vacancies may be filled in the same manner at any time. If none is appointed, the judge of the probate court shall appoint at any regular term, on the application of any landowner.

85-1605 Application for survey

Every owner of land, any portion of which lies in any district, though the remainder lies in an adjoining district or an adjoining county, who desires the lines around his entire tract to be surveyed and marked anew, shall apply to the processioners of said district to appoint a day when a majority of them, with the county surveyor, will trace and mark the said lines. Ten days' written notice of the time of such running and marking shall be given to all the owners of adjoining lands if resident within this State; and the processioners shall not proceed to run and mark such lines until satisfactory evidence of the service of such notice shall be produced to them.

85-1606 Surveyor's duty; plat; certification; evidence

It shall be the duty of the county surveyor, with the processioners, taking all due precaution to arrive at the true lines, to trace out and plainly mark the same. The surveyor shall make out and certify a plat of the same, and deliver a copy thereof to the applicant; and in all future disputes arising in reference to the boundary lines of such tract, with any owner of adjoining lands, having due notice of such processioning, such plat and the lines so marked shall be prima facie correct, and such plat, certified as aforesaid, shall be admissible in evidence without further proof.

85-1602 General reputation, when evidence

General reputation in the neighborhood shall be evidence as to ancient landmarks of more than 30 years' standing; and acquiescence for seven years, by acts or declarations of adjoining landowners, shall establish a dividing line.

85-1603 Adverse possession

Where actual possession has been had, under a claim of right, for more than seven years, such claim shall be respected, and the line so marked as not to interfere with such possession.

85-1609 Protest to judge of the probate court; appeal to superior court; verdict and judgment

Any owner of adjoining lands, who may be dissatisfied with the lines as run and marked by the processioners and surveyor, may file his protest thereto with the judge of the probate court within 30 days after the processioners have filed their returns, specifying therein the lines objected to, and true lines as claimed by him; and it shall be the duty of the judge of the probate court to return all the papers, including the plat made by the surveyor, with said protest, to the clerk of the superior court of the county or counties where the disputed land lies (copies being sent to the adjoining counties); and it shall be the duty of the clerk to enter the same on the issue docket, as other causes, to be tried in the same manner and under the same rules as other cases. The verdict of the jury and the judgment of the court shall be framed to meet the issue tried and decided: Provided, that it shall not be necessary to run any lines between adjoining landowners except the lines in dispute.

The returns of the processioners have been held to be *prima facie* evidence (67 SE 2d 195, 85 SE 2d 808, and 148 SE 2d 74); thus the burden of proving otherwise rests on the protestor. Where a party has been in actual possession of land for more than 7 years, under claim of right, the processioners shall respect that claim (19 SE 2d 538). Under this ruling the processioners may fix a boundary line, straight or crooked, in accordance with either the writings or possession. At no time can they set a new line unrelated to either the writings or occupancy (94 SE 824). Although processioners may mark out only established land lines, an established land line may be one established by acquiescence for a period of 7 years, evidenced by acts or declarations of adjoiners (151 SE 834).

7-3 Kentucky

The establishment laws of Kentucky requires the appointment of processioners (four, with any two as a quorum) who have the authority to monument lines on the request of landowners or the owner. Differences exist between the laws of Georgia and Kentucky, but the aim is the same: establishing and marking boundary lines by a legal process. Section 73.070 of the Kentucky law requires the use of the magnetic meridian as follows:

> (1) Every survey shall be made by horizontal measurement.
> (2) In resurveying lands, the surveyor must execute the survey by the magnetic meridian, but he shall certify and show in his plat the degree of variation in the magnetic needle from the true meridian at the periods of the original survey and of the resurvey, if it can be done.

As with Georgia, the statute of limitations for adverse possession is 7 years.

7-4 Connecticut

The statutes of Connecticut, Section 47-34, provide:

> When the boundaries of land between adjoining proprietors have been lost or become uncertain and they cannot agree to establish the same, one or more of them may bring a complaint to the superior court for the county in which such lands or a portion of them are situated; said court may, upon such complaint, order such lost and uncertain bounds to be erected and established and may appoint a committee of not more than three disinterested property owners, who shall give notice to all parties interested in such lands to appear before them and, having been sworn, shall inquire into the facts and erect and establish such lost and uncertain bounds and may employ a surveyor to assist therein and shall report the facts in their doing to the court.

After presenting the report to the court, the court may accept, reject, or listen to appeals. If the report is accepted and filed, it is binding on all the parties.

It is interesting to note that provisions are not made for the appointment of a surveyor on the committee—only property owners.

7-5 Illinois

Chapter 133, Section 12 of the Illinois law was amended in 1969 to read as follows:

Surveys shall become legal surveys in the following manner:

Any landowner desiring to establish the location of the line between his land and that of an adjoining landowner may do so as follows:

(a) He shall procure a land surveyor licensed in this state to locate the line in question and shall compensate such land surveyor.

(b) The land surveyor shall notify the owner or owners of adjoining lands that he is going to make the survey, which notice shall be given by registered or certified mail at least twenty [20] days before the survey is started.

(c) Whenever all the owners of the adjoining lands shall consent in writing, the notice shall not be necessary.

(d) The lines and corners shall be properly marked, monumented by durable material with letters and figures establishing such lines and corners, referenced, and tied to corners shown in the corner record book in the office of the county surveyor, or to corners shown on a plat recorded in the plat books in the office of the county recorder.

(e) The land surveyor shall present to the county surveyor for entry in the legal survey record book, a plat of such legal survey, together with proof of notice to the adjoining landowner or landowners, or waiver of such notice. Notice shall be given by the land surveyor to adjoining landowners by registered or certified mail within ten [10] days after filing of the survey.

(f) The lines, as herein located and established, shall be binding upon all landowners affected, their heirs and assigns, unless an appeal is taken as provided for by 1 R. S. 1852, ch. 103, Section 8 [17-3-58-5], as amended. The right to appeal shall commence when the plat of the legal survey is recorded by the county surveyor in the legal survey record book. [Acts 1969, ch. 96, § 3, p. 215.]

The surveyor cannot decide who has title to land; he merely locates boundaries. A court trial to determine the effect of the original act produced this response from the court:

> Where a bill was filed under the provision of the act of 1867 for the resurvey and establishing of corners and lines in place of those made by government surveyors, which had become obliterated, the line thereby established must be taken and held to be the true and correct line, as between parties and privies to the decree.
>
> —*Ambrose* v. *Raley* 58 Ill. 506.

7-6 Colorado

The Colorado statutes call for the court to appoint one or more disinterested surveyors to locate the lost, destroyed, or disputed corners and boundaries. The location can be in accordance with either writings or acquiescence as stated in the following statutes.

38-44-101. When action may be brought. When one or more owners of land, the corners and boundaries of which are lost, destroyed, or in dispute, desire to have the same established, they may bring an action in the district court of the county where such lost, disputed, or destroyed corners or boundaries or parts thereof are situated against the owners of the other tracts which would be affected by the determination or establishment thereof, to have such corners or boundaries ascertained and permanently established. If any public road is likely to be affected thereby, the proper county shall be made a party defendant.

38-44-103. Pleadings—trial of issues. The action shall be a civil action, and the only necessary pleadings therein shall be the petition of plaintiff describing the land involved and, insofar as may be, the interest of the respective party and asking that certain corners and boundaries therein described, as accurately as may be, be established. Either the plaintiff or the defendant, by proper plea, may put in issue the fact that certain alleged boundaries or corners are true ones or that such have been recognized and acquiesced in by the parties or their grantors for a period of twenty consecutive years, which issue may be tried before a commission appointed in the discretion of the court.

33-44-106. Hearing. At the time and in the manner specified in the order of court, the commission shall proceed to locate said boundaries and corners and for that purpose may take the testimony of witnesses as to where the true boundaries and corners are located; and, when so ascertained, the commission shall mark the same by erecting or putting down permanent and fixed monuments at all corners so located. In its report to the court, the commission shall file a map or plat showing all monuments, lines, and any other evidences or witness marks that will more nearly identify the corners *and, if that issue is presented, shall also take testimony as to whether the boundaries or corners alleged to have been recognized and acquiesced in for twenty years or more have in fact been recognized and acquiesced in. If it finds affirmatively on such issue, it shall incorporate the same into the report to the court.* (Emphasis added)

38-44-109. Corners and boundaries established. The corners and boundaries finally established by the court in such proceedings, or an appeal therefrom, shall be binding upon all the parties, their heirs and assigns, as the corners and boundaries which have been lost, destroyed, or in dispute; but if it is found that the boundaries and corners alleged to have been recognized and acquiesced in for twenty years have been so recognized and acquiesced in, such recognized boundaries and corners shall be permanently established.

In an appeal from a trial (Forristall v. *Ansley*, 170 Colo. 391) the appeals court specifically approved the findings of the commissioner based on 20 years' recognition and acquiescence. In other words, the surveyor appointed to the commission is expected to take 20 years' recognition and acquiescence into consideration in his findings.

7-7 Missouri

In a 1900 Missouri Supreme Court case the following was found:

The eighth instruction (of the court) merely directed the jury that if they found that Hearrel, the former county surveyor, had found and fixed the center of the section as the statute of this state directs, the point of intersection found by him was the true and legal center corner of the section. Certainly this is the law. It is of the utmost importance that a corner thus fixed by law, and as to which there is no suggestion of dereliction of duty in its establishment, should remain a permanent corner.

The surveys of the lands of this state were made in the first place by the surveyors of the United States. The corners and monuments fixed by them are binding upon juries and courts alike. The subdivision of sections is provided by our statutes, and where this has been done by county surveyors, and interior corners fixed, these, also, are binding, in the absence of proof that such officers have violated the law in their methods.

—*Granby Mining & Smelting Co. v. Davis,* 57 SW 126.

7-8 Arkansas

In Arkansas the law provides that a certified copy of the survey of the county surveyor is *prima facie* evidence. In a litigation that followed the following was found:

> The statute does not make it the county surveyor's business, nor invest him with any general powers to determine boundaries between individuals. He is a public convenience, not a general arbiter, although his acts in the line of his duty may make a prima facie case against individuals. It is taken as proof in the absence of better.
> —*Smith v. Leach,* 44 Ark. 287.

> Appellant first argues that "The certificate of survey by the county surveyor, Mr. Coleman, was filed for record and is prima facie correct and the court should have so held." It is true that a certified copy of the record of the county surveyor is prima facie evidence of the correctness of a boundary line as it appears from the survey. However, this is a rebuttable presumption and any duly qualified surveyor may testify as to its correctness.
> —*Mason v. Peck,* 239 Ark. 208.

7-9 Arbitration

Boundaries in dispute may be settled by arbitration between the disputing parties. If both agree to accept the results of the arbitrator or arbitrators, in advance, the findings will be binding on the parties. At law this is considered as an adjudication by a method of the parties' own choice and should not be set aside because of errors; but misconduct, gross mistake, fraud, or partiality may be cause to set the award aside.

In some states, such as Illinois, the statute law specifically provides for an arbitration method in land disputes. Usually such arbitration is limited to boundary location and cannot generally be resorted to when determining ownership matters. Since the surveyor does not ordinarily decide ownership matters, only location of boundaries, arbitration comes within the scope of his activity, and he is usually the one eminently qualified for boundary arbitration.

7-10 Establishment by the County Surveyor

In some states the statute provides that the county surveyor, on application of any property owner, shall locate corners. Fees are as prescribed by law and are generally too low to be practical. In addition, many county surveyors are elected without regard to the qualification of the man elected. In general, the practice of a county surveyor's doing private surveys is becoming obsolete, and the private practitioner determines property boundaries.

In some states provisions are made that the reports of the county surveyor are *prima facie* evidence of the facts presented. In others regular establishment proceedings are set up with provisions for notice to adjoiners and appeal. If uncontested for a specified period of time, in some states, the findings of the county surveyor become final.

7–11 Agreement by Formal Documents

Two parties can always agree on a boundary line and consummate their agreement by a written contract.

7–12 Summary

The surveyor must always be aware of the fact that at some time in the past a land boundary may have been fixed in position by some establishment procedure sanctioned by law, and if the law has been complied with, the boundary so fixed is correct regardless of the later evidence found. Usually, the establishment procedure is by court approval but not always. In some states, unless appealed to the courts, the findings of the establishment authority is final.

When a commissioner, processioner, or surveyor is given the authority to commence establishment procedures, each is usually told to consider both writings and possession of a duration of the statute of limitations: and the findings can be based on either consideration.

In all establishment procedures a written report must be filed. It is the obligation of the surveyor to know where such reports are filed, and the surveyor should examine the reports for possible establishment in his area of practice. In some instances the statutes have been changed; it is the law that was in effect at the time of establishment that is controlling.

CHAPTER 8

Procedures for Locating Written Title Boundaries

8–1 Introduction

The procedures used to locate property described by a written conveyance is the subject of this chapter. Resurveys, location of tracts described by metes and bounds, location of lots and blocks, or the location of parcels described for any land right are included; original surveys made to create new land divisions are treated in Chapter 13.

The purpose of this chapter is twofold: first, to relate procedures on how to locate boundaries already described; second, to discuss the duties and obligations of the surveyor who locates described boundaries.

Most clients, when they present a survey problem, feel that the surveyor's first move is to rush into the field and make a survey. This is far from the truth, since documentary and map evidence must be examined and research conducted. A complete analysis of the problem is essential. In some instances, two-thirds of the time may be spent in the preparation of fieldwork and office completion and only one-third in fieldwork. If the area being surveyed has been stripped of monuments by the usual causes of obliteration, considerable time must be spent questioning old residents and others acquainted with the historical background of the area. The discovery of a proper point from which to start measurements may require days, whereas the act of measuring after the reference monuments have been located may only take a few hours. Logical procedures to follow in solving these problems are outlined in this chapter.

8–2 Boundaries Defined by Written Documents

Writings used to describe an ownership (deed, will, etc.) are indispensable evidence; in this country transfer of real estate title must be in writing. The legal description of a parcel—that is, the writings competent to convey title to the parcel—can be considered as an indispensable guide giving the surveyor necessary information about the location of the client's property. A legal description is seldom complete; it generally will refer to other documents such as maps, surveys, adjoiners, monuments, and prior conveyances in the chain of title. With all of the information available, the legal description and all reference documents, either directly called for or implied, plus all adjoiner surveys, the surveyor then goes on the ground and locates the parcel in accordance with the best-available evidence as discussed in Chapter 2.

In 1946 the ACSM in its *Technical Standards for Property Surveys* made the following recommendation: every parcel of land whose boundaries are surveyed by a licensed surveyor should be made conformable with the record boundaries of such land. This policy, if followed, precludes setting monuments along lines whose standing depends solely on an unwritten title right.

As previously stated, deeds valid when formed are not made invalid because of loss of evidence. Something must be used to locate the deed on the ground. Provided better evidence does not exist, reputation of existing occupancy can be used as the best evidence of where the land should be located. The *Technical Standards for Property Surveys* does not intend to exclude possession lines from being used to indicate where the original writings were located; it merely means that if the possession lines are not in agreement with the writings, the surveyor should disclose the situation, not hide it by setting monuments in a line based solely on possession. An obligation exists to reveal these discrepancies.

Locating land described by a written conveyance is a great challenge to the surveyor; it is detective work; it is the process of discovering what transpired in the past; it is the understanding of past events, evidence, and history; it is the final conclusions from the facts discovered; it is a professional service to the public.

Many of the surveyor's acts are based on judgment, reasoning, and conclusions; these are not subject to standardized rules or applications. But if all surveyors followed systematic procedures and uncovered all the evidence, one would arrive at the same conclusions as would another surveyor using equal care.

8-3 Arrangement of Subject Matter and Systematic Procedures in Conducting the Survey

The functions of property surveyors are described by loosely used terms rather than precise phrases. Because of this, clarification of these terms, along with an explanation of the surveyor's authority to perform property surveys, is deemed necessary. Although it can be said that no two surveying problems are the same, it is possible to group the majority of situations into a few classifications and systematize the necessary procedures in such a way that nothing is overlooked, and the client is assured of the professional service he deserves. Basically, all questions of claims fall into three catagories: (*a*) questions of title (the quality of title or who owns the land), (*b*) question of where title lines are located, and (*c*) a combination of the two previous questions. The steps outlined in this chapter will not fit all situations, nor will they perfectly fit any one specific survey, but they will provide a guide for general situations.

Beginning with Section 8-8 the steps are grouped into five phases: contact, research, fieldwork, compilation of evidence, and presentation. Intermingled in the discussion of these steps are included the duties and obligations of the surveyor. The importance of each step will shift with different types of surveys; and, in some, certain steps may be omitted and new ones added. It might be advisable to introduce into cost-accounting procedures these or similar divisions so that cost estimates of future operations can be predicted from past experiences.

NATURE OF LOCATION SURVEYS

8-4 Definition of Terms and Surveyor's Functions

The surveyor, when locating the perimeter called for in an existing conveyance, may do any or all of the following functions:

1. A resurvey.
2. A survey in accordance with instructions in a written description.
3. A possession survey relative to written title lines.

Although defined here, the subject of original surveys is more thoroughly treated in Chapter 13.

DEFINITION. *Original surveys are a consideration of a conveyance and are made for the purpose of identifying land on the ground prior to or as a consideration of conveyancing.*

Original survey, when properly used, means the survey called for or presumed to have been made at the time a parcel or parcels were created. The essential character of an original survey is this: except where a senior right or a lawful fee right obtained by possession is interfered with, the location of the survey as set on the ground is legally correct. An original survey creates boundaries. The purpose of a *resurvey* is to determine where the footsteps of the original surveyor were located.

In some parts of the United States, especially in the Northeast, the determination of whether an original survey was or was not made is difficult; early records of land surveys have been lost or destroyed or are deficient in explanations. In such areas, where land was sold by a plan, the basic presumption is that an original survey was made even though direct proof may be lacking. In other areas, for example, California, direct proof that an original survey was made or was not made is almost always available; a need for such presumption is remote.

For sectionalized lands the last General Land Office survey prior to the government's disposing of the land is the original survey. Some townships may have several original surveys within their boundaries. At the present time, when a subdivision is created, most states require a survey prior to the sale of any lot; that survey is an original survey. Boundaries can be created in two ways: by a survey from which a description is prepared or by a written description without benefit of survey. If a survey is made for a metes and bounds description and if the survey is called for in the description, the survey called for is an original survey. Metes and bounds surveys made after the formation of a conveyance, in most areas, are not considered original surveys. In Maine and New Hampshire monuments set soon after a conveyance have been elevated to the status of an original monument by the following court cases:

> Where a monument does not exist at the time a deed is made, and the parties afterwards erect such a monument, with intent to conform to the deed, such monument will control.
> —*Lerned* v. *Morrell*, 2 N.H. 197.

> If a conveyance of land refers for its boundaries to monuments not actually existing at the time, and the parties afterwards deliberately erect the monuments as and for those itended, they will be bound by them in the same manner as if erected before the conveyance.
> —*Proprietors of the Kennebec Purchase* v. *Tiffany*, 1 Me. 219.

> Monuments, whether they embrace more or less land than the deed, are controlling whether the deed is given to conform to them or they are erected to conform to the deed.
> —*Bemis* v. *Bradley*, 126 Me. 462.

In each of these three cases quoted in which monuments were set after the formation of the conveyance, the setting of monuments controlled only when done by the parties to the deed. The court's theory was that where the monuments were set by the parties to the deed, it showed their original intent. Since later purchasers were not a party to the original deed, their acts do not disclose the original intent of the parties to the deed and thus are inferior in the evidentuary chain.

Free survey is used to designate a survey of property that is arbitrarily selected by the owner and is "free" in the sense that the lines of the survey are not fixed by a former conveyance. The terms "original surveys" and "free surveys" as used herein are often synonymous. Although the land surveyor rarely performs a wholly free or wholly original survey, he often creates, within subdivisions, lots that are independent of any former survey or conveyance and are free or are original surveys. If a surveyor creates a new lot that does not touch the boundary of a new subdivision and the survey is a consideration of the subdivision, or if he carves out a parcel of land wholly within an ownership and describes the parcel by a metes and bounds description that calls for the survey, he is performing an original free survey. Most original surveys are dependent on a resurvey to establish a record point of beginning. Original surveys today are more often called "subdivision surveys."

The term *resurvey*, in a sense, is reserved to mean the relocation of lines marked during an original survey. In a broad sense, a resurvey can mean the relocation of any former survey irrespective of whether the former survey was an original survey or not. For example, a surveyor locates the called-for mean high-tide line on a certain day, and at a later date a new survey could be for the purpose of (1) locating the present mean high-tide line (not likely to be in the same place as in the first survey because of erosion and accretion) or (2) to resurvey the former surveyor's lines. Neither would be a resurvey of an original survey as defined above.

PRINCIPLE. *Resurveys of original surveys are for the purpose of restoring the original surveyor's lines in the same position he marked them.*

As generally used, "resurvey," without qualifications, has this connotation.

Location of land described by a lot number within a recorded subdivision is generally a resurvey, since practically all laws regulating land divisions by plats require presurveying. There are exceptions where the land platted was not originally surveyed.

A description of land to be surveyed reads, "Beginning at an oak tree located 33 feet northerly of the intersection of Foss and South Grade Roads; thence N 22°01'E, 207.01 feet; thence N 89°07'E, 242.01 feet; thence ... etc." The surveyor in monumenting these lines is not performing a resurvey (no original survey was called for) but is performing a survey as based on the written record. No suitable term at present exists that adequately describes this function; herein, it will be called *surveys based on the record*.

Metes and bounds survey, as used by surveyors, is a broad term that can mean any of three survey functions performed in conjunction with metes and bounds descriptions. The phrase would include (1) a survey of the previously described land, (2) original surveys being made for the purpose of preparing a metes and bounds description, and (3) if a metes and bounds description calls for an original survey, a later location of the land described would be a resurvey as well as a metes and bounds survey.

A metes and bounds description can be written without the benefit of a survey, and any survey made to locate this already-described land would be a "survey based on the record."

Possession surveys are to locate land lawfully gained or lost by acts of possession or to show the relationship between title lines and possession lines.

Overlaps and gaps: when discussing defects found in titles, the terms "hiatus," "compound hiatus," "confusion," "point of confusion," "area of confusion," and "gore" are used to express overlaps, gaps, and indefinite ownership areas between adjoiners.

Hiatus, according to *Webster's New International Dictionary*, means "an opening; a gap, a space where something is wanting." In title problems the meaning should be confined to a gap rather than overlap (Fig. 8-1).

Compound hiatus means multiple gaps on the same adjoiner.

Confusion, at law and with respect to land, is the intermixture of the land (paper titles alone or paper titles and possession) of two or more owners so that their respective portions can no longer be distinguished. This term includes the concept of "uncertainty" that may be either a gap or an overlap.

Area of confusion is enclosed by the limits (width and length) where confusion exists.

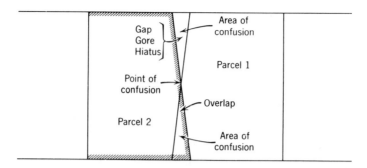

Figure 8-1 Title defects.

Compound confusion exists where one title has confusion with several adjoiners.

Point of confusion is the place where confusion commences or, as often used, the point of change from a gap to an overlap. See Fig. 8-1.

A *gore* is a small triangular piece of land. The term, originally applied in New England, usually refers to a narrow, sharp-pointed, triangular parcel of land, more often occurring as a gap than an overlap.

An *overlap* is an extension of a written title over and beyond another written title. On the ground there may be only one line of possession, but the writings may be in conflict so as to create a "paper" overlap. Ordinarily, the overlap belongs to the senior title holder, although possession may dictate otherwise. "Encroachment" is a preferred term to use where possession overlaps paper title. Technically, ownership of the same interest cannot overlap; someone has ownership, and someone else has what appears to be a right of ownership (color of title).

8-5 Legal Authority for Regulating Property Surveys

Original surveys of property are for the purpose of creating new parcels of land prior to conveyancing; they are properly regulated by legislative action. How land may be subdivided, how it must be monumented, what measurement accuracies must be maintained, and what use may be made of the land are all questions within the police power of the jurisdiction and may be covered by regulations.

The purpose of resurveys is to relocate the lines described in an existing conveyance. How this shall be done is a question of interpretation of the meaning of a conveyance. Courts interpret documents. Although the legislature may regulate how a conveyance shall be formed, once it is formed, the courts interpret, in the light of the laws existing as of the date of the deed, where the location shall be. *All surveys based on the record* are of this nature; the courts interpret their meaning and location.

The legislature may properly regulate who may perform resurveys, what kind of monuments must be set on resurveys and how survey plats must be made and delivered, but it is not proper for them to legislate how the court is to interpret conveyances or what evidence the court must use.

PRINCIPLE. *Original surveys to divide land are or may be regulated by legislative action; once a conveyance is made, the location of the land described is interpreted by the courts.*

When subdividing or platting new parcels of land, the surveyor obeys the statutory laws, but when the surveyor is locating the land described in a conveyance, he looks to court rules and past decisions to tell him how to interpret the writings and where he should locate the boundaries.

8-6 Property Surveyor's Authority

PRINCIPLE. *With the possible exception of establishment proceedings in accordance with law, the property surveyor has no judicial authority; when resurveying for clients, the force of the property surveyor's authority is derived from reputation and respect.*

In a resurvey no one is forced to accept the surveyor's findings. The surveyor's monuments are given force only by the consent of landowners or the court.

After a surveyor has monumented land boundaries, the proprietor will view the results and accept or reject them. On occasion possession will vary from marked lines, and the owner will refuse to accept the results. Sometimes the adjoiner objects and withholds his consent. If a court trial ensues, the court will declare whether the survey results are final or not. But in no instance can the property surveyor enforce his opinion on others; he must convince the owners or the court that he is right.

Without trust in a property surveyor there is no reason for his existence; the property surveyor is hired by people because they believe him to be capable of being correct and just. If the people believe that a surveyor is dishonest, his findings will not be approved and his efforts will be for naught. Property surveyors depend on the public, or the courts, to approve their monuments; they are arbitrators, and they serve the public.

8-7 Fact Finding and Conclusions

PRINCIPLE. *First, the surveyor finds all the facts about a property; in this sense he is a fact finder. Second, the surveyor comes to conclusions from the facts; it is the conclusions the surveyor makes that mark him as a professional.*

Frequently, the surveyor is referred to as a fact finder of information pertaining to property-line locations. However, this is not the limit of his responsibility and duty. After the surveyor has gathered all the facts, he must come to a conclusion about where deed lines exist and also about whether ownership lines might have changed because of prolonged possession as discussed in Chapter 3.

Finding the facts about a particular property is often delegated to a subordinate or to office help; deriving a conclusion from the facts is a professional responsibility. The collection of maps, adjoiner deeds, and measurements represents a necessary preliminary step prior to the final conclusions, but this is not what elevates the surveyor above the crowd. The surveyor's final correct conclusion and the ability to say what is right and what is wrong separates him from others. Although research is essential and no surveyor can come to a correct conclusion without all the facts, surveyors should not be thought of solely as fact finders.

The responsibility for fact finding can be divided into four parts:

1. Facts furnished by the client.
2. Search of the written records (deeds, adjoiners, maps, etc.).
3. Fieldwork, search for monuments, locating possession, and making measurements.
4. Seeking testimony and information from old residents and other surveyors.

CONTACT WITH CLIENTS

8-8 Initial Contact with Client

As a matter of professional standing, professionals do not actively solicit clients in the manner of business people; they should not advertise or seek recognition in a self-laudatory fashion. Standing in a community, knowledge, prestige, and personal integrity are the foundations for success. Obtaining clients is a result of referral, personal contact, and discreet notices of services rendered.

The client may become aware of his need for a survey from various outside contacts. He may have had similar experiences in the past and realizes that this is the proper course of action, or he may be advised by his attorney that a survey is essential to settle a title problem. Occasionally, the client learns of surveying services from descriptive literature found in abstractor's offices, real estate offices, title insurance offices, and the like. Much public education is needed to inform property owners when they need a property survey and that after an adequate survey has been made it can be likened to purchasing a life insurance policy on one's property.

Referrals by previous clients are the most common means of client contact. A property owner will preferably engage a surveyor he has seen work in his community rather than a complete stranger.

8-9 Initial Conference with Client

The initial conference is generally an inquiry by the client about what services the surveyor can render. For such inquiry a fee is not appropriate, since this should be considered an opportunity to educate the public in the need for expert help.

It is rare indeed when a client knows precisely what he wants or needs, although he may think so. The property owner sometimes contacts the surveyor when he should more properly seek legal counsel. The usual statement of the prospective client is, "I need my property surveyed." The surveyor then (1) tries to reach an agreement with the client about the extent and kind of work to be done, (2) presents his estimate of cost, and (3) tries to arrive at an agreement.

8-10 Contracts with Clients

An agreement between a client and the surveyor arising from an "offer" by the client to provide "consideration" for the surveyor to locate a particular parcel of land and the "acceptance" by the surveyor to do so for the consideration is a *contract*, provided the agreement is possible and legal. The law will enforce contracts, both written or implied. Not only will the law enforce performance of a duty required by a contract but also a certain quality of work (called *implied warranty*) will be upheld. Most liability suits stem from work of inferior quality. Figure 8-2 shows one type of contract form used.

8-11 Agreements with Clients

Liability may result from a failure to fulfill an agreement; hence the surveyor should only agree to do those things that he is capable of and is competent to perform. A client may request any one of the following items:

1. A survey of conveyance lines according to a particular written document.
2. A survey to divide property into smaller parcels.
3. A survey to locate a particular line of possession relative to known monuments within a given area.
4. The survey and location of a client's property rights (either written or unwritten).
5. A survey for construction or topography purposes.

At this point, it should be made clear to a client that the survey will only locate written deed lines and that deed lines are not necessarily "ownership lines." Sometimes a client thinks a surveyor is capable of determining ownership lines; thus when a client requests a survey of his property, the land surveyor, to avoid future embarrassment, should condition his acceptance on

```
            The following is a confirmation of work ordered to be performed. If any of
            the information shown hereon is not in accordance with your understanding,
            please advise us immediately. We will not be responsible for any errors
            or misunderstanding which may arise from a lack of proper notification.

                            (NAME AND ADDRESS OF SURVEYOR)

                                  W O R K   O R D E R

     Taken by:_____                    Job No._____
                                                         Date_____

     Ordered by:                              Bill to:

     _____                 _____
            (Name)                                   (Name)

     _____                 _____
           (Address)                                 (Address)

     _____                 _____
            (Phone)                                  (Phone)
                                              (Requests to bill someone other than the
                                              undersigned must be accompanied by an
                                              authorization)

     Location (Street, City, or County)_____
     _____

     Title Ins. Co._____    Escrow No._____
     Documents Received_____
     Description of Work_____
     _____
     _____

     Sketch if Necessary



     Starting Date Promised_____ Completion Date Estimated_____
     Basis of Charge_____

     Bills are rendered monthly for the work done in the preceding month, and are due
        and payable within ten days after presentation.
     Client agrees to pay all reasonable collection expense and legal interest.
     Cancellation of this order presupposes payment for work already completed.
     Corners marked or stakes set are not to be used for construction until confirmation
        of work actually performed is received from this office.


                                         _____
                                         Signature of Responsible Party
        CONFIRMING COPY
```

California Council of R. E. & L. S.

Figure 8–2 Contract.

the premise that the survey is to be made in accordance with a deed to be furnished by the client.

When a client requests a property survey, it is logical to assume that he means to ask, "Do my possession lines agree with my written deed?" The client already knows where his fences are, so why would he want them located? Unless it can be shown by reasonable evidence that fences represent monuments of original survey lines (see Section 2-52) surveyors do not survey from fences. If, by some circumstances, the client does want his fences located, especially for unwritten title action, the surveyor must be certain that the client understands that the survey is not a location of his written conveyance lines. If this is not made clear, the surveyor will surely be embarrassed in the event of a later boundary dispute.

Sound business practice requires that the request for the survey and the specific offer be put in writing to reduce or eliminate future problems or misunderstandings.

RESEARCH OF RECORDS AND DOCUMENTS

8-12 Research Responsibilities

PRINCIPLE. *As a minimum a property surveyor who decides to make a survey from a written conveyance assumes the responsibilities of obtaining copies of (1) all adjoiner conveyances called for in the legal description furnished him, (2) all maps called for, (3) all recorded adjoining surveys, (4) public agency maps that are kept in such a form that they are available, and (5) government township plats and field notes where called for.*

In some areas the responsibility for research goes beyond that just named, especially in searching for adjoiner deeds. In areas abutter's deeds are readily available from a title company or abstractor; in others they are not.

It has been said that a surveyor can make an accurate survey but not a correct survey without research. Lines and positions can be accurately located with respect to other lines and monuments, but if the lines and monuments used to measure from are the incorrect ones, the accurately located lines are without meaning or effect.

The objective of research is to gather all written evidence (including plats) pertaining to evidence of title and evidence of monument locations. Although fieldwork follows the search in the records, evidence discovered in the field sometimes suggests additional places to search.

Each state, county, or local government has its own peculiar manner of storing and filing property-location data. A worthy project for a surveyor organization is the establishment of a local plat library.

The surveyor, as an expert, is held responsible to examine maps, writings, or commonly known sources of written information that might disclose evidence of property location. Most survey offices have large investments in a plat library, and often all the research necessary is within private files.

Where and how a surveyor does his research is so variable that it would be impossible to describe all places to look. This variability of sources of information is one reason why the experienced elder surveyor of a community has an advantageous position; this surveyor knows where to find data. In addition, he will have considerable amounts of reference materials and previous survey plats in his files.

In some sections of the country legal descriptions have deteriorated to the point where it becomes necessary to read former deeds in the chain of title to determine what was originally conveyed. Whereas the first conveyance may give clear instructions about how to locate a line, later conveyances performed by the uninformed may show the wording altered to "bounded on the east by Caleb Brown." When looking at Caleb Brown's deed it says "bounded on the west by Alfred Bryant." Neither tells how to locate the bound. Without adequate record research in such situations the surveyor does not know what monumentation he should be looking for; he is stumbling in the dark. Generally, the question is not whether research should be undertaken but how much research should or needs to be done. Without the research, a correct location cannot be made.

Generally speaking, the county surveyor or city engineer's records are the most fruitful source of property-location information, although not always. Road-survey plats, field books with reference ties, sewer surveys, water-line surveys, and numerous others often show the location of former monuments.

The courthouse and the recorder's office have the reports of court decrees, public records, and probate information. The state highway division, public utilities, tax assessor's office, railroads, special commissions, abstract offices, and title companies, to mention a few, are good places in which to search for information.

Original government survey notes concerning sections are readily available. Most original notes are now filed in state offices, and copies may be found in many counties.

The client's neighbors may have information of previous surveys; the adjoiner's deed and his plats and writings can disclose valuable evidence. Not all survey plats are recorded, and, generally, they can only be discovered by local inquiry.

In some parts of the country the surveyors have joined collectively in professional groups that, among other things, share information of surveys in their area. From a card index or master map the individual surveyor may be able to ascertain who has notes on surveys in a particular area, and he can borrow or examine them.

8-13 Documents Used for Property Surveys

PRINCIPLE. *The decision of which document shall be used as the basis for a survey is made by the surveyor.*

The client informs the surveyor what he wants surveyed; the surveyor then accepts, rejects, or negotiates what will be done. Ordinarily, the surveyor places on the client the burden of furnishing a description on which the survey is to be based. In some areas of the United States the surveyor searches the records and determines for himself the contents of the client's deed.

Not all documents on which a survey can be based are of equal dignity or value. A deed to property rarely recites all easements of record; a survey based on a deed can result in property monuments being located in the streets. Tax deeds are notorious for error; they are notices of taxes, not descriptions of property.

It is the obligation of the surveyor to inform the client which document should be used or relied on. Property surveyors are to protect the client's interests; if they agree to survey from a document that is potentially faulty, without notifying the client about the dangers, a truly professional service is not being rendered.

Ordinarily, the documents preferred are in this descending order of importance: (1) abstractors' opinion or title insurance policy, (2) deed, (3) mortgage or loan descriptions. Tax deeds should only be used as a last resort.

8-14 Ownership and Location of Land

PRINCIPLE. *The property surveyor does not decide who owns land or rights in land; he merely locates land; and, except for special agreements with respect to unwritten rights, he only locates land in accordance with written descriptions.*

Ownership of land is complex and falls within legal concepts. To determine true ownership, investigations are made into mortgages, trust deeds, valid signatures, fraud, insanity, payment of taxes, special assessments, heirs, possession, and numerous other things. In America the property surveyor has never been given the authority to decide who has ownership to land and in all probability will never be given that authority.

The legality of a document is a matter to be referred to an attorney or title company. If a client has paid for an abstractor's opinion or obtained title insurance, the description in the insurance policy or in the abstractor's opinion is guaranteed by others, and unless there is an obvious error, the surveyor usually accepts the description at its face value.

8-15 Location of Easements

PRINCIPLE. *Surveyors should not agree to locate all existing easements pertaining to a property; the surveyor should merely agree to locate easements in accordance with descriptions furnished and those that are visible or of public notice.*

Easements necessary for the enjoyment of a property may be automatically transferred, whether mentioned in a conveyance or not. Title insurance policies guarantee to define every "recorded" easement as of the date of the policy, but easements may be created after the title policy is issued. The property surveyor should not by any verbal or written statement lead a client to believe that no easements other than those located exist or that the written deed is being guaranteed; the location of a particular written description is being certified to.

When an easement, though not originally belonging to land, has become appurtenant to it by grant or prescription, a conveyance of the land will carry with it

such easement, whether mentioned in the deed or not, though it may not be necessary to the enjoyment of the land by the grantee.
—*Cole* v. *Bradbury*, 29 A 1097, 86 Me. 38.

Where an easement has become appurtenant to an estate, a conveyance thereof carries with it the easement belonging to it, whether mentioned in the deed or not, though it is not necessary to the enjoyment of the estate of the grantee.
—*Spaulding* v. *Abbot*, 55 N.H. 423.

FIELDWORK

8-16 Planning for Field Work

It is unfortunate, but often true, that the licensed surveyor does not do the fieldwork; he is usually too occupied with computations, platting, contacting new clients, and tending to general business activities. The surveyor must devise means of instructing his subordinate so that he can be confident that instructions will be followed. These instructions usually consist of (1) specifying which monuments will be searched for or, in the event these monuments cannot be found, alternate monuments, (2) what measurements will be made, and (3) specifying the accuracy required. In addition, the surveyor sometimes tells the subordinate what to do with the evidence and measurements uncovered.

The field party is furnished with all the maps, plats, topography sheets, aerial photographs, deeds, road surveys, and all information found while researching; or he is furnished a composite map, prepared in the office, showing the results of the research. Office closure computations are made, where it is possible to do so from the record data, *prior to* going into the field and are furnished to the field men. No lot or block in a subdivision, wherein the subdivision map gives complete dimensional data, should be surveyed without this necessary step. Map blunders or errors can often be resolved before fieldwork, and, if they cannot, the field man will be prepared to look for the errors.

A study of aerial photographs will often save considerable time and materials for the field party. A stereoscopic examination of the photographs may indicate obstacles to normal measuring procedure, such as extreme slopes, nonintervisibility between points, density of vegetation, and possible search areas for property corner locations. The necessary control traverse can be laid out on paper with due consideration to the required accuracy, strength of figure, and desired location of control points.

The surveyor can plan ahead by determining from the photographs and plats whose property he will need to enter to perform the fieldwork. Evidence of ancient lines (not visible on the ground) may be apparent on the photograph to a trained individual, and the field crew can be instructed where to intensify the search for monuments.

Timing is important. A survey in the business district may require Sunday work. Surveys next to school yards are best made after hours, and measurements anywhere are best performed when least populated. Other considerations such as the position of the sun, direction of the wind, stage of the tide, and other things must be taken into account in the effective planning of the fieldwork.

Metes and bounds descriptions require special attention, since the responsible surveyor must interpret the meaning of the description and transmit his interpretation to the men before they go in search of evidence. Instructions about final conclusions cannot be given until all evidence is gathered, but instructions in the order of importance of deed elements can be given. In the following deed, furnished the party chief, the phrases italicized there and the written instructions given are the key to determining what to do.

> Hector MacKinnon and Sarah Jane MacKinnon, his wife, both of San Diego County, California for and in consideration of the sum of Twelve Hundred and Fifty Dollars do hereby grant to J. B. Reichard of San Diego County, California, all that real property situated in said County of San Diego, State of California, bounded and described as follows: *Beginning at a post* 35.95 chains west of the northeast corner of the southeast quarter of Section (26) twenty-six, township 13 south, range 4 west S. B. M.; thence S 7°55'E, 18.41 chains *to post*; S 82°05'W, 13.02 chains *to post*; N 7°55'W, 20.09 chains *to post*; east 13.30 chains *to place of beginning* and containing (25) twenty-five acres *as surveyed April 30th 1893 by Berry McLarsen.*

Instructions to field party:

1. Inquire as to the whereabouts of the parties named in the deed and whether any living relatives exists. If found, inquire of them where the original posts were located.
2. Search for *posts* as set by McLarsen.
3. Look for possession evidence that might prove original survey position.
4. Locate east quarter corner of Section 26.
5. If survey of McLarsen is not discoverable from the posts or possession evidence, make a solar observation at the east quarter corner and run *due west* 35.95 chains. Stake property.
6. Record measurements are the last resort. Only use them in the absence of other acceptable evidence. The deed as written will not close mathematically as shown on the traverse sheet furnished.
7. If McLarsen's survey lines are discoverable, tie his lines to the east quarter corner.

8-17 Field Evidence

Collection of evidence in the field consists of (1) finding monuments or other memorials and (2) tying these together with measurements. Sometimes this step and the completion step (setting final monuments from conclusions drawn) are performed simultaneously, but ordinarily the surveyor determines all of the available evidence before drawing conclusions and setting final monuments.

8-18 Search for Monuments

As previously pointed out, all surveys must start from a monument or monuments; there are no exceptions. If one were to analyze the cause for two surveyors coming to different conclusions about the location of a point, the lack of discovery of correct monuments would often be the underlying cause. The importance of monuments stems from previously stated principles and is summarized as follows.

PRINCIPLE. *Except where a senior right is interfered with, monuments, either natural, artificial, record, or legal, if called for in a conveyance and if found undisturbed, indicate the intent of the original parties. If called-for monuments cannot be found or if they are found disturbed, their former position can be identified by competent witness testimony or acceptable physical indicators of boundaries.*

This principle emphasizes the first command for field search: find the monuments called for or locate the former position of the monuments. As stated in Chapter 2, this can be done (1) by the discovery of the monument itself or accessories to the monument, (2) by other monuments set to perpetuate the position of the original monument and supported by writings, (3) by competent witness evidence, and sometimes (4) by reputation and uncalled-for monuments.

The field search is a search for evidence. The searching field man needs to have knowledge of what he is to look for and an understanding of the significance of the things found; therefore, knowledge of the fundamentals of evidence (Chapter 2) is essential. Also, a field man needs to possess good "horse sense" to know where to look and to recognize evidence when it is found.

A bit of rust may mark the former position of an iron pipe; a scar on a tree may indicate an old hack; rotted wood may determine the spot of a former wooden stake; fences, hedgerows, ditches and other boundary barriers indicate likely places to search; signs of cultivation or tree rows may indicate the direction of lines. The success or failure of a surveyor is often in direct relation to his ability to discover evidence; the surveyor is a detective.

Technically, uncalled-for monuments, except as noted in Chapter 2, do not control a deed, but uncalled-for monuments are one of the best sources of evidence used by the surveyor. The position of these monuments indicates where owners think their lines ought to be, and, although they may not be paper-title lines, they may identify unwritten ownership lines. The sheer weight of evidence of numerous monuments, all properly interrelated, may prove a location. Sometimes the only evidence is uncalled-for monuments.

Any conveyance, sufficient as of the day of its execution, cannot be made insufficient because of destruction of evidence of monuments. It is the duty of the surveyor to hunt, probe, interview witnesses, spade the earth, and examine trees, fences, and other objects until he has found the best-available evidence from which to make his survey. No rule of time exists; it may take a few minutes, or it may take a week. Within certain limits, a person can predict how long it will take to measure a given distance, but it is impossible to determine how long it will take to find a monument. This is the uncertainty of property surveying and the reason many surveys are not susceptible to exact cost estimates. The following principle is self-evident:

PRINCIPLE. *The surveyor hunts and searches in the field until he has found the best available evidence upon which to base his survey. Time is not a consideration.*

8-19 Uncalled-For Monuments

In locating legal descriptions within the United States there are probably more uncalled-for monuments found on the ground than there are monuments described. Until recently very few states required the surveyor to record what he set in replacement of original monument locations. In those states where widespread loss of evidence of original monument positions exists, the acceptance of monuments that appear to have been set by a surveyor is quite common, although no chain of history of the monument may exist. In some states the reputation of a monument is the best-available evidence, and it becomes controlling for that reason alone. In those states where there has been an active professional surveyor's group, laws have been enacted to require filing public records of discovered monument positions, as in California in 1930 and in Colorado in 1963. Until such time as all states adopt a recording policy for set monuments, surveyors will be working in the dark. See Section 2-57 for the discussion of uncalled-for monuments.

8-20 Importance of Possession

Possession lines cannot be ignored. If possession lines do not agree with written deed lines, the relationship of the written lines to that of possession must be shown; possession lines may be ownership lines. Fences, buildings, roads, and other evidence of possession that were constructed on some information or some assumed knowledge of the actual line may serve as proper memorials for the boundary.

PRINCIPLE. *Possession may memorialize original survey lines.*

There are two distinct and different types of possession. Possession representing original survey lines as marked and surveyed is treated in a vastly different way from possession not in accordance with original survey lines. See Section 2-52.

If after an original survey is made, fences and improvements are constructed in accordance with the original survey, the fences and improvements stand as monuments representing the location of the original surveyor's lines. When all the original monuments have disappeared, the only remaining evidence of the location of the original survey may be the lines of possession. Possession in accordance with an owner's idea of where his lines ought to be, or possession in accordance with an erroneous survey, or possession in accordance with any survey made after the original survey is not possession representing where the original surveyor marked his original lines.

Where a survey is called for, the lines and monuments set by the original surveyor are conclusive control for resurveys. If possession monuments the original lines, it is a controlling factor in locating the lines of the original surveyor. But if possession is not related to the original surveys, it is then treated as an encroachment.

Perhaps the greatest value of possession evidence is as an indicator of where to search; lines of possession usually follow property lines, and most monuments are found at the termination of possession.

8-21 Testimony

Testimony about where monuments used to be, especially in rural areas, is of value, provided the rules of evidence are followed. Human memory quite often outlives the physical monument but, on the other hand, may be subject to the flaws of memory. In some states the surveyor has the authority to take statements of living witnesses under oath (see Section 11-6). In the absence of this right, a witness can always be taken before a notary public. The oath serves to deter a person from later changing his testimony or to refresh his memory, and it may be of value in the event of death.

8-22 Measurements

At least one monument, even though it may only be the line of possession, must be discovered before measurements can commence, but after the discovery of the first monument, the process of searching and measuring are often done simultaneously. Measurements indicate likely places to search, and early in the search for monuments, it is sometimes advantageous to make quick approximate measurements until such time as a suitable starting monument is found. But usually it is a waste of time to make approximate measurements, find the monument, and then remeasure precisely. It is often better to measure carefully to find the monument and then tie in the monument and correct the measurement to the true line by a computation.

8-23 Uncertainty of Position Caused by Measurements

PRINCIPLE. *A found undisturbed, original monument, expressing the intent of the parties of the conveyance, fixes a point that, as between the parties, has no error in position. All restored monuments, established by measurements, have some error in position.*

A discovered, called-for, and undisturbed, original monument accurately determines position. Irrespective of measurements from distant points, the position is unalterable, fixed, and has no error of location. If measurements to the point differ from the record, the measurements are in error; the position occupied is correct.

All nonoriginal monuments set by measurements from fixed points have some error in position. Although the error may be insignificant, it may be present, and the primary concern of the surveyor is to find starting points that are free from position errors; these are original monuments or the spot formerly occupied by an original monument.

8-24 Permissible Uncertainty of Measurements

PRINCIPLE. *The magnitude of permissible uncertainty of measurements is determined by court interpretation.*

The courts are the final authority on interpreting where boundaries called for in old conveyances should be located. And this authority includes the right to decide how much measurement uncertainty can be tolerated. The opinion of surveyors, as experts, may influence the courts' thinking in trial cases, and it is always proper for surveyors to express an opinion on measurement accuracies, but surveyors do not have the final say.

Permissible uncertainty of measurements has varied throughout the years. In the early days of this country, with the use of the compass, directional certainty was poor, and errors of 20 feet in a mile of line were not uncommon. Today, an error of 1 foot in a mile is ordinary accuracy. In years to come, an error of 0.10 foot in a mile may be considered excessive for property location. The court adjusts its thinking as measurement technology improves; accuracy standards for resurveys change with the development of new equipment.

A deed reading "N 10° E, 100 feet to an iron pin" will, in the event the pin is lost, have the former pin's position restored exactly (as closely as the surveyor can measure) "N 10°00'00"E, 100.00 feet." This is a court rule: the rule does not say the distance may be within so many minutes of the direction or within so many feet of 100.00 feet. The original measurements to the pin were certainly in error to some extent, yet this does not alter the rule to restore the lost pin at exactly 100.00 feet.

In a regular section of land, the section corners and the exterior quarter corners were set by the government surveyor. In the event that the quarter corner of a regular section was lost, the court declares that it must be reset "exactly" on a straight line connecting the section corners and "exactly" midway. When the corner was originally set, the government would accept as sufficient, a position within one chain of that specified. By law in sectionalized land the government could set rules of accuracy on how close a monument need be set to its true position. But this law in no way affects the courts' prerogative to specify how the corner is to be relocated in the event that it is lost. The court does not say the corner shall be relocated halfway, plus or minus one chain; it says "exactly" halfway. The government recognized this as early as 1903 in the *Manual of Surveying Instructions for the Survey of the Public Lands:*

Surveyors who have been United States deputies should bear in mind that in their private capacity they must act under somewhat different rules of law from those governing original surveys, and should carefully distinguish between the provisions of the statute which guide a Government deputy and those which apply to retracement of lines once surveyed. The failure to observe this distinction has been prolific of erroneous work and injustice to landowners.

PRINCIPLE. *The error of measurement originally permitted when tying original monuments together has no relationship to the accuracy required to reestablish an original lost monument position.*

In restoring a lost monument (not an obliterated monument) to its former reported position, the law in specifying restoration procedures assumes that the original reported measurements are free from error. In locating property in accordance with an existing conveyance, the surveyor must be as accurate as is necessary to protect his client's interests. If damages occur, the court will decide what the responsibility of the surveyor was.

One thing is certain: an error of measurement must be proved, and that can only be proved by subsequent measurements. The limits of accuracy required cannot extend beyond that which can be proved by measurement procedures. If surveyors cannot measure more accurately than ±0.03 feet in a mile, the courts certainly cannot assess damages for an error of 0.03 feet.

PRINCIPLE. *In the absence of the owner's specifying unusually high accuracy, it is presumed that the surveyor will work to an accuracy consistent with the purpose that the survey is to accomplish or standards of the profession in that locality.*

All surveys of written conveyances should, as a minimum, be just as accurate as that of a reasonably skilled prudent land surveyor under similar circumstances. Obviously, if the land surveyor were told that a 14-story building were to be erected adjacent to the property line and his liability would be high, he would exercise considerably more care than he would for a farm survey. Chapter 10 discusses survey accuracies and standards. If a client requests a property-line survey of low accuracy, the surveyor should refuse, since he has a moral obligation to the adjoiner as discussed in Chapter 5. The property surveyor should not reduce accuracy merely because he has reduced his fee. The courts have repeatedly held that cost is not a factor in determining liability; in fact a surveyor can be held liable for damages resulting from a free survey.

Registered or licensed surveyors always have the responsibility to instruct their field men about the accuracy required and on the procedures to be followed. Failure of a subordinate to attain the specified measurement accuracy is never an excuse to escape liability; surveyors must see that their instructions are followed.

Accuracy of position in resurveys or surveys based on the record is difficult to define. It not only depends on attainable measurement accuracies but also on legal interpretation of conveyance words. A deed reading "the west half of lot two" as shown in Fig. 8-3 is difficult to interpret. The uncertainty in position arises from legal interpretation, and unless the surveyor can resolve correctly the question of what is the direction of the easterly line of the westerly half, he may have large position errors.

Surveyors have proportional measurement problems. Suppose that a rectangular block, as originally monumented, contained 12 lots each 50×100 feet. Six lots faced one street. On discovery of the original block corner monuments, the surveyor must divide the frontage into six equal parts. Many types of measuring devices could be used to do this. Monuments set apart one-sixth of the length of the block and on a straight line connecting original monuments would be correct in position. A tape too long or too short could be used to divide the block into proportionate parts, but the true unit length of each lot would be unknown. In any proportionate measurement problem, it can happen that the monument as set will be in its correct position, yet the measurements will be inaccurate as compared to a defined unit.

Assuming that the surveyor correctly interprets a deed and monuments the line described, he should perform his measurements with sufficient certainty to prevent damages to the client. The usual reason that an owner wants a survey is that he wants assurance that when he erects improvements they will be on his land.

Judges are apt to look upon errors in terms of positive measurements. If a building were proved to be encroaching on the adjacent property by 1 inch, the fact that the surveyor had closed with a relative error of 1/20,000 or even 1/50,000 would probably fail to impress the judge; the point that the building encroaches 1 inch is all that counts.

8-25 Preservation of Discovered Evidence

Physical evidence, particularly monuments, must be noted and described in field books or in some other manner to preserve evidence of their location and being. Photographs are well suited for this purpose, especially when taken systematically and supplemented with a sketch. In the event of court litigation,

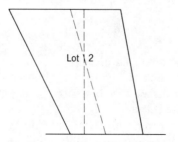

Figure 8-3 West half of Lot 2.

the surveyor must be prepared to prove his location of the property. Field notes and other records of field evidence are presented to support the surveyor's opinion of property location.

8-26 Field Notes

Field notes are a record of evidence discovered and an account of what was done in the field; they are shorthand notes of an entire project.

Ordinarily, a sketch to an approximate scale best presents the results of a property survey. Written notes in long hand, although effective at the moment, are usually difficult to interpret at later dates.

A checklist of some specifics to include in field notes follows:

1. Date, name, and address of client.
2. Names of party personnel and position and duty.
3. Condition of wind and observed temperatures.
4. Equipment used, with serial numbers.
5. Legal name of parcel being surveyed (book and page recording, lot and block, subdivision, etc.).
6. Work order number and purpose of survey.
7. North arrow with origin of bearings.
8. Description of all monuments found, not found, and set.
9. Measurements as made (slope distance and vertical angles, temperature, taping, horizontal angles, etc.).
10. Corrected distances and angles.
11. Description of all monuments set and ties taken.
12. Relation of possession to survey lines.
13. Outline of parcel surveyed with colored pencil.
14. Sketch of parcel staked.
15. Oaths of witness evidence (when applicable).
16. Names and addresses of adjoiners, old residents, and so on.
17. Reference to any records used or called for.

8-27 Assembly of Data

After the field search is completed and the discovered monuments have been tied together by measurements, office work such as computations and analysis of results is necessary. Finished plats are often prepared at this point and used for the final staking of the survey.

All evidence discovered must be considered during the analysis. In most cases, the field man can be of great help in deciding where new monuments are to be placed, what is to be shown on the plat, and what calls should be used if a new description is to be constructed.

Whether used or not, all data should become a permanent part of the file, with sufficient notation and cross-referencing to be of benefit to future surveys.

COMPILATION OF EVIDENCE

8-28 Computations

Computations serve several purposes: (1) to check on the presence of blunders and the quality of measurements, (2) to determine where final monuments belong, and (3) to calculate area.

If a perimeter is surveyed, the figure must close mathematically. As previously stated, a mathematical closure never proves the absence of errors; it provides some assurance against the presence of blunders and excessive errors. Systematic errors, such as the incorrect length of tape, cannot be determined by computations.

Even though area is the least-important consideration in weighing evidence, it and frontage are of prime importance to the landowner. In the matter of liability, surveyors have probably paid damages more frequently for errors in area computations than from other sources. Blunders in computing areas must be avoided!

Computation sheets and computer tapes are preserved and filed with other materials of the survey. They are evidence to properly express the field measurements and should be free of blunders.

8-29 Platting

PRINCIPLE. *Every property survey should result in the delivery of a plat, whether recorded or not.*

A plat is the result and summation of the survey; it is documentary evidence that the client is entitled to.

Every land survey requires a map properly drawn, to a convenient scale, showing all the information developed by the survey. A given land survey also requires a proper caption, proper dimension and bearings or angles, and references to all deeds, and other matters of record pertinent to such survey, including monuments found and set.[1]

Customarily, the original drawing belongs to the surveyor; the client receives copies. Blueprints or blue-line copies of the results of surveys cannot be easily altered; copies are often more certain evidence.

A plat is a special form of a map and must be subject to cartographic

---

[1] *Technical Standards for Property Surveys*, American Congress on Surveying and Mapping, 1946.

240 / PROCEDURES FOR LOCATING WRITTEN TITLE BOUNDARIES

discipline. The plat should show land boundaries with all data for description and identification. Such a plat functions as a clearer and more comprehensive description of the title than does a field inspection (see Chapter 14).

8-30 Conclusions

After a surveyor has conducted all the research, has completed the field search, is satisfied that no other evidence can be found, has made his computations, and has platted the positions of known monuments and facts, he is ready to come to a conclusion about the validity of found monuments, where new monuments should be set, and where property lines belong. This requires skill and judgment. The property surveyor's faculty for concluding and expressing an opinion on the location of property lines places him above the technician.

To arrive at a proper conclusion it is essential to completely understand the problem at hand and the object of the survey.

PRESENTATION OF RESULTS

8-31 Final Conference with the Client

Although this step is not always necessary, it is advisable to explain to the client what has been done, why it was done, and to take him around and show him each corner located or established. In the written report given to the client, it should be noted that the client was shown the corners.

In the event that possession disagrees with title lines, it may be necessary to explain the situation to the client and find out if he wants his record corners set. If possession has been long continued, it may be desirable to maintain the status quo and not start a neighborhood squabble by setting monuments.

Title discrepancies, brought to light by a title search, may require early conferences before proceeding further. It might be advisable for the client to seek the counsel of an attorney to correct serious title defects.

During the final conference, time should be taken to explain the value of monuments and why they should be preserved.

8-32 Completing the Survey

The final step in a property survey is the presentation of a plat, certificate of survey (Chapter 14), a surveyor's report, and a statement of cost. This is a formal step. Maps and plats presented should be properly titled, signed, sealed, certified, in accordance with statutes or regulations, and delivered. If the surveyor wants to limit his liability, he has the opportunity to do so at this time. Warnings can be given that certain title defects may give future troubles, that certain encroachments may have ripened into a fee title, or that other matters may have adversely affected the land. If a letter is submitted to the client and a copy filed with the job, future disputes can be averted. As a protection for the surveyor it is suggested that only delivered plats be sealed and numbered.

8-33 Forms of Surveying Results

Several of the states require, by statute, a record of survey that must be filed and recorded. In most cases any survey that leads up to or involves a land division requires a plat that must be recorded. Even in the absence of statutory requirement, the surveyor should present the client with results in a form that will not only serve the immediate need but also be beneficial to all interested parties in the future.

A written statement accompanying the plat will tell of (1) the monuments searched for, discovered, and used for basis of measurements; (2) records and documents examined for information; and (3) conclusions, results, and recommendations pertaining to possible unwritten title rights.

Other items such as photographs of the monuments, measurement data, and monument ties may be presented to the client.

8-34 The Surveyor's Report

When detailed explanations are to accompany the plat, a report is sometimes in order. The report allows the surveyor to record, at length, differences and discrepancies found, resolutions of problems (with reasons for making specific decisions), recommendations for unresolved problems, and recommendations for further investigations or actions needed by the client, surveyor, or attorney. The report is an ideal device for limiting liability to specific responsibilities, thus reducing the scope of implied warranties.

CHAPTER 9

Apportionment Procedures for Land and Water Boundaries

9-1 Introduction

Situations arise where several parties have clear rights to a portion of a parcel of land, yet the division line between adjoiners is not clearly spelled out. In numerous court cases involving such situations, the court decisions handed down help the surveyor determine where boundaries belong. The discussion on how to divide land areas between adjoiners who have rights to some of the area will follow this sequence: (1) apportionment of the beds of vacated streets, (2) apportionment of accretions or relictions adjoining waters, (3) apportionment of discovered excess or deficiency, (4) monumenting the dividing line between fractional conveyances, and (5) wills and proportional rights. As with so many legal points, court decisions may vary from state to state.

VACATED STREETS

PRINCIPLE. *In private conveyance of land abutting a street or highway where the fee to the street or highway belongs to the abutting owner at the time of the conveyance, it is presumed that the conveyance goes to the centerline of the street or highway.*

This rule is of universal recognition, but it must be remembered that the fee to the street has to be in the party making the conveyance. At the time the Dutch controlled New York, dedicated streets gave fee to the Crown; now when such streets are vacated, title rests in the state. The conveyance of part of the street cannot extend beyond what the grantor owned or had interest in.

The reason for the foregoing rule is stated in the following court cases:

> It is favorable to the general public interest that the fee in all roads should be vested either exclusively in the owner of the adjacent land on one side of the road, or in him as to one half of the road, and as to the other half, in the proprietor of the land on the opposite side of the road. This is much better than that the fee in long and narrow strips or gores of land scattered all over the country, and occupied or intended to be occupied with roads, should belong to persons other than the adjacent owners. In the main, the fee in such property under such detached ownership would be and forever continue unproductive and valueless. True it is that the fee in a road

or in one half the breadth of land occupied by a road is generally not of much value to an adjacent proprietor, but it goes to enlarge his holding and probably enhances somewhat the value of his estate, when a detached ownership would usually leave it of no value whatever. At all events, this much may be asserted confidently, that as the fee in roads has to reside somewhere, it is more desirable that it should be in the owners of the adjacent lands than elsewhere. And detached ownership being less desirable, or not desirable at all, any actual intention to establish it in a particular instance, or in the great mass of instances, is less likely to exist than is an opposite intention.

—Johnson v. *Arnold*, 91 Ga. 659, 18 SE 370.

It seems to be the universally recognized rule that the conveyance of land bordering upon a public highway conveys title to the center of the highway, subject to its use by the public, whether it is so expressed in the deed or not; and where a conveyance or bond to convey designates the public highway, (or street) as one of the boundaries of the tract, it will, in the absence of languge showing a contrary intention, be construed as including the highway itself to the center or middle thereof.

—Williams v. *Johnson*, 149 Ky. 409, 149 SW 821.

The grantor, by apt and fitting words, may exclude this presumption of fee passing to the center of the road and reserve the entire road to himself, subject to the easement of the public (*Cottle* v. *Young*, 59 Me. 105).

The question, then, whether, in a conveyance of land abutting upon a highway, the highway is included and passes to the grantee, or whether it is excluded and does not pass, becomes in all cases a matter of construction and intention merely, from the language used by the parties, and such surrounding circumstances as are proper to be taken into the account in ascertaining the intentions of the parties,—keeping always in view the legal presumption that the parties intended to include the highway, and that the burden is upon the party who assumes to show that the party intended the contrary.

—Buck v. *Squiers*, 22 Vt. 484.

If a statute exists wherein the dedicated street passes fee title to the public agency, the adjoiner cannot claim fee title to the street.

Where a statutory dedication of a street is made, the fee passes to the municipality, and a subsequent purchaser of abutting property "is estopped by the solemn act of his grantor, from claiming title to the center of the street."

—Helm v. *Webster*, 85 Ill. 116.

When there is ambiguity or doubt about the intent of the grantor, the presumption will prevail (*Van Winkle* v. *Van Winkle*, 184 N.Y. 193), and in the absence of fraud or ambiguity, parol evidence is not admissible to rebut the presumption (41 Ind. App. 647 and 90 Ky. 426).

A conveyance of land by lot, block, and map wherein the map shows that the property abuts the highway gives to the lot owner the fee to the centerline of the highway, although the map by its lines shows the lot as extending only to the sideline of the highway. However, in two jurisdictions the courts have held the opposite opinion (Maine and Maryland). Thus the actual application is dependent on case holding in a particular jurisdiction.

> But the lots are conveyed by a plan, and by the copy of the plan, furnished to us, it does not appear that the lot last mentioned embraces any part of the way. The southern line of the lot separates it from the way, whether that is regarded as the line of the lot or the line of the way, or both; it clearly delineates the limits of each, and a conveyance of the lot by the plan does not carry the fee to the center of the way, for in order to have that effect, the grant must extend beyond the southern line of the lot as laid down on the plan. The boundary of a lot by a wall or fence would limit the grantee to it, although it might also be the boundary of a road. The same rule of construction must apply to a line on a plan.
> —*Southerland v. Jackson*, 32 Me. 80.

In general a call for a bearing and distance to a highway does not exclude the presumption of going to the centerline, but a call to the sideline and along the sideline probably limits the conveyance to the sideline.

> Where a survey gives the dimensions and quantity of the land conveyed exclusive of the public way, it does not operate to destroy the presumption that the fee to the roadbed was conveyed; for the reason that such dimensions and quantity of the usable land is ordinarily deemed by the purchaser of paramount importance in determining its availability for the uses designed by him.
> —*Van Winkle v. Van Winkle*, 95 N.Y. App. 605.

> The fact that the description only brings the lot to the edge of the street can make no difference, for the description which thus brings the lot to the edge of the street must be merely understood as specifying the land that the purchaser may hold and use as exclusively his own, and as defining the line at which the public easement begins; the purchaser owning, subject to that easement, to the center of the street.
> —*Schneider v. Jacob*, 86 Ky. 101.

> The three lots were described as a whole, and not separately, and as described formed a parallelogram, the corresponding sides of which were, respectively, 100 feet and 140 feet. The northern boundary of the lots was described as the southern boundary of Avenue A, the eastern boundary as the western boundary line of Seventeenth street of such city, the southern boundary as the northern margin line of a "20-foot alley," and the western boundary described as a straight line 140 feet long, and parallel with the eastern boundary and Seventeenth street. It will be observed that this property abuts Avenue A on the north, Seventeenth street on the east, and an alley on the south, but extends no farther than the southern margin of Avenue A, the western margin of Seventeenth street, nor the northern margin of the alley; thus rebutting what might otherwise be a presumption that the deed passed the fee, subject to the public easement of right of way to the center of the avenue, street, and alley, which were named as three of the boundaries of the property. While the avenue, street, and alley constitute three of the four boundaries, there is no presumption in this case, because of these recitals, that it was contemplated or intended to pass title to any part of the soil of such thoroughfares.
> —*Tuskegee Land & Secur. Co. v. Birmingham Realty Co.*, 161 Ala. 542.

An exception to mentioning the sideline of a street to prevent conveyance of the fee to the street occurs in two jurisdictions. In *Woodman* v. *Spencer* (54 N. H. 507) the description "on the easterly side of the road" conveyed to the centerline. *Cox* v. *Freedley* (33 Pa. 124) is the other jurisdiction.

Bounded by, on, along, or over a street does not exclude an interpretation to the centerline of the street. A description of "easterly on highway" gives title to the centerline (*Chatham* v. *Brainerd*, 11 Conn. 60). "Thence with said road to starting point" conveys fee to the centerline (*Johnson* v. *Arnold*, 91 Ga. 659). "In the line of a street" goes to the centerline (*Healey* v. *Kelly*, 24 R.I. 581). On the other hand a deed bounded by "the east line of Clyde street" excluded Clyde street (*Wallace* v. *Free*, 50 N.Y. 644).

In general, failure to mention a street does not exclude the intent to convey to the centerline of the street.

For further details on this subject, the reader is referred to *American Law Reports* (annotated, 2ALR 6, 1919).

PRINCIPLE. *When a street is dedicated for one particular purpose or use, the fee reverts back to the original owner or his legal successors or heirs once that purpose is terminated.*

At times streets or alleyways are dedicated for unique special purposes such as trash collection, law enforcement, or fire protection. If the specific purpose is identified in the dedication, the full fee and usage of the land reverts to the original owner or his successors once the roadway is no longer used for that specific purpose.

9-3 Apportionment of Vacated Streets

From the previous discussion it is apparent that in most states whenever a subdivision is made with dedication of streets, the adjoiner has a fee right to a portion of the street and the fee rights usually extend to the centerline of the street. If there is a statute requiring the dedicator to give a fee right to a public entity, then the adjoiner to the street acquires no fee rights in the street.

Uniformity of opinion on how to extend the sidelines of a parcel into a vacated street does not exist. This discussion is to point out methods that have been used.

In a few states, the fee rights in the bed of streets must be shown on the map at the time the subdivider is preparing a subdivision map. In such cases there is no cause to invoke a common-law rule for distribution of the bed; such states are excluded from this discussion.

"Reversion rights" is used to describe the portion of the street that a lot has coming to it in the event the street is vacated. Since the adjoining lot has always had a fee right in the street, the term "reversion" is rather misleading; how can you get back what you already have? Probably the real meaning is that "the fee owner has his land back unburdened or unencumbered by a public easement." Because "reversion rights" has been widely used, the term's use is continued herein.

246 / APPORTIONMENT PROCEDURES FOR BOUNDARIES

PRINCIPLE. *Within the interior of a subdivision, reversion rights extend to the centerline of the street; uniformity of opinion does not exist about the direction of the sidelines of each parcel. Along the boundary of a subdivision, reversion rights extend to the limits of the subdivider's fee rights in the street that may be beyond or short of the centerline.*

In determining the direction of a parcel's sidelines in streets, three methods have been used.

METHOD 1. *The sidelines of lots are extended.*

Most states have rejected this method; it was adopted in the state of Washington as per the court case quoted next. Since it causes many inequities and indeterminate situations as shown in Fig. 9-1, it is not recommended.

> We are of the opinion, however, that we must take the statute providing for the reversion of property in vacated streets as solely controlling in the matter. It provides that the vacated portion shall belong to the abutting property owners, one-half to each. Relators obtained their tide land lots with the streets dedicated between them and subject to the vacation statute. The question to be determined then is, how is the vacated portion to be apportioned, one-half to each? We think these lands must be alloted the same portion of the vacated streets that they would be in other platted lands; the lateral lot lines to be extended straight from their former property line to the center line of the vacated street.
>
> —*State* v. *Patterson*, 173 P 186, Wash.

METHOD 2. *The new frontage on the centerline of the street is to be in the same proportion as the old frontage on the sideline of the street.*

This method is commonly used in the apportionment of accretions, and in two states it has been applied at the centerline of streets (*Showalter* v. *Southern Kansas Ry. Co.* 49 Kan. 421; and *Blackwell & SW Ry. Co.* v. *Gist*, 18 Okla. 516). Fig. 9-2 shows a hypothetical solution to a problem. Unfortunately, there is no starting and ending point that fits every situation; in Fig. 9-2 it was assumed that street intersections marked the beginning and end. Suppose there were a long curved road with intersections 10 miles apart; where would you

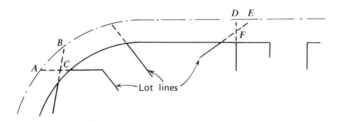

Figure 9–1 The sidelines of lots are extended to the centerline of the street to apportion the street. This method is not recommended, except for the state of Washington, where the court has approved it. Who owns areas ABC and DEF?

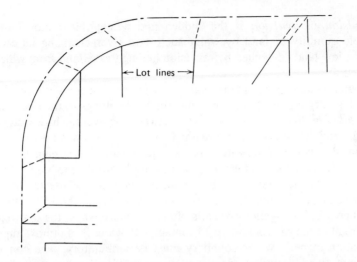

Figure 9-2 The sidelines of lots are extended in a direction so that each lot has the same proportionate length on the centerline as it had on the sideline. This method has been approved in Kansas and Oklahoma. The problem with the method is that the beginning and end of apportionment are difficult to determine in some circumstances.

commence apportionment? In those states that have approved this method, it is probably best applied to subdivisions wherein the apportionment begins and ends at the subdivision boundaries.

METHOD 3. *The direction of the sidelines extend at 90° to the centerline of the street or on a radial line on curves.*

This method is logical and easy to apply; it is recommended, except in those states where some other method has been approved.

APPORTIONMENT OF WATER BOUNDARIES

9-4 Introduction and Definitions

By definition, *riparian* "pertains to living on the bank of a river, of a lake, or of a tidewater." *Riparian rights* pertains to the abutter's rights to the soil under the water, the water or use of water, and accretions or erosions. What the surveyor does to apportion soil rights is the subject of this chapter.

Accretion or *alluvion* is the act of growing to a thing, and along waters it is the gradual and imperceptible growth of land by natural causes. In common law the riparian owner has a right to all accretions that attach to his land.

Erosion is the removal of lands by the action of waters. By common law lands eroded away are permanently lost to the riparian owner. If land reappears by accretions, the land reappearing is governed by the rules of accretions.

Revulsion or *avulsion* is the sudden and perceptible removal of a considerable quantity of land by water such as a river changing its course. By common law lands detached by revulsion belong to the land from which it was detached.

Reliction is an increase in land caused by the permanent withdrawal of water as in a lake drying up. By common law the land belongs to the riparian owner.

Thread of a stream is a line halfway between the average high water mark of the right and left banks of a stream; however, in a few states it is halfway between the average low water marks. The thread of a stream is sometimes called the *median line* and is not to be confused with the *thalweg*, which is the deepest part of the stream, or, better stated, the place where the last drop of water will flow. The division line between states along the Mississippi is the main channel of navigation which is often, but not always, the thalweg.

The right of the riparian owner to contact with water is of utmost importance; the riparian owner's water boundary must be ambulatory. The concept of a permanent fixed boundary does not apply to riparian property lines. The boundary is where it is at a given moment of time. Sometimes the determination of the location of a water boundary, as along a sandy ocean front, is meaningless; with a change in seasons from summer to winter the boundary may vary a small amount to several hundred feet. Regardless of the disadvantage of an ambulatory boundary, the necessity of contact with water is more important.

Navigable Waters. The U.S. Government, at the time of formation of the Constitution, was granted lands of the public domain, and as the public domain was disposed of, the beds of all navigable waters were reserved. As new states were formed as a condition of statehood, such beds passed to the states. In the 13 original states, Texas, and Hawaii the beds of navigable rivers were never vested in the Federal Government. In the public domain states, in theory, the beds of non-navigable waterways passed to the upland owner.

Under federal rules the test for navigability is defined in court cases as follows:

> Rivers must be regarded as public navigable rivers in law which are navigable in fact. They are navigable in fact when they are used, or are susceptible of being used, in their ordinary conditions, as highways for commerce, over which trade and travel are or may be conducted in the customary modes of trade and travel on water.
> —Daniel Ball, 77 U.S. 557, 1870.

> A legal inference of navigability does not arise from the actions of surveyors in running meander lines along the banks of the river. Those officers are not clothed with the power to settle the questions of navigability.
> —*Oklahoma v. Texas*, 258 U.S. 574, 1922.

Most unfortunately, some states do not recognize the federal rule, as, for example, Minnesota. In Minnesota the court proclaims that any stream is navigable when it will float a canoe of the lightest draft down a waterway at flood time! To date this concept has not been tested in federal court. When the state has final jurisdiction (as in rights to fish), the state rule will probably be final.

Some states, such as Georgia, adopted the "bosum rule." Section 85-1303 of the code of Georgia reads:

> A navigable stream is one capable of bearing upon its bosum, either for the whole or part of the year, boats loaded with freight in regular course of trade. The mere rafting of timber or transporting of wood in small boats shall not make the stream navigable.

In states such as Georgia the question of navigability thus becomes a question of fact to be supported by the presentation of sufficient evidence, either pro or con.

Once a state has acquired the bed of navigable waters, it can dispose of the bed or let others use it in accordance with law. In some states the beds have been sold, given away, or even leased. For example, by colonial Massachusetts' law the upland owner was given the bed to the low water mark of the sea but not more than 100 rods (applies to New Hampshire and Maine). In California, by constitutional provision, the bed of navigable, nontidal rivers down to the low water mark belongs to the riparian owner; however, the state now contends that the beds of all navigable waters are held in trust for all the people, and the beds cannot be alienated under any conditions.

The riparian owner has a right of access to navigable waters. Since the state owns the bed, it has control over pier rights. Where the state has given the right to construct piers to the bulkhead line and has failed to specify the direction of the pier, several litigations have resulted.

9-5 Navigable Waters

The general rule is that the riparian owner has title to the bed of non-navigable waters (rivers and lakes) adjoining his land. The direction (described later) of the dividing line between adjoining owners has been the subject of many litigations.

When a person deeds land, he can retain the bed of non-navigable waters, but unless he specifically excludes the bed in a conveyance, the bed passes whether mentioned or not. In general, it is presumed that the grantee acquires the bed of a watercourse owned by the grantor. When the reference is to the watercourse such as "along Bear Creek," "along the river," "to the creek," or "up the river" without other words of exclusion and where the grantor owns the bed, the boundary line is placed in the thread of the river. In cases it has been held that when the deed says "to the south bank of the creek and then along the south bank" or "to the low water mark and then along the low water mark" the boundary is to the line mentioned. In Kentucky (*Hough* v. *Ohio River Sand Co.*, 288 SW 2d 655, 1956) a landowner's title was limited to the low water mark, but the general description said the "northwest part" of the entire tract; the court held the conveyance went to the thread because of the contrary statement. In the event of ambiguous language the general rule is that the conveyance goes to the centerline of the creek (see 78 ALR 3d 604).

Section 85-1305 of the Georgia code attempted to clarify this problem by the following:

> The owner of a stream not navigable is entitled to the same exclusive possession thereof as he has of any other part of his land, and the legislature has no power to compel or interfere with him in its lawful use, for the benefit of those above or below him on the stream, except to restrain nuisances.

The legislature then attempted to identify the difference between accretion and avulsion in Georgia code 85-1302:

> The beds of streams not navigable belong to the owner of the adjacent land; if the stream of water is the dividing line, each owner is entitled to the thread or center of the main current; if the current changes gradually, the line follows the current; if from any cause it takes a new channel, the original line, if capable of identification, remains the boundary. Gradual accretions of land on either side accrue to the owner.

Here one can see the possibility of ambiguity by the courts in the application of the statute. Yet the surveyor's specific responsibility is one of the collection of evidence relative to the "how" of the stream change and the subsequent interpretation of the evidence. The surveyor may be required to rely on the collection and interpretation of old and new aerial photographs, core borings of trees, and soil samples of soil borings from the area in question.

9-6 Area of Accretions and Relictions

Along rivers or lakes accretions or relictions belong to the riparian upland owner; the problem of the surveyor is to determine the direction of the division line between adjoiners. With very rare exceptions, the direction of the dividing line is not on the prolongation of the property line; it is on some equal basis, whereby each owner is given equality of access to the water.

Before accretions or relictions can be divided between adjoiners, at least two survey lines must be determined. For adjoining navigable waters, the first line to determine is the location of the present boundary between the owner of the bed (usually the state) and the riparian owner. The second line is the boundary between the owner of the bed and the riparian owner at some time in the past. If there has been no inland erosion on the riparian owner's land, the line is usually where it was located as of the date of the deed's formation. If there has been erosion, the line is the boundary existing at the most distant inland erosion. In most states the boundary line between the upland riparian owner and the owner of the bed is the average high water mark, although there are minor exceptions.

By federal rule the limits of the bed of tidewaters is the mean high tide line where it exists at any moment of time, and the *average high tide* is defined as the average of all high tides over a period of 18.6 years, the period of time it takes the sun and moon to complete one cycle. The average high tide varies from place to place, depending on the shape of the shore, the average weather, and the latitude; it is not a uniform elevation. The government maintains tide stations along the U.S. shores, and average readings can be obtained for those points. Since the average high tide varies little within reasonable distances, it is generally assumed that it has a constant elevation. This assumption can be significantly in error in long, narrow bays, especially where a river is pouring in a large volume of water.

Where there is wave action and a sandy beach, the determination of the mean high tide line at any moment of time is not very important; the next storm will probably alter it. Around bays or protected shorelines, the location of the mean high tide is rather static.

In California, by court decree, the boundary line is the neap high tide line, that is, the average of all tides when the moon is halfway between new and full. The difference between this and the federal definition is about a half foot in elevation.

In a few states an attempt was made to establish the line of vegetation as the dividing line. In Washington state Mrs. Hughes, who acquired her land prior to statehood, sued, and the federal court said:

> The question for decision is whether federal or state law controls the ownership of land, called accretion, gradually deposited by the ocean on adjoining upland property conveyed by the United States prior to statehood. We hold that this question is governed by federal, not state, law and that under federal law Hughes, who traces her title to a federal grant prior to statehood, is the owner of these accretions.
>
> —*Hughes* v. *State of Washington*, 389 U.S. 290.

This has not settled the question of the division line for those who acquired their land after statehood.

When surveyors are called on to testify relative to areas of accretions or relictions, especially in tidal areas, courts and clients are prone to expect the surveyor to locate the land in question relative to such features as mean high tide, an so on, yet the courts have not legally determined exactly what benchmark or reference is to be used. A surveyor is capable of measuring from fixed identifiable reference points, but until the courts of the various jurisdictions identify the parameters of measurements, any surveyor who assumes the responsibility of locating these lands is assuming a great responsibility, for the courts have not unanimously agreed about elevation, time sequence, and the like, and in fact some courts and individuals have sought such identifying features as salinity of water, vegetation level, and species of trees and even grasses to determine tide.

Although the location of the waterline between the upland owner and the federal interest is fixed by court decision, this in no way means that the same rule regulates the relationship of the state to its citizens. Whereas many of the states with ocean frontage have adopted the federal rule of mean high tide line as the division line between the upland owner and the state, several have not, and some are contrary (California is neap high tide; Washington is the line of vegetation; Maine, New Hampshire, and Massachusetts are the low water mark but not more than 100 rods). Unless the state supreme court has approved the mean high tide as the dividing line, the surveyor has no authority to assume that it will be.

The boundary of the bed of rivers is not easy to determine. Rivers run downhill, and the division between the upland and the bed is a gradient line, especially so at high water. The best discussion of how to determine the gradient boundary is found in the *Texas Law Review* (volume 30, 1952, p. 305) by Arthur D. Kidder who was the expert in the Red River case between Texas, Oklahoma, and the Indians.

Artificial fills of land by humankind are not accretions; the boundary line between the upland owner and the owner of the bed is where it was before commencing the fill. In the San Francisco Bay area, the mud flats were filled with trash and rubbish. Since the weight of the overburden compressed the mud, excavation to determine the elevation of the line between mud and fill will not give the true elevation of the mud prior to the fill.

In California the courts have ruled that the state owns to the average neap high tide line at the last natural conditions. In the Bay of San Diego in 1916 a flood took out the Sweetwater dam and caused large depositions of mud in the bay; thus the last natural conditions date way back. The federal rule is that the average high water is as it exists at any moment of time regardless of the natural condition of accretions. Thus if a jetty is constructed and it causes accretions, under California rules the accretions belong to the state; under federal rules they belong to the riparian owner.

In Massachusetts the state owns the bed of great ponds, that is, fresh water lakes of more than 10 acres. The 10 acres are determined at the average high water mark. Lakes nearing the 10-acre mark may be difficult to determine, since weather cycles do occur. The first 10 years may be below the average rainfall and the next 10 may be above average. In some cases it takes a court to determine the final status of who owns the bed.

9-7 Apportionment of Accretions

This portion of the discussion assumes that the beginning and end of accretions has been determined and the problem is to apportion the area between adjoiners. Almost never has the prolongation of property lines been approved. The most commonly approved methods are as follows.

Proportional Shore Line Method.

PRINCIPLE. *Riparian owners are to have the same proportion of frontage on the new shoreline as they had on the original shoreline.*

Figure 9-3 shows application to a specific case. The old shoreline is divided into ownerships A, B, C, D, E, and F. The new shoreline is divided into parts A',

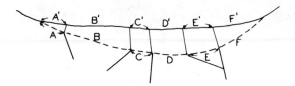

Figure 9–3 Proportional shoreline method. The new frontage has the same proportional linear distances as the old shoreline. $A/A' = B/B'$, and so on.

B', C', D', E', and F' in such a way that $A/A' = B/B' = C/C' = D/D' = E/E' = F/F'$. Except in states where there has been a ruling to the contrary, this method is recommended.

Colonial Method. In Maine subdivision owners of small lots had their riparian rights to the low water mark of the ocean divided by a principle outlined in the case of *Emerson* v. *Taylor* (9 Me. 35). In Fig. 9-4 straight lines are drawn from one lot corner to another as they exist on the shoreline (at mean high tide). Perpendiculars are erected at the end of each line. The angle between the perpendiculars (angle *ABC*, angle *BCD*, etc.) is measured and the property line is set on the bisection of the angles.

Round Lake Pie Method. An approximate round lake has its bed divided much as you would cut up a pie. The center point of a lake is selected (selecting it is not easy), and lines are radiated out from that point to each original shoreline point as shown in Fig. 9-5.

Long Lake Method. The long lake method of dividing the bed of a lake is illustrated in Fig. 9-6. The problem of dividing the bed of a long lake usually arises in disputes over ice-cutting privileges (see *Romney* v. *County of*

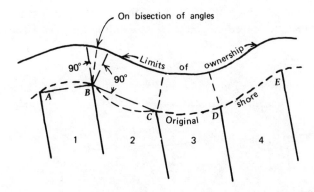

Figure 9–4 Colonial method. At the original shoreline angle *ABC* is measured; the property line between Lots 1 and 2 is on the bisection of angle *ABC* as shown.

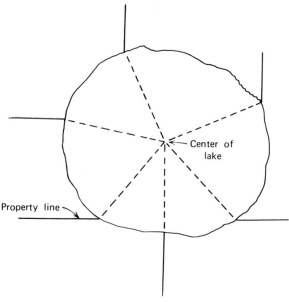

Figure 9–5 Round lake pie method. A point is selected in the center of a round lake, and lines are run from that point to each property corner at the shoreline. This establishes ownership rights in the bed.

Stearns, 130 Minn. 176). Each end is treated as a pie and the in-between lines are determined as being at 90° to the thread of the lake.

Area Apportionment Method. Judges are likely to apportion land in accordance with its best use. Sometimes use for crops is more important than frontage. In one such case in Louisiana area was used.

9-8 Seaward Limits of States

Each state generally assumed that its seaward limit was the limit of the extent of the United States and leased oil rights to various parties. The U.S. Supreme

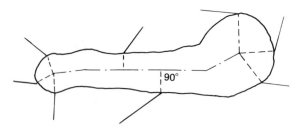

Figure 9–6 Long lake method. Each end of the lake is treated as a round lake, and the centerline of the lake is constructed in between. Property lines are at 90° to the lake's centerline.

Court decided by court opinion that the limit was the mean low water mark. Following this, by an act of Congress, the rights to the beds of lakes (great lakes in particular) and the ocean were given to the states. Along the oceans the limits of state ownership were set at 3 geographical miles (1 minute of arc at the equator or 6,076.10 feet) from the low water mark used on the navigation charts. On the East Coast it is the average low water mark, and on the West Coast it is the average low-low water mark. An exception was made for the Gulf of Mexico states of Spanish origin (Texas and Florida), wherein the limit was 9 geographical miles. Historic bays (Monterey Bay in California) belong to the states.

At the present time in some of the eastern seaboard states litigation is being tried to determine if the upland owner's rights extend to the line of vegetation or the mean high-tide line.

EXCESS AND DEFICIENCY

9-9 Apportionment of Excess or Deficiency in Subdivisions

Apportionment of discovered discrepancies is always a rule of last resort; if a found surplus or deficiency exists, the rule is to discover where it belongs and give it to that parcel. Only when several parties have equal rights should apportionment be considered.

If parcels of land are granted in sequence, the senior party gets what is described in the conveyance; the junior party receives a remainder. Most frequently when parcels are created by metes and bounds descriptions each parcel is created in sequence; however, this is not always true. In Hawaii the original deeds went out in sequence to different parties. In a court case the court ruled that all the parties acquired their land at the same time; merely because the court was slow in giving some of the parties their deed did not alter the fact that they all acquired a right to the land at the same time. Each party was entitled to a proportion of any excess or deficiency on the basis of simultaneous creation.

Within a subdivision, wherein numerous lots are all created at the moment a map is filed, each lot has proportional rights to excess or deficiency between discovered original monument positions. Since original monument positions (where they existed when set) cannot be disturbed, proportional rights must end at an original monument position.

The thinking of the court is well illustrated in the following case:

> These proceedings are unlike a case where a deed for a certain portion of land is made to one person, designating specific boundaries of the land conveyed, and afterward a deed is made to still another person for another portion of the land, designating other specific boundaries, but the boundaries of the two portions of the land overlapping each other so as to appear to convey a strip of the land to both persons; for the title in such a case for all of the portion of land first conveyed would vest in the first grantee before the second deed was executed, and the grantor would

have no power to again convey any part of the property which he had previously conveyed to the first grantee. In this case all the parties are claiming under the same instrument, the same proceedings, the same decree, and the titles of the allottees under the partition proceedings all vested at one and the same time and by the same instrument. The right of everyone to his full quota of land is therefore equal to the right of every other one. It is evident that the partition proceedings are erroneous; for they apparently show that the land partitioned is about 2 chains longer, north and south, than it actually is; but it must not be supposed that this mistake or error occurred solely in the measurement of any one particular allotment. It should be supposed that the error occurred during the measurement of the entire length of the entire tract of land from north to south; and when the land is ascertained to be about 2 chains shorter than the partition proceedings show it to be, each allotment should bear its proportionate share of the loss. This would certainly be inequitable for one of the allottees to lose all, and the others to lose nothing.

—McAlpin v. Reicheneker, 27 Kan. 257.

Since apportionment is a rule of last resort, the first part of the discussion will be when not to apply apportionment.

PRINCIPLE. *An error or mistake occurring on a map is not apportioned.*

In Fig. 9-7 the number 45 has been obviously transposed to 54; the error is corrected and apportionment does not apply. Usually, a mathematical error can be proved by a closure.

PRINCIPLE. *Along the boundary of a subdivision, the original subdivider most frequently erred in locating the subdivision boundary line; if that can be shown, the lots adjoining the boundary receive the discrepancy.*

The original location of lots cannot be disturbed. If the original owner (or surveyor) located lots on the adjoiner's land, the loss of land, because of the adjoiner's rights, cannot be divided among others (see Fig. 9-8).

				450' measured					
45	45	45	45	54	45	45	45	45	45
1	2	3	4	5	6	7	8	9	10
20	19	18	17	16	15	14	13	12	11
45	45	45	45	45	45	45	45	45	45

Figure 9-7 The 45 in Lot 5 has been obviously transposed to 54. The block measures 450 feet as given on the original map. When such an error occurs and it can be proved where it exists, the usual procedure is to place the error where it occurs.

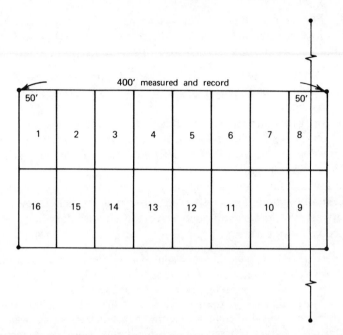

Figure 9-8 Solid circles indicate found original control points. Lots 8 and 9 must suffer the entire loss of land because of a senior right; proration is not applicable.

PRINCIPLE. *In those states where the remnant rule is applied (Minnesota and New Jersey), the end lot with irregular dimensions receives all the excess or deficiency.*

In Minnesota (*Barrett* v. *Perkins*, 113 Minn. 480) and New Jersey (*Baldwin* v. *Shannon*, 43 N.J.L. 597) the courts have specifically ruled that where lots are regular in character (say, even 25-foot width) and there is an end lot of irregular dimension (say, 31.82 feet), all discovered excess or deficiency must be put in the irregular lot as based on the theory that whatever was left over was supposed to go in the irregular lot. In most states the method has been rejected.

PRINCIPLE. *In New York where a lot is described by both a subdivision lot number and by a metes and bounds description, the metes and bounds description is controlling.*

The rule is ill advised, since it can lead to many problems; however, in New York it is the law (*Mechler* v. *Dehn*, 196 N.Y.S. 462).

PRINCIPLE. *Excess or deficiency cannot be apportioned beyond an original monument position.*

In the usual situation where streets are improved with sidewalks and paving, the streets as built are presumed to be correctly located and any discovered excess or deficiency within one block is not distributed into other blocks.

PRINCIPLE. *For excess or deficiency occurring within a subdivision block wherein all lots have equal standing, the discrepancy is normally divided among the lots in proportion to each lot's linear dimensions.*

Figure 2-11 shows how it is normally done. It is presumed that the original measurer erred with a long or short tape, and each lot was originally in error by a proportionate amount.

In the majority of states when a street is dedicated, it is given its exact width irrespective of found shortage or surplus in a subdivision. In Utah in a reported case (*Coop v. Lowe Co.*, 263 P 485) deficiency was divided between the lots and the street, a rare exception.

In the following Ohio case of 1963 the common-law rules for the majority of states are clearly stated. The court recognized that the deficiency had to be divided between the lots existing between fixed streets; the alley and streets were to receive full measure.

> Said lots are part of Davis' Fourth Addition to the Village of Mineral City, Ohio which was received for record in the Recorder's Office of Tuscarawas County, Ohio on January 2, 1890 as shown by Plat Book 1B at page 16 in said office.
>
> Subject to qualifications, the general rule is that where a tract of land is subdivided and is found to contain either more or less than the aggregate amount called for in the survey of tracts within it, the proper course is to apportion the excess or deficiency among the several tracts. See 11 C.J.S. Boundaries § 124, page 737.
>
> In 8 Amer. Jur. Page 796, Section 71, it is stated that "if after a tract of land has been subdivided into parts or lots, and title thereto has become vested in different persons, it is discovered that the original tract contained either more or less than the area assigned to it in a plan or prior deed, the excess should be divided among, or the deficiency borne by all of the smaller tracts or lots in proportion to their areas." See also Marsh v. Stephenson, 7 Ohio St. 264.
>
> In 7 Ohio Jur. (2d) Page 615, Sec. 10, it is stated that "if there is a shortage on the plat, the rule is well settled in this state that it will be divided as nearly as possible prorata between the parties." See also Cincinnati, S. & C. R. Co. v. Tuttle, 18 Ohio Cir. Ct.R. 630, 7 Ohio Cir. Dec. 63.
>
> In Nilson Bros., Inc. v. Kahn et al., 314 Ill. 275, 145 N.E. 340 it is stated in the second proposition of the syllabus that "[i]n the absence of any agreement or question of title by adverse possession, where a block has been platted into lots and lots sold, a shortage in block will be prorated among the several lots."
>
> Although it is the general rule that where land is subdivided by a plat, the shortage shall be apportioned among all of the lots, such apportionment will only be made in the absence of facts showing that equity does not require the application of a different rule. See Sellers v. Reed, 46 Tex. 377.
>
> It would be impractical, if not impossible to take all of the land included in said Davis' Fourth Addition to said Village of Mineral City, Ohio and pro-rate this alleged deficiency among all of the lots contained therein. To do so would affect the location of all existing streets and alleys in the said addition.
>
> In Hillside Cotton Mills v. Bartley et al., 156 Ga. 271, 119 S.E. 404, the court held that "where there is a shortage in the actual land subdivided and platted into

lots and blocks with intervening streets, each block should, if possible, be treated as distinct, and the shortage therein should be distributed among the lot owners, except so far as possession has fixed the limits."

This court is of the opinion that this is a proper rule to apply in the instant case under the evidence adduced.

For the purposes of this case, the court will treat the lots herein before enumerated that are bounded by an un-named alley on the south, by Logan St. on the north, by Fair St. on the west and by Clay St. on the east as a single block.

The recorded plat of said Davis' Fourth Addition to said Village of Mineral City, Ohio shows that an alley 16½ feet wide, between lots 230, 231, 232 and 233, and lots 242, 243, 244 and 245 was donated to the village by the acceptance of said Plat. Lots 230, 231, 232 and 233 were intended to be 99 feet long, and lots 242, 243, 244 and 245 are shown to be 92 feet long with the exception of lot 242 which was intended to be 92 feet long on its north line and 95.75 feet long on its south line. To recognize the length of these lots as shown by the plat as actual would leave a space varying from 3.7 feet on the north to 8.2 feet on the south for the alley.

The alley not having been vacated, and having been donated to the village for public use by the proprietor before any lots were sold, the court must fix its width as 16½ feet. Any shortage must be apportioned as nearly as possible and practical pro rata among the lots in question.

—*Labus* v. *Jones*, 197 NE 2d 244, Ohio, 1963.

DIVISION LINES BETWEEN FRACTIONAL PARTS

9-10 Determining the Division Line between Fractional Parts

At common law, with the possible exception of wills and proceedings in partition, the rule is this: a fractional part of a whole parcel conveys a fractional part of the area, not the value. This rule can, of course, be altered by statute as is true for sectionalized lands.

When a fractional part is conveyed and the common law applies, no ambiguity about the area conveyed exists, but there is ambiguity about the direction and location of the division line between the fractional parts. If the easterly and the westerly halves of a lot are conveyed, an infinite number of division lines can be drawn that will divide the lot into two equal areas. In one California case an owner of the east half erected an irregular fence enclosing the desirable portion of the parcel while the owner of the west half was absent. Since the fence enclosed exactly half the area and since the boundary was indeterminate, the court refused to order the fence to be moved to a straight line or to be parallel with the easterly boundary. Such is an unusual situation; in several states the courts have ruled on a definite procedure for establishing the division line.

In sectionalized lands where federal law is applicable, definite statutes state exactly how each fractional part shall be divided, and it is not based on equal areas. Federal statute always applies when land is owned by the federal government or when land is under the jurisdiction of federal law. The moment a state is created or at the moment land passes from the federal government to a private party within a state, cases are tried in state courts, and state law applies. In most instances, the states adopt federal laws as controlling; however, this is not a requirement, and in some instances, they have not adopted federal laws as controlling. In Fig. 9-9 the southwest quarter originally contained 179 acres and by federal rule the east half would contain 80 acres and the west half 99 acres. One party acquired the entire southwest quarter and while under state jurisdiction sold off the east half. Under these circumstances, does the state common law or the federal law apply? In one California case (also in Michigan and Minnesota) the court adopted the state common law and gave an opinion as follows:

> Words used in a conveyance are to be given their ordinary and popular meaning, unless used in a technical sense, or having a special meaning, or the context shows that they are used in a different sense. The word "half" has a plain, common, and natural meaning, and when used in describing lands is to be understood literally. There is nothing uncertain or equivocal in the term, and if used without qualification in the deed to respondent must be given its literal significance as one of two equal parts into which anything may be divided, and to have conveyed to him the east half of the quarter section in quantity and acreage. (*Jones* v. *Pashby*, 62 Mich. 614, [29 N.W. 374]; *Cogan* v. *Cook*, 22 Minn. 137.)
>
> —*Wood* v. *Mandrilla*, 167 Calif. 608.

This California situation *cannot* apply to those cases where the parcel was divided into the east and west halves by the Federal Government or when the

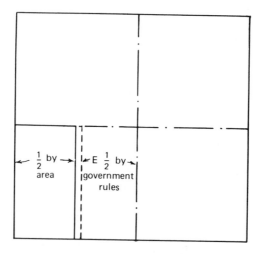

Figure 9-9 This section had excess in the west half, and by the U.S. government of determining the west half of the southwest quarter, it contained 99 acres. The east half contained 80 acres. The owner of the entire southwest quarter sold the east half, and the California court ruled that half of the area was sold.

division was made while under Federal Government jurisdiction. In each of the foregoing California cases, one owner had erected a fence and had possession. In effect, the court did not rule against possession. In California government-checking agencies for subdivision maps and for records of surveys have insisted that federal rules be followed, and practically all surveyors have followed federal rules. Unless there is possession indicating that adjoiners have agreed on some other mode of a division line between them, it is recommended that California surveyors follow federal rules. By far the majority of other states do.

It is conceivable that two separate parcels could be conveyed by patent by the Federal Government—that is, the west half of the southwest quarter and the east half of the southwest quarter as shown in Fig. 9-9. After the patents were issued, one buyer acquires the entire southwest quarter. He then conveys the east and the west halves under state law. The patents from the government were not of equal area, but in some states the next conveyances by the party owning the southwest quarter would be of equal area. Thus the same descriptions could have different meanings, depending on the sequence of jurisdiction. This type of possibility places an added responsibility on the surveyor to examine and explore the complete background of a parcel prior to the preparation of property descriptions by proportions.

Whereas technically the division line between fractional parts of parcels created under state laws is indeterminate, many of the states have handed down decisions fixing the method of determining the division line. The following, as based on such decisions, is recommended.

PRINCIPLE. *Make the dividing line straight.*

It is doubtful that many states would adopt the California notion wherein the purchaser had the right to establish a fence in any crooked direction to suit his whims. The division line was straight in all court cases reviewed, except that one.

PRINCIPLE. *If a lot is divided into fractional parts and there are abutting streets, unless the intent is otherwise, divide the area in fractional parts exclusive of the streets.*

Except where there is a contrary intent, the foregoing is the general rule and should be followed. To indicate one court's opinion on what constitutes a contrary intent, the case of *Ferris* v. *Emmons* is given next. One party acquired the northeast quarter of Block 195 and the other acquired the northwest quarter. Each block contained 40 acres "measured from the centerline of the streets." The streets on the west and east side of the block were not of the same width; half the street on one end was 15 feet greater than at the other end. If the area of the two parcels were computed from the centerline of the streets, the division line between the two parcels would differ by $7\frac{1}{2}$ feet from that when the areas were computed from the sidelines (see Fig. 9-10). On determining the intent of the original conveyance, the court examined the original map and quoted its opinion as follows:

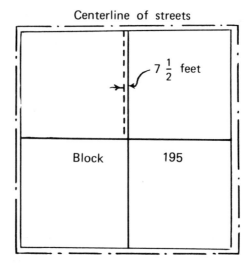

Figure 9-10 The half street on the east side is 15 feet wider than the half street on the west side. If measurements go from the sideline of the streets, the property line shifts 7½ feet for the northeast quarter.

The sole question presented for adjudication is whether the center line, which divides block 195 into an east and a west half, is to be determined by including or excluding the streets on the east and west sides of the block. Appellant's contention that the streets are to be excluded in fixing the division line between parcels and in ascertaining the net acreage covered by each conveyance, finds some support in the cases which hold that, however clearly it may appear that the owner of a parcel of land holds title to the center of an adjoining street, subject to the public easement, and that the boundary of the parcel is technically, therefore, the center of the street, in view of the fact that the owner of such parcel of land has no right to the possession or occupancy of any portion of the public street, it will ordinarily be presumed that the parcel does not include such portion of the street. Earl v. Dutour, 181 Cal. 58, 183 P. 438, 6 A.L.R. 1163; Joens v. Baunbach, 193 Cal. 567, 226 P. 400.

This principle does not apply to the present case, for the record is replete with evidence sustaining the findings of the court below and irrefutably indicating that the common grantor intended, by the respective deeds to the predecessors in interest of the parties hereto, to convey fractional parts of block 195 as measured from the center line of the adjoining streets. That respondent is the owner of the strip of land in dispute is established by the evidence. The deed to respondent's predecessor, delivered in the year 1888, describes the property covered therein as "The North East quarter (N.E. ¼) of Block One Hundred ninety-five (195) of the Pomona Tract, according to map of said Tract duly recorded in Book 3, pages 96 and 97 of Miscellaneous Records of Los Angeles County, California, estimated to contain Ten (10) acres of land." This map is in evidence, and materially assists in deciding the question presented by the appeal. A careful study and understanding of it discloses that the net acreage (viz., that acreage devoted solely to private use and occupancy) of the several fractional parts of the forty-acre blocks of the tract, of which block 195 is one, adjacent to the streets of the tract, is noticeably less than the acreage of corresponding fractional parts of forty-acre blocks not abutting the streets; thus clearly indicating that, for purposes of dividing the several forty-acre blocks of the tract into fractional parts, all measurements are to be taken from the center line, rather than from the side lines, of the streets. This is borne out by the legend appearing on the face of the map, which discloses that the tract is subdivided into blocks of three sizes; the largest blocks containing forty acres each, the second containing ten acres each, and the smallest containing two and one-half acres each, "measuring from the center of the streets," the ten-acre blocks being divided into "8 equal lots Net including streets," and the two and one-half acre blocks being divided into lots "on the same basis, similar lots in the same blocks all being of the same width." Reading the several deeds here involved in conjunction with the map to which they refer in describing the property transferred, it is immediately apparent that fractional parts of block 195, as well as the fractional parts of all other forty-acre blocks in the tract, are to be measured from the center lines of the adjoining streets. In addition, and by way of explanation of what was intended by the parties to the original conveyances when they used the word "block" in the instrument of transfer, the respondent introduced a substantial amount of parol evidence, both as to the common reputation and custom in the community of Pomona, prior to the institution of this action, as to the meaning of the word "block," as used in the map above referred to and the several conveyances made pursuant thereto by the common grantor. The wintnesses testifying in this respect stated that it was commonly understood that "the blocks extend to the center of the streets," and that "the man who owns the property on the side of the street would have less acreage than the man who was on the side where there was no street, where he had the fractional part of the block."

—*Ferris* v. *Emmons*, 6 P 2d 950.

Where owners have not established the division line between adjoining fractional parts conveyances, the following procedure is recommended. In the minds of most people the easterly half indicates parallelism and so does the westerly half.

PRINCIPLES. *If the original parcel was in the shape of a rectangle or parallelogram, make the division line on the average bearing of the two sides.*

If one deed reads "the east half of a lot" and the other reads "the lot except the east half," make the division line parallel with the east line.

If a nonsectionalized land deed reads "the southwest quarter or the southwest 10 acres," make the parcel a parallelogram in the southwest corner of the whole parcel.

If the descriptions call for the east and the west halves and if there is no logical reason for making the division line parallel with either the east or west boundaries, make the division line north and south.

Usually north is relative to the map referred to (magnetic or relative to found monuments, with quoted bearings between them). In the absence of inferences, make north astronomic north.

As was pointed out, the division line between two halves under state common law is not definitely described. The courts are reluctant to disturb an existing possession that encloses the correct area, and in California the courts have refused to hear such a case. The foregoing principles are recommended procedure for an equitable solution when no posession exists. When one party has correctly established his correct area by an enclosure, the survey will find that in most states it is not wise for him to disturb this possession, and if both adjoiners assist in the erection of a fence or barrier, this arrangement certainly should not be disturbed by a contrary opinion.

In determining the true interpretation of "half," a confusion has existed for decades. Does half mean half of the area, or does it mean half as determined by dividing specific lines in half as is done when dividing a quarter of a section in half? Different applications of methods apply at the federal and state levels.

WILLS AND PROPORTIONAL RIGHTS

9-11 Wills and Proceedings in Partition

In a will where heirs are given fractional parts of land, the area is divided by acreage. In some instances each party is given an area such as the east 10 acres and the west 10 acres. In the event of discovered excess or deficiency, the discrepancy is divided equally between the parties. In court cases wills are usually divided by value, not necessarily by acreage. By the time a surveyor receives a request for a survey of land deeded in a will, the direction of the dividing line is usually defined; the problem is to adjust the line to give equal areas. In court cases where the surveyor is part of the proceedings in partition, he may be called on to divide an area by value.

Of course, in a will, where one party is designated to receive a remainder, division of excess or deficiency does not exist.

REFERENCES

Brown, Curtis M., *Boundary Control and Legal Principles*, 2d Ed., John Wiley & Sons, New York, 1969.

Manual of Surveying Instructions, 1973, U.S. Department of the Interior, U.S. Government Printing Office, Washington, D.C., 1973.

Shalowitz, Aaron L., *Shore and Sea Boundaries*, U.S. Government Printing Office, Washington, D.C., 1964.

CHAPTER 10

Measurements, Errors, and Computations

10-1 Introduction

Of the important forms of evidence pertinent to boundary location, measurements are the specialty of the property surveyors and are necessary for determining quantity and location of title ownership. The expert property surveyor is skilled in making such observations and competent in their evaluation.

In court cases seldom does the opposing attorney ask questions on measurement accuracy or procedures, probably because he has insufficient background to know how to form questions. This does not mean that the surveyor need not be prepared to answer extensive questions in this area; sometime, someplace, there is the exceptional attorney who is extra diligent or who has sought adequate survey knowledge and can quickly bring out shortcomings of the less-qualified surveyor. Several states have adopted by statute the fact that the surveyor is to follow the procedure of *The Manual of Instructions for the Survey of the Public Lands of the United States* in resurveys. Asked by an attorney in court, the question "Did you use a 2 pole chain in your survey?" causes problems, since the statute of the United States which requires the use of a 2 pole chain has never been repealed. The answer to the question is no, based on that fact that the manual authorizes the use of other measuring devices. A background of knowledge of past and present measurement methods is needed.

The aim of this chapter is (1) to discuss the dependability and accuracy of existing measuring devices, including the most modern and old, (2) to analyze errors and uncertainties inherent in measuring devices and procedures, and (3) to define what accuracy is expected of a professional surveyor when locating property lines—that is, to make a statement of allowable uncertainty of measurements.

Techniques of measurements, how to operate a transit, how to hold a plumb bob, and how to drive a pipe with a sledge hammer are not treated in this book. It is assumed that the professional surveyor has developed sufficient skills in such matters. The treatment here is more from the point of view of how measurements are used to prove property locations rather than how to make measurements.

Many excellent books and articles have been produced in the geometrical theory and practice of surveying, and some of these references are cited at the close of this chapter. This chapter does not duplicate the material that is so well expressed elsewhere, and the reader who desires additional information on these subjects may consult some of the references available to him.

Units of length and direction are means of expression. The surveyor should be aware of the uncertainties that exist in a given measurement, how these may be reduced, and by what means the most probable values can be obtained.

The problem of measurements is more serious with the property surveyor than it is with other surveyors. Those working on topographic maps, construction layout, and engineering quantities must fill a present need; the property surveyor must produce measurements not only for an immediate purpose but also for measurement calls that will be on record for many years to come.

10-2 Types of Measurements

The principal measurement problem in property surveying is one of determining position. For the most part, this position is projected on a horizontal plane, although there are occasions when altitudes or surface distance must also be considered.

To obtain a horizontal position, distance and direction are required. From these distances and directions all area quantities, shape, location, and other geometric data are derived.

10-3 Distance

When referring to distance, it is now a horizontal distance relative to some actually defined unit. In early U. S. history in some states surface measurement was standard practice. Property surveying in the United States has been subject to many different units of length, usually of historical significance or application. The most common unit in use today is the *foot*, which is defined as one-third of a yard, the standard unit for property surveys in the United States. The yard is standardized by the U.S. Bureau of Standards, and the official comparison of this length is kept in Washington, D.C. The *yard* is defined as being equal to 0.9144 meter, or 1 foot=1200/3937 meter. In 1959 a new definition of comparison of the foot–meter relationship was agreed on: 1 inch=2.54 centimeters exactly. At the time it was agreed that the U.S. survey foot would remain the same. The difference between the two definitions only amounts to two parts in a million, which can hardly be of significance to the land surveyors, although it may be of significance in extended geodetic control.

The surveyor must not assume his own value for the foot, nor must he take the manufacturer's word that the length of the tape is consistent with the standard value.

The official unit of measurement of BLM is the chain, consisting of 66 feet as established by statute.

10-4 Standardization of Tapes and Distance-Measuring Devices

PRINCIPLE. *It is the responsibility of each surveyor to know the length of his tape or distance-measuring device under specified field conditions.*

The surveyor may determine the length of his tape by sending it to the U.S. Bureau of Standards, Washington, D.C. The Bureau, in the laboratory, compares the length with a bench measure and issues a certificate or a report stating the values of the length for certain conditions of temperature, support, and pull. Tapes conforming to the specifications for standard steel tapes will be certified by the Bureau, and a precision seal showing the year of standardization will be placed on the tape. A tape not conforming to the specifications will be tested by the Bureau and a report will be issued. Several firms manufacture calibrated tapes and provide similar certificates.

A steel tape is considered as standard when it has been calibrated by the Bureau and found to conform to the following specifications:

1. It shall be made of a single piece of metal ribbon.
2. None of the graduations shall be on pieces of solder or on sleeves attached to the tape or wire loops.
3. The graduations must not be dependent on spring balances, tension handles, or other attachments liable to be detached or changed in shape.
4. The error in the total length of the tape, when supported horizontally throughout its length at the standard temperature of 68°F. and of standard tension, shall not be more than 0.1 inch per 100 feet.

The standard tension is 10 pounds for tapes 25 to 100 feet or from 10 to 30 meters in length and 20 pounds for tapes longer than 100 feet or 30 meters.[1]

Several universities have means of determining the standard length of tapes and provide this service in the public interest. The comparison may be made on two monuments set with a specific distance inscribed on the tablets. It is advisable for the surveyor not to conduct such a comparison himself but to have some official of the university or the one in charge of such comparison monuments to supervise the calibration and attest to such or witness same in writing.

The surveyor who has an extensive practice and employs several tapes would benefit by having one tape reserved solely for comparison. This tape may be certified by the Bureau and can be kept on a reel and used only from time to time to check the length of the tapes employed on the job. The standard tape should be recertified periodically. During the manufacturing process internal strain develops, and while setting on the shelf, the relaxation of the strain will cause

---

[1] Lewis V. Judson, *Calibration of Line Standards of Length and Measuring Tapes at the National Bureau of Standards*, NBS Monograph 15, Washington, D.C., 1960.

the tape to change in length. In one case a lovar tape expanded almost ¼ inch in a 4-year period.

If a surveyor is going to court and is to testify based on a distance measured with an electronic measuring device, he should be prepared to prove that he has frequently standardized his equipment. Most important is to prove that the instrument is consistent within given limits under variable conditions of temperature and atmospheric humidity.

To check consistency it is necessary to have a standardized base line. Several state associations and some universities in cooperation with NGS have established base lines for the standardization and testing of electronic distance measurement (EDM) equipment.

10-5 Units of Length

Although the official standard of length in the United States is the meter, it is probably the least used in property surveys. The units used to convey property are those that belong historically with the origin of the title. The English mile formed the basis for property granted by the British Crown. In the areas that were once under Spanish or Mexican sovereignty, the vara was the prevalent unit, and in Texas it still is the unit used by the state. Customary units of length, according to the Mendenhall Order of April 5, 1893, are to be derived from the United States Prototype Meter. Table 10-1 lists some of these units and their equivalents.

How long is a unit of length? Throughout history, humankind has used units of measurement to record the intended length of a line, but unfortunately the definition of the length of the unit used was not always preserved in history. The English *rood* (rod or perch as now known) was once defined as the total length of the left feet of 16 men, tall and short, as they came from church on Sunday.

PRINCIPLE. *For any conveyance of real property, the length of the unit of measurement is that used as of the date of the deed.*

Values of measurement for the same unit sometimes varied from locality to locality. In the year 1919, the legislature of Texas standardized the vara to be $33\frac{1}{3}$ inches. This, in no way, is retroactive, and conveyances made prior to 1919 in Texas are not bound by the legislative act. When retracing original survey lines, unless the surveyor uses the same chain length as was used originally, he will not reach the same point as the original survey and will not be following the original surveyor's footsteps.

PRINCIPLE. *Every measurement of distance is subject to some errors, either known or unknown.*

Perfect measurement of land boundaries does not exist. Distances, whether made by a tape, electronic distance equipment, or the application of trigonometry, are subject to some errors. The individual should strive to conduct his measurement so that it will be as close to the standard as possible. Many of the

Table 10–1 Units of Length[a]

Units	Inches	Links	Feet	Yards	Rods	Chains	Miles	Meters
1 inch	1	0.126263	0.0833333	0.0277778	0.005050	0.001262	0.000015783	0.02540005
1 link	7.92	1	0.66	0.22	0.04	0.01	0.000125	0.2011684
1 foot	12	1.515152	1	0.333333	0.060606	0.015151	0.000189394	0.3048006
1 yard	36	4.54545	3	1	0.181818	0.045454	0.000568182	0.9144018
1 rod	198	25	16.5	5.5	1	0.25	0.003125	5.029210
1 chain	792	100	66	22	4	1	0.0125	20.11684
1 mile	63,360	8000	5280	1760	320	80	1	1609.3472
1 meter	39.37	4.970960	3.280833	1.0936111	0.198838	0.049710	0.000621370	1

[a] Lewis V. Judson, *Units and Systems of Weights and Measures*, National Bureau of Standards Circular 570, U.S. Government Printing Office, Washington, D.C., 1956, pp. 22–23.

applied corrections to tapes and other equipment are dependent on a formula developed in laboratory conditions and applied in field conditions that may be different. Every applied correction has some error.

Some surveyors, engineers, and courts accept a distance of 100.00 feet as being absolute, when in reality it is only a distance measured under conditions that were unique at that time only and could not be repeated.

10-6 Methods of Determing Distance

Far back in the Egyptian culture, the problem of distance was solved by comparison. The rope stretchers (Fig. 6-1) compared the length of rope with a distance measured on the ground. Bars and rods have also been used for this comparison. As recently as 1900, bars were used for base-line measurements by the U.S.C. and G.S., and land surveys were measured with poles.

The invention of the surveyor's chain by Edmund Gunter, an English astronomer in 1620, was an improvement over the ropes and poles used at that time. This device (Fig. 6–3) became most common in use and was the means for laying off most of the public domain in this country. The chain was made in various lengths, such as 66 feet, 33 feet, and 100 feet; and some chains were made too long intentionally to compensate for errors. All were bulky to handle, and usually the chain would be dragged on the ground, and the length would be in error because of slope, kinks, bad alignment, and other conditions.

Steel ribbon tapes came into popular use at about the beginning of the twentieth century. The use of these tapes reduced the errors in measurement and permitted the surveying team to measure distances in a shorter time with better accuracy.

Steel tapes are obtainable in almost any conceivable length from a few feet to a thousand. They are graduated in various units such as feet, meters, yards, and chains. A common tape used by property surveyors is the 100-foot band chain.

Good quality taping can be performed by two men. The tape may be either suspended from the ends or fully supported, provided its length is known for the conditions prevailing. The use of taping pins is a great help in marking the end points of the tape lengths as well as a means for keeping track of the number of tape lengths.

Whereas surveyors today commonly use electronic distance measuring devices, it must be remembered that the distances established for most conveyances were based on old measurement methods. Accuracies obtainable in former times have no relationship to accuracies attainable today. For those situations where taping was not practical, tachymetric methods were sometimes used. Stadia, the best-known tachymetric method, was used more than any other, and the accuracy of this method, if carefully done, was not more than 1 part in 300. The subtense bar had limited use after the development of optical angle measuring instruments capable of 1-second measurements.Tachymetric methods found popularity in Europe, with some instruments giving precision closures of 1 on 5000 under controlled conditions.

More details on the tachymetric surveys may be found in many of the popular textbooks on surveying as well as in the journals published by the technical societies.

Many surveys were made in early times with a wheel or "odometer" giving surface distance. Such devices are useful for special purposes today. Good results depend on a smooth rolling surface, which is not often obtainable.

In recent years, electromagnetic and electro-optical devices have come into use for measuring distances. Whereas these devices are of great value in making remeasurements today, they were not available in former times. Sometimes in retracing former surveys, it is better to use the equipment available at the time, especially when retracing a survey originally made with a compass. Local magnetic variations are not detected in modern directional devices, and errors of slope measurement do not show up. The object in a resurvey is to "retrace the footsteps of the original surveyor," and in some instances, such cannot be done by the use of modern equipment that skips part of the ground traversing. Marked trees and corners can be easily overlooked. After a survey is retraced with older equipment, the surveyor should return and accurately determine the position of found or set monuments; otherwise, he is merely accomplishing a refinement of the earlier survey.

10-7 The Early Determination of North

Definition of direction must be relative to two fixed points, and from the facts known about the earth today, the direction formed by the two poles is most logical to use as a reference. In early history there was much debate about whether the earth was round or flat. Earlier people, especially seafaring men, did observe that certain stars remained relatively fixed and others seemed to rotate around them.

Polaris, from which many surveyors determine north, was not always the North Star. At the time of the pyramids, Thuban (Alpha Draconis) was within 4° of the celestial pole, and Polaris was 25° away. From about 1800 B.C. to 1000 A.D. no bright star existed near the north pole. Today, Polaris is less than a degree from the pole, and it will reach within 0°35' before it travels away from the pole. Some 10,000 to 12,000 years from now Vega will be in a position to be considered the North Star.

Substantial proof exists that the north star of 2900 B.C. was used to construct the Great Pyramid's entrance passage, because it was so built that Thuban at lower culmination (3°42' below the pole when pointing due north) could be seen from the bottom of the passage. The base of the entrance is only 0°03'06" west of north! Other evidence indicates the pole star, Thuban, was probably in use 5000 B.C.

10-8 Direction

Distance alone will not define the position of a point; but with direction location is identified. Direction is determined by angular or circular measure in a horizontal plane. In the United States the sexagesimal system of angular measure, which divides the circle into 360 degrees, each degree into 60 minutes, and each minute into 60 seconds is in common use. The angular measure affects position in direct proportion to the distance. As an example, 1 minute subtends an arc of about three-tenths of a foot at a thousand feet. The accuracy of directions is dependent on the instrument used, the procedure employed, and the length of sight.

The discovery of the earth's magnetic field and the invention of the compass came before Columbus' time. The use of magnetite for determining north was known in the twelfth century, and the Chinese knew of the properties of lodestone about 1000 B.C.

Early navigators in Western Europe were blessed with an almost zero declination, an advantage that no longer is in effect. A belief arose that the compass always pointed north, and during Columbus' voyage, his men were frightened by the discovery of the compass deviating from true north. In 1635 Henry Gillibrand discovered changes in the magnetic needle, and it was in 1722 that daily compass variations were discovered. The first isogonic chart appeared in the nineteenth century.

The magnetic field of the earth is not constant; it is subject to annual changes, daily variations, and magnetic storms. At present, no certain method exists to prove what the magnetic declination was on a particular date in the past without positive observations on that date, and it is even more difficult to predict what future declination will be. Directional calls found in old deeds and based on magnetic observations are difficult to interpret. This one phase causes inexperienced surveyors agony in attempting to determine the correlation between old bearings and the present reading. Many surveyors who conduct retracements are woefully ignorant of magnetic declination and its application to retracements.

10-9 Reckoning

Directions must be referred to a meridian. Often the meridian is assumed along one line purely for the purpose of providing relative bearings for each of the sides. The U.S. Rectangular System was intended to be oriented to astronomic north, although later surveys reveal considerable deviation therefrom.

"True" north may be defined by one of three meridians: astronomic (north as defined by the celestial poles); geodetic north based on a reference spheroid (generally Clarke's); or geographic north. Although these differ very little (only a few seconds), true north in property surveying is generally considered as that determined from star observations.

Once the direction is reckoned with respect to a meridian, all other sides may be related by angle measurements. Metes and bounds descriptions are sometimes oriented by carefully reading the compass needle on one of the sides and then relating the other sides by the deflected angles measured. When the survey is retraced years later, the magnetic direction of any line can be determined, within a limited accuracy, by applying the change in magnetic declination for the elapsed period of years.

Most of the public lands were surveyed with the compass. Today, such methods are not generally satisfactory, although many surveyors still use the compass needle for retracing old lines and for a basis of orienting the figure. Some states have and do prohibit the use of a compass in land surveying.

In those areas where grid systems are employed, grid north or grid south may be the reference meridian. This system has many advantages, since it is plane rectangular, that is, the meridian lines are parallel and can be easily corrected to astronomic north. A more complete discussion of rectangular systems will be found later in this chapter.

10-10 Methods of Observing Directions

The method of observing magnetic bearings for each side of a property has passed into obsolescence and, in fact, is prohibited by law in some areas. Besides the annular and secular changes and the danger of local attraction, compass needle readings can, at best, be read only to the nearest quarter degree.

Although alidades of some sort have been in use for a long time, the modern transit is a development of the last century. The terms "transit" and "theodolite" are not well defined by popular usage. The instrument's telescope must turn about the horizontal axis to be properly termed a *transit*. The term *theodolite* is generally associated with the more precise instruments, although it is possible to have a theodolite that will transit and a transit that is a theodolite. Most surveyors today use a theodolite.

Instruments and methods of observation are classed either as "repeating" or "directional." By alternating the use of the upper and lower motions, the value of an angle may be repeated or "run up" several times so that small fractions can be measured even when smaller than the smallest division on the instrument. *By repeating an angle more precision can be obtained than is possible by a single reading.*

Direction instruments have no lower motion and are usually of higher precision than repeating instruments. The angles may be derived from the difference in the observed directions of two lines. *A number of readings with a direction instrument will improve the accuracy but not the precision.*

The quality of direction should be consistent with the quality of the distance. It would be ridiculous to use a magnetic compass with a carefully taped distance and absurd to read angles to a second when the distances were determined by stadia. As in all survey operations, the observer should provide for a check against blunders when reading angles. This may be realized easily by "doubling" the deflection or angle to the right, even though the single reading will satisfy the precision requirements.

Systematic readings will minimize blunders and increase efficiency. Many systems for turning angles are described in the various textbooks on surveying. The surveyor should decide on the procedure that best suits his needs, and he should follow this method in an orderly fashion.

The vernier transit, which has been the workhorse for property surveyors for so many years, is now generally replaced by optical reading transits for many purposes. The optical transit, usually of European design, has many advantages over the usual vernier transit and, when put to use, will produce a higher precision for less cost. See Table 10-2 for angular units.

10-11 Meridian Determination from Celestial Observations

For many centuries the stars have been the guide for humankind to locate direction. The fact that the earth spins on its axis causes the stars to appear to rotate about a point in the sky (the north celestial pole), which point has become known as true north. One star, Polaris, is in close proximity to the north pole and is a convenient star for observers in the middle latitudes of the Northern Hemisphere. All stars apparently cross the observer's meridian each day. If the bearing of the star is computed for the time the observation was made, the meridian can be located.

Many sections of the international boundary of the United States, most of the state boundaries, much of the county boundaries, and most of the U.S. Rectangular System are defined in terms of "true" bearings. Although the monuments set to perpetuate their location are controlling, the intent was to orient many of these lines with the cardinal direction. In most states, surveys that are described by metes and bounds will be oriented with astronomic north if another system is not named or implied.[2] For the surveys that are based on other meridians, it will often be more efficient to initiate, terminate, or check directions with a star observation.

With present-day surveying equipment and time pieces all star observations are relatively simple to perform in the field, and with programmable computers of the electronic type, computations are no problem. In performing subdivision surveys and any survey to create a new parcel, directions should be based on either a sun or polaris observation, or, if the state has established a plane coordinate system with sufficient monument density, that should be used.

10-12 Measurements from Photographs

Photogrammetry is almost as old as photography. In the broadest sense, it is the science and art of making measurements from photographs. The intent of this brief discussion is not to present the theory of the science or the technique of the art but to point out some of the proved and accepted applications that may be made in property surveys by knowledgeable surveyors.

The practical beginning of metrical photogrammetry in the United States was in the middle 1920s, nearly a generation after the fine start made in Europe. It is

---

[2]*Richfield Oil Corp.* v. *Crawford*, 249, P 2d 600.

Table 10–2 Table of angular units

Unit	Degrees	Minutes	Seconds	Grads	Hours	Radians
Degrees	1	0.01667	0.0002778	0.90	15	57.2958
Minutes	60	1	0.0166667	54	900	3437.75
Seconds	3,600	60	1	3,240	54,000	206.265
Grads	1.1111	0.01852	0.0003086	1	16.7	63.66
Hours	0.0667	0.00111	0.00001852	0.060	1	3.820
Radians	0.01745	0.0002909	0.000004848	0.01571	0.2618	1

not known when these measurements were first made for cadastral surveys, but as early as 1938, property damages caused by forest fires were surveyed with the aid of aerial photographs, and property lines were located for reservoirs in the Tennessee Valley Authority. By the close of World War II, the use of aerial photographs for property-line location was common.

Many experiments and practical applications of modern photogrammetric principles to boundary location have been performed in Europe. The earliest known work occurred in Italy in 1931. Reallotment surveys in Germany made use of coordinates of property corners obtained by photogrammetric methods. The accuracy was quite satisfactory, and over a large-scale operation with high-density ground-control monuments, it was successful for property registration purposes. The project required the identification of all points on the photos using small targets. Similar projects were done in the Netherlands, Sweden, and France. A recent land-registry program in Austria made use of the teleprinter and electronic computer, undoubtedly the most automatic property survey performed up to that time.

In recent years there have been several noteworthy experiments in the United States. The U.S. Forest Service conducted a survey to locate sections of forest lands for timber management in Missouri and the Lake Tahoe area. Together with ground control, the measurements were made photogrammetrically from good quality aerial photographs. In a test area in Utah, BLM subdivided several townships with the use of photogrammetric methods. Picture points served to locate the position of the corner, with plane tables used in making the actual establishment. This test was well within the required accuracy stated in the *Manual of Instructions*, but the costs were 50 percent above that of the usual conventional methods. In Minnesota the center of sections and 1/16 corners were set by photogrammetric methods.

The present American practice uses the tool of photogrammetry in many ways such as pipeline location, "paper" surveys for highways, and subdivision planning. In all cases, final location and measurements are on the ground.

10-13 Photogrammetric Measurements

Before distance, direction, or position can be determined from photographs, identification of points on the pictures is essential. The corners may be either marked before the photography or identified afterward by field inspection. When the corners are marked with a target there is danger of its becoming moved, destroyed, or confused with some other object. The practice of premarking points is quite prevalent in Europe, where small colored cards are centered over the monuments, or other devices are used that will cause them to contrast well on the photograph. Lime or plastic strips spread on the ground are sometimes used for premarking points. When the identification is left until after the photographs have been taken, some additional cost may result.

The discovery of accepted property corners is usually the biggest task for the property surveyor, but after the discovery of the corners the geometric location of these points may not present a great problem. It is possible to determine these positions photogrammetrically if extremely precise methods are used. The photography needed is of the highest order, at a very large scale, and taken under ideal conditions. A considerable amount of ground control is necessary to orient the photogrammetric models. With high-precision plotters measurements have been made from photographs with a certainty of less than 1 inch that would satisfy the requirements of many property surveys. Whether the courts will accept such measurements as proper evidence of property title has not been tested.

Such a modern method of surveying could be applied to the layout of an urban subdivision, but as yet it has proved too costly for other than topography plotting. It is anticipated that in the near future photogrammetric methods will be used more extensively. The future of photogrammetry as a tool for surveyors is unlimited.

10-14 Availability of Photographs

The United States has been completely covered by aerial photographs. In some areas, especially around large cities, the coverage has been repeated several times. Most of these photographs are available to the surveyor at a small cost. Such governmental agencies as the Geological Survey, Production and Marketing Administration, Tennessee Valley Authority, Corps of Engineers, Agricultural Stabilization and Conservation Service, Forest Service, and others keep on file many photographs of various ages and scales and have means of supplying prints to individuals. Many states, counties, cities, and private companies have photographs that have been taken in connection with some special purpose, and these can be purchased for a small charge.

Some practicing surveyors take their own photographs. It is not always essential to use an aerial camera; a reliable press camera with a large negative size will be satisfactory for many uses. Using a small chartered airplane and a conventional press camera, an entire township can be photographed for about $250. The information obtained may be worth several thousand dollars.

10-15 Survey Computations

Within the last few years an unbelievable development in the science of computations has occurred. Electronic calculators take raw data, compute the error of closure, adjust the traverse, and calculate the area in a matter of minutes, whereas it formerly took hours. Most of the drudgery and errors of office work have been eliminated.

In the early days of surveying, textbooks all had logarithm tables in the appendix. At one time computation sheets were stored with every job. It is now quicker to recompute than it is to try to understand old computation sheets with sine, cosine, bearing, and distance all handwritten. Because of the difficulty and tedious nature of computations in early times, the cause of errors can be understood; proof by computations was often omitted.

Mechanical calculators came into common use by 1930; electric calculators were common by 1950; electronic calculators capable of solving for the sine, cosine, and tangent became common by the 1970s. Most computations prior to 1920, if done, were generally by logarithms.

The only purpose for calculating an error of closure is to determine the precision of the survey—that is, the relationship of angular measurements and distance measurements to each other and then to calculate area.

An error of closure does not indicate the accuracy of property corners. Assumptions are made when a traverse is balanced that any errors are distributed uniformly in each course, yet this is not necessarily true. No one is precluded from placing any corrections in those lines in which the error most probably occurred. The selection of the proper lines requires experience and a knowledge of the theory of errors and adequate field experience.

Once adjusted lines are computed, then area can be determined. The calculated area of a balanced, closed traverse is determined by the method used. No two methods of area computation will give the same results and may vary in the first or second decimal place.

10-16 Consistency of Significant Figures

The number of significant figures in a numerical expression is the number of digits in the expression, minus any zeros whose sole purpose is to indicate the position of the decimal point. Built into the expression of any numerical value is an error of half the magnitude of the unit of that last significant figure (e.g., 1432 stands for all those values between 1431.5 and 1432.5; therefore, the error in expressing the value is ±0.5).

Value	Significant Figures	Error of Expression	Relative Error
0.023	2	±0.0005	1/46
172	3	±0.5	1/344
84.15	4	±0.005	1/16,830
3956	4	±0.5	1/7,912
1.320	4	±0.0005	1/2640
32,100	3 to 5	±50 to ±0.5	1/642 to 1/64,200

Significant Figures	Relative Error
1	1/2 to 1/18
2	1/20 to 1/198
3	1/200 to 1/1,998
4	1/2,000 to 1/19,998
5	1/20,000 to 1/199,998

The error in the expression of a measured quantity should be consistent with the error in the measurement itself. Thus, if a distance was measured and recorded as 14,319.42 feet but the error in the measurement was 0.5 foot, the last two figures are meaningless, and the expression merits only five significant figures instead of the seven shown.

The expression of computed results should be consistent with the expression of the values from which they were derived. This consistency can be investigated by applying the theory of propagation of accidental errors.

Product of Two or More Values. For $U = xyz$, then

$$\frac{\sigma U}{U} = \pm \sqrt{\left(\frac{\sigma x}{x}\right)^2 + \left(\frac{\sigma y}{y}\right)^2 + \left(\frac{\sigma z}{z}\right)^2}$$

This shows that the relative error of the product can never be smaller than the relative error of any one of the factors. To provide a working guide, this statement is generalized into the following rule: *The number of significant figures in any product cannot be greater than the least number of significant figures in any one factor.*

EXAMPLE 1. $A = ab$ $\quad a = 4.1$ feet $\quad\quad b = 74.0$ feet
$\quad\quad\quad\quad\quad\quad\quad\quad \sigma a = \pm 0.05$ foot $\quad \sigma b = \pm 0.05$ foot

$$A = 4.1 \times 74.0 = 303.40 \text{ feet}^2$$

Now

$$\frac{\alpha A}{A} = \pm \sqrt{\left(\frac{0.05}{4.1}\right)^2 + \left(\frac{0.05}{74.0}\right)^2} = \pm\sqrt{0.00015355} = \pm 0.0124$$

or 1/81

$\sigma A = \pm 0.0124 \times 303.40 = \pm 3.76$, or about 4 in the third significant figure. To be consistent, $A = 300$ feet^2 (two significant figures).

Sum of Two or More Values. For $U = x + y + z$,

$$\sigma U = \pm \sqrt{(\sigma x)^2 + (\sigma y)^2 + (\sigma z)^2}$$

This shows that the error of a sum can never be smaller than the error of any one of the terms. To provide a working guide this statement is generalized into the following rule: *the number of significant figures in any sum shall be such that the error in expressing the sum is equal to the largest error in expressing any one term.*

EXAMPLE 2. $U = a + b + c + d$

$$a = 4321.6$$
$$\sigma a = \pm 0.05$$

$$b = 42.043 \quad \sigma b = \pm 0.0005$$
$$c = 31.72 \quad \sigma c = \pm 0.005$$
$$d = 91.0 \quad \sigma d = \pm 0.05$$

$$\sigma U = \pm \sqrt{(0.05)^2 + (0.0005)^2 + (0.005)^2 + (0.05)^2}$$
$$= \pm \sqrt{0.0025 + 0.00000025 + 0.000025 + 0.0025}$$
$$= \pm \sqrt{0.00502525} = \pm 0.07 \text{ (larger than any one term)}$$

From the rule, $\sigma U = \pm 0.05$, which approximates $\sigma U = \pm 0.07$. Now $U = 4321.6 + 42.043 + 31.72 + 91.0 = 4486.363$, properly rounded off to 4486.4.

This last example shows that by rounding off the computed sum to a more consistent value, a shift of 0.037 is made. This is commonly termed the *rounding-off error*. If this computed value is used to compute another value, the rounding-off error will be introduced as a systematic error into the computation. Thus rounding off should not be done until the final result is obtained. For most computations this statement can be modified by saying: *one significant figure more than is necessary should be maintained until the final result is obtained*.

Consistency between Linear and Angular Measurements. There is a generalization using the following particular example. (See Fig. 10-1.) The position of a point B from point A may be defined in terms of a distance D and an angle θ from some reference line.

An error in the angle ($\sigma\theta$) would produce a shift in the position of the point B to B'. An error in the distance (σD) would produce a shift in the position of the point from B' to B''. To be consistent,

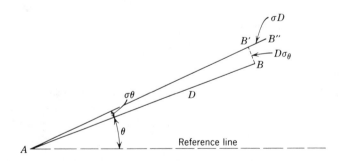

Figure 10–1 Relationship between angles and distances.

$$BB' = B'B''$$

therefore

$$\sigma D = D\sigma\theta \quad \text{or} \quad \frac{\sigma D}{D} = \sigma\theta$$

The relative error in distance = *the angular error in radians*

Relative Error	Angular Error
1/57	1°
1/690	5'
1/3,440	1'
1/10,300	20"
1/206,300	1"

ERRORS IN MEASUREMENT

10-17 Precision and Accuracy

Precision refers to the degree of refinement by which a quantity is determined. *Accuracy* is a measure of how close to the true or exact value the determination is. In some circumstances, the number of decimal places is a measure of the precision. It is not to be construed that higher precision will insure more accurate results. Consider the example that follows in which three angles of a triangle were measured first by a sextant and next with a theodolite.

Sextant	Theodolite
64°40'	64°41'22"
42°25'	42°24'52"
72°55'	72°53'40"
180°00'	179°59'54"

The three angles measured by the sextant total to the required 180°00', whereas the angles measured with the theodolite fail to close by 6". Obviously, the quantities measured by the theodolite are more precise, but those determined by the sextant are quite accurate *within their limit of expression*.

Precision without accuracy is meaningless, and better accuracies can be obtained only through increased skill and more refined equipment. Precision is achieved by employing equipment of high quality with skilled techniques under the most favorable conditions. Each must be compatible with the others, or the effort is for naught.

In most cases, the true accuracy cannot be ascertained, but by careful analysis of the errors consistent with the precision used, an estimate of the accuracy can be ascertained.

10-18 Classification of Errors

An *error* may be defined as the difference between the true value and the observed quantity. Errors must not be confused with *blunders*, which also cause differences in the derived values, but are illegitimate and must be removed.

In general, all errors may be classed into two groups: "systematic" and "accidental." *Systematic errors* are those that have predictable quantities and signs. With sufficient collateral evidence the effect of these errors may be removed by correction. A typical example of a systematic error is the thermal expansion of a steel tape due to the change in temperature. If the temperature, coefficient of expansion for the tape, and the length are known, the correction can be applied mathematically or made physically. It should be noted that systematic errors behave in a predictable way; the magnitude and direction of these errors are functions of determinable variables.

Accidental errors may be divided into two types: "random" and "constant." The *random errors* have equal chances of being plus or minus. An example of this is the setting of a chaining pin at the end of the tape. Say that the chainman can set the pin within ±0.005 foot. Therefore, each time the pin is set the chances are equal of its being advanced or set back by the 0.005 foot.

A *constant accidental error* is one whose quantity is unknown but which will have its effect in the same direction throughout the operation. Alignment error is one such example. It is impossible to determine how much error is introduced by reason of the tape's not being in a straight line, but the tape will always have the effect of producing too large an observation; therefore the effect is constant in direction. The calibration error of the tape is another fine example of constant error. If a certain tape is certified to be 100.00 long, its length is uncertain by ±0.005 foot at least. We do not know if it is too long or too short, but, whichever, it will be the same every time it is used, and the error will accumulate in direct proportion to the number of opportunities.

Even though systematic errors can be corrected for, there will always be some accidental error accompanying each systematic correction. In the preceding example, the temperature correction is not complete because of uncertainties in reading the thermometer. Many of these "residual" errors are small or even insignificant, but often they must be considered when writing specifications.

Blunders and mistakes have to be avoided, or the work must be redone. Such acts as misreading the tape, transposing numbers, dropping a tape length cannot be corrected for and have no place in professional caliber surveying.

10-19 Theoretical Uncertainty (tu)

All surveyors realize that no quantity can be measured to the absolute true value and that each stated measurement has some inherent doubt. It is essential that each measurement be accompanied with a statement of its reliability. In control surveys, such as are conducted by the governmental agencies, the customary method of expressing the quality of the measurement is by stating the *relative error of closure*. These ratios are computed by dividing the indicated closure discrepancy by the total distances. For this, the familiar expressions such as 1 in 5000 and 1 in 10,000 are used.

A survey that has been conducted to locate the title ownership of real property must be certain to a degree that any conflicts arising over encroachments, and the like, cannot be attributed to the poor quality of the survey. Closures of traverses must be good, certainly, but it is possible for a traverse, displaying a small closure discrepancy, to contain excessive uncertainties. The concept of relative error of closure is not very meaningful in this circumstance, for it would permit a larger uncertainty in a survey of a large lot than in one of half size. It is, perhaps, better for the surveyor to examine his methods, procedure, and equipment to estimate the uncertainty in derived positions of the points that his survey has passed through. This expression of quality stated with the quantity will be of great help to the client, the court, retracing surveyors, and all who have interest in the property in question. As an example, a property line may be expressed as "659.74 feet, ±0.08." This would indicate that the true length will probably lie between 659.66 feet and 659.82 feet, and the most likely value is 659.74. If a suit between the coterminous owners were to result over an encroachment of 1 inch, it would be obvious that this much uncertainty exists in the survey itself.

The value ± 0.08 cannot be determined by casual guess. It must be calculated from sound theoretical analysis. The surveyor has, of course, corrected his distances for all known systematic errors and has stated the uncertainty in the accumulated results of accidental errors only. For many of the accidental errors, it is quite difficult to ascertain what the magnitude is, whereas for others, the effect of the error is quite obvious. A number of the errors will be so insignificant that their effect will be overshadowed by the others.

It is likely that the random errors will accumulate as the square root of the opportunities or better expressed by the equation

$$E = \pm e \sqrt{n}$$

where E is the total effect of a particular error (e) and n is the number of opportunities. Constant errors accumulate in direct proportion to the number of opportunities, or

$$E = \pm e \cdot n$$

When the total error has been calculated for each of the contributing factors, the resultant is found by the expression

$$E_T = \pm \sqrt{E_1^2 + E_2^2 + E_3^2 + \cdots}$$

where E_T is the final resulting error in the observation and $E_1, E_2, \ldots,$ are the full effect of each of the contributing errors.

The value E_T may be expressed in linear units, rotation, or whatever is appropriate to the situation. This value may also be termed the *theoretical uncertainty* (*tu*). This value has very little to do with the closure, except that the experienced discrepancy should fall within this expected range. *It is just as likely for a closure to be zero as for it to be any other number within the expected range.*

Relative error of closure is dependent on the length of the survey as well as the quality with which it was executed. If need be, the surveyor can express the theoretical uncertainty as a ratio to the perimeter and have a *relative uncertainty* (*ru*).

10-20 Errors in Taped Distances

The surveyor in practice must be familiar with the various corrections necessary to remove the effect of systematic errors. In general, these involve only two problems: length and alignment. Length is affected by temperature, tension, and the standard dimension of the tape. Alignment error can be both in vertical and horizontal planes. Vertical alignment or slope can be corrected mathematically or by leveling the ends of the tape. Alignment errors, although accidental in character, have a systematic effect; and, if proper values of the amount of malalignment can be determined, a correction can be applied to the observed value: $C_A = -(\sigma^2 n/2L)$, where n is the number of sides, L is the length of tape, and σ is the standard deviation of the alignment error. When the tape is allowed to sag, there is a shortening that is a form of vertical malalignment.

Corrections to the length of tape for temperature, elastic stretch, and sag can be taken from the family of curves in Fig. 10–2 and Fig. 10–3. More complete information on corrections to taped distances will be found in any standard textbook on surveying.

Accidental errors in taped distances occur also in alignment and length. For each correction applied to remove a systematic error there will be a "residual accidental error." Uncertainties in determining the temperature, inherent error in the standardized length, and the like, all have accidental errors of either the random or constant class. Human limitations account for some of the accidental errors in distance measurements such as estimating the fraction, setting the pin, and like things. To calculate the uncertainty of the distance, a careful analysis must be made of all the contributing factors. These elements will include the following:

1. Calibrated length.
2. Variation in temperature.
3. Uncertainty in tension.
4. Error in determining slope.
5. Error in setting the pins.
6. Fraction and end reading.
7. Horizontal alignment.

Others errors may be present, but for the purpose of the following example, it is considered that those listed are the only ones that have a significant effect. After

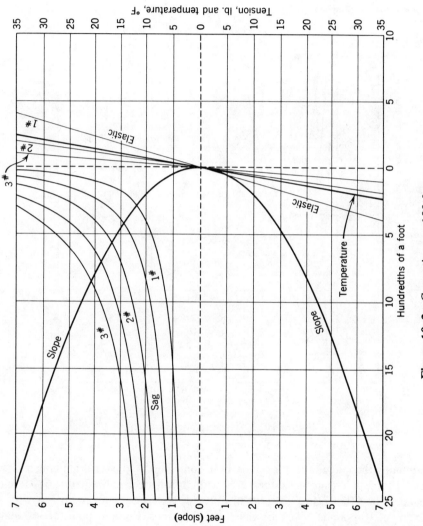

Figure 10–2 Corrections to 100-foot tape.

286 / MEASUREMENTS, ERRORS, AND COMPUTATIONS

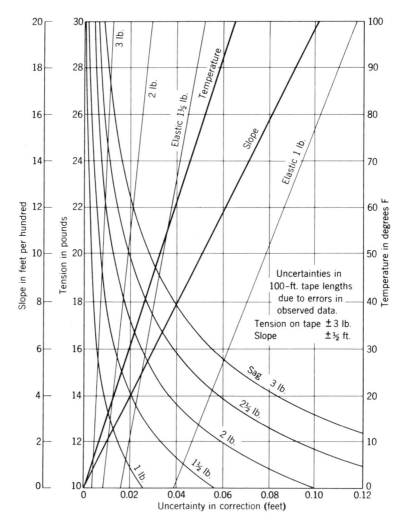

Figure 10-3 Uncertainties in taping.

estimating the magnitude of each of these contributors, the resultant uncertainty can be computed as follows: tape used was carbon steel, calibrated at 68° F and found to be 100.02 feet long when fully supported on a level surface with a tension of 10 pounds. The tape weighed 1.8 pounds. For a distance of 360 feet, the tape was fully supported on the ground, and the tension of 10 pounds was exerted. For the remaining distance the tape was suspended from the end points; 25 pounds of tension was applied, and the plumb bobs were used to mark the end points. The ground was sloping 4¼ percent generally. The average temperature was observed to be 92° F. The computation of uncertainty is as follows:

Error	Magnitude	Frequency	Total	Total Squared
1	0.005	6.6	0.03	0.0009
2	0.02	1	0.02	0.0004
3	0.003	6.6	0.01	0.0001
4	0.02	7	0.05	0.0025
5	0.005	6	0.01	0.0001
6	0.005	8	0.01	0.0001
7	0.016	7	0.04	0.0016
				0.006

$$E_T = \sqrt{0.006} = \pm 0.08 \text{ foot for } 659.74 \text{ feet}$$

10-21 Errors in Observed Directions

If directions are derived from angle measurements at each of the sides, the new side will inherit the uncertainty of the previous one. When directions are measured directly such as with the magnetic compass, each side is independent of the others and the carry-over is not in effect. Unfortunately, the errors in reading the compass are so large that this method is not worthy of consideration here.

Accidental errors associated with the measurement of angles involve such considerations as pointing, reading the instrument, centering over the station, centering of the signal over the sighted station, instrumental errors, and others. It is seldom that systematic errors are involved in this operation, and residuals need not be considered. Of the accidental errors, those that are random are most prevalent and lend themselves conveniently to error analysis.

Careful and skilled operation can minimize the effect of pointing and centering over the station so that their contribution is insignificant. Under ordinary situations, the instrumental error can be minimized by an appropriate program of observations. It must be emphasized that these factors cannot always be ignored, but a typical situation would require consideration of only the pointing error, the inherent error, and the reading error. The following example is for an angle measurement made on a backsight 300 feet away and a foresight 659.74 feet distant. The instrument is a half-minute vernier transit with insignificant instrumental errors. The angles were turned once and doubled. A plumb bob string was used on the short backsight, and a range pole on the foresight. In this example, the errors of instrument centering, instrument adjustment, pointing, and the like, are considered insignificant. The point occupied is the fourth station from the initial corner.

Reading Uncertainty. As the instrument must be set on zero, the single and double angles observed, three readings are involved, but only the initial and the final are used for deriving the angles. The limitation of the instrument is $\pm 15''$. The full effect of the reading error therefore is

$$\frac{\pm 15'' \sqrt{2}}{2} = \pm 11''$$

Signal Error. On the short sight, the plumb bob can be held to within ± 0.005 feet of the station, and this error contributes $0.005/300 = 0.000017$ radians or about $\pm 4''$. On the foresight, the range pole is centered within ± 0.05 feet and introduces $0.05/660 = 0.000076$ radians or $\pm 17''$ error.

The result of these three uncertainties is the square root of the sum of the squares or $\sqrt{11^2 + 4^2 + 17^2} = \pm 21''$, the uncertainty in the angle measurement. If this had been the uncertainty for each of the preceding four sides, the inherited uncertainty is $\pm 21 \sqrt{4} = \pm 42''$. The uncertainty of the direction of the foresight will then be $\sqrt{42^2 + 21^2} = \pm 48''$.

10-22 Reliability of Meridian Observations

Observations taken on the stars for azimuth determination are subject to errors, and the results will be uncertain to some extent. The contributing errors will appear in the instrumentation, ambient conditions, and the attending observations.

Instrumentation. When extreme vertical angles are observed, some of the instrument adjustments become quite critical. If observations are being taken at $20°$ or more above the horizon, the manufacturer's manual and a good textbook on surveying should be consulted. Maladjustments of the line of sight, horizontal axis, and plate bubbles each contribute to the uncertainty in the horizontal angle reading. A typical reading taken on Polaris (altitude of $40°$) may be uncertain by $\pm 0.3'$. This combined with the pointing, centering, targeting, and reading errors may result in an uncertainty of $\pm 0.5'$ for a typical Polaris observation.

When altitude is observed, it is dependent on the horizon as determined by spirit bubbles or a pendulum. On vernier transits, the vertical angle can be read to the nearest minute only, and there is no practical way of doubling the angle. The uncertainty in the observed vertical angle of a star with a vernier transit is about $\pm 1'$ and with the optical transits about $\pm 20''$. See Fig. 10-4.

Ambient Conditions. Temperature and barometric pressure observations are necessary to correct for atmospheric refraction. The uncertainties in temperature and pressure will not be significant, but it must be realized that the corrections are empirical only and will not completely remove the error because of refraction. It is for this reason that low-altitude observations are uncertain by several minutes when below $10°$.

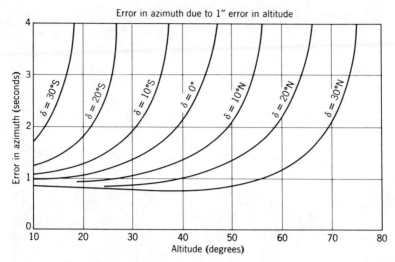

Figure 10-4 Errors due to altitude.

Attending Observations. Time, latitude, longitude, declination, and right ascension are not ordinarily a result of field measurements but are derived from other observations and sources.

When the azimuth is derived from an expression involving the hour angle, time must be considered. Under some conditions, standard time to the nearest minute will be sufficient, and in others a fraction of a second is necessary (Fig. 10-5). With a good timepiece it is possible to observe time with an error of not more than $\pm 1^s$. If a chronometer and stopwatches are used, the error may be reduced to as little as $1/10^s$. The timepiece should be checked against the Bureau of Standards time signal, WWV, the Dominion Observatory Time Signal (Canadian), CHU, or some other reliable standard time.

Latitude can be determined by star observation, or in the United States it is more practical to scale this value from a reliable map. Latitude with an uncertainty of not more than $\pm 1''$ can be scaled easily from a 7.5' Topographic Series Map published by the USGS. See Fig. 10-6 for the effect of the error in latitude on the resulting azimuth.

Longitude may be scaled from a map in the same manner as latitude, and the error effects the hour angle in the same manner as observed time but in different units, since there are 15 degrees for each hour. A condition that would require time accuracy to $\pm 10^s$ would permit a longitude error to $\pm 2.5'$.

Declination and right ascension are taken from the ephemeral tables and have no significant errors.

The derivation of the azimuth of a star will involve in every case the declination and a combination of (1) latitude and altitude, (2) hour angle and

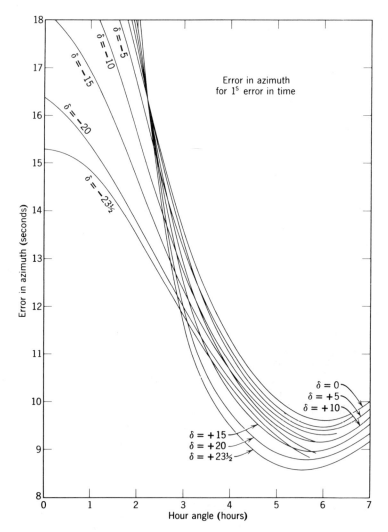

Figure 10-5 Errors due to time.

latitude, or (3) hour angle and altitude. Figure 10-8 illustrates the uncertainty in azimuth with these various combinations under various conditions. These factors, together with the instrumentation errors (Fig. 10-7), must be considered in deciding on the star, time of observation, and choice of expression. There is no reason why the surveyor cannot obtain an azimuth reliable to ±0.3′ with ordinary equipment by observing either the sun or Polaris.

10-23 Errors in Position

The error in position is, of course, the combined effect of the uncertainty in distance and in direction. When rectangular coordinates are employed, it is

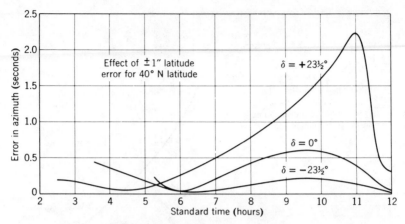

Figure 10–6 Errors due to latitude.

most convenient to express the uncertainty in terms of latitudes and departures. It will be recalled from elementary surveying that the latitude of a line is found by D cosine ϕ, where D is the length of the line, and ϕ is the azimuth of the line. The departure is found by the expression D sine ϕ. Therrefore, the error in latitude can be determined by differentiating the foregoing expressions, first in respect to σD and then in respect to $\sigma \phi$. Using the errors found in the previous examples for distance and direction and taking, for example, a bearing of N 60° E, the following error in position is derived.

Latitude (N)

$$\sigma N = \cos 60° \; \sigma D = 0.50 \times 0.08 = \pm 0.04 \text{ foot}$$

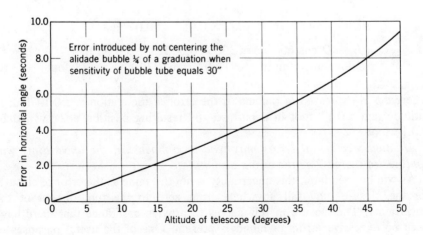

Figure 10–7 Trunnion error.

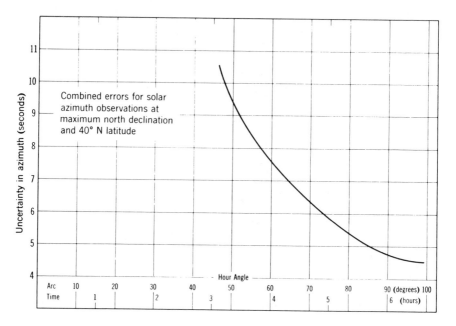

Figure 10–8 Combined errors.

$$\sigma N = -D \sin 60° \text{ sf} = 572 \times 0.00023 = 0.13 \text{ foot}$$
$$\text{resulting uncertainty} = \sqrt{(0.04)^2 + (0.13)^2} = 0.14 \text{ foot}$$

Departure (E)

$$\sigma E = \sin 60° \ \sigma D = 0.87 \times 0.08 = \pm 0.07 \text{ foot}$$
$$\sigma E = D \cos 60° \ \sigma\phi = 330 \times 0.00023 = \pm 0.08 \text{ foot}$$
$$\text{resulting uncertainty} = \sqrt{(0.07)^2 + (0.08)^2} = \pm 0.11 \text{ foot}$$

where $\sigma\phi$ is expressed in radians. If the error at the station is ±0.14 foot in latitude and ±0.11 foot in departure, the resulting position error would be ±0.18 foot.

Although accidental errors may tend to compensate, the uncertainty will continue to increase as the errors accumulate.

As will be seen later, this uncertainty is not the same as the error of closure, for the closure may fall anywhere from zero to the full amount of the uncertainty. The closure merely serves to provide assurance that there have been no excessive errors or blunders present. One of the useful purposes to which this method of analysis can be put is in preparing specifications (Sections

10-27 to 10-35). If it is required that the position of a point be certain within a given tolerance, this type of calculation can aid the surveyor in planning the method of taping, program of observations, how refined the temperature measurements should be, and the like.

10-24 Errors in Altitudes

Many of the systematic errors in leveling can be eliminated by proper techniques. The instrumental error and the effect of curvature and refraction can usually be eliminated by balancing the distances to the rod stations. Systematic errors in the rod have no serious effect, except when the difference of elevation is great.

For all but the more precise surveys, such as are conducted for geodetic control, corrections are unnecessary.

There are many accidental errors in leveling, but the two that usually have the most significant effect are the reading of the rod and the centering of the bubble. Both of these are random accidental errors and occur each time the rod is sighted and are a function of distance. Assuming that the rod can be read to the nearest ± 0.005 foot when the distance is 200 feet and that the spirit bubble can be centered to one-fourth of a division, the following is the calculation of the uncertainties of a level line extending for 2 miles:

1. If the bubble tube is graduated so that the divisions are 20" apart, the uncertainty in centering the bubble will introduce 20"/4 or ± 5" error in the line of sight. In 200 feet this will amount to about ± 0.005 foot on the rod.
2. The rod reading, as stated previously, has an error of ± 0.005 foot per rod reading. If the rod is read only once for each backsight and foresight, this error, and the one due to centering the bubble, has as many opportunities to occur as the number of rod readings.
3. If the line is 2 miles long, about 53 rod readings would be taken, and total uncertainty (due to the two causes under consideration would be $\pm 0.005 \sqrt{2} \sqrt{53} = \pm 0.051$ foot.

Because only two factors were considered in this analysis, the actual uncertainty would be slightly greater, and proper allowance should be made.

Closure of level circuits will not indicate the amount of uncertainty, but it will provide a certain amount of assurance that the line is free of blunder.

10-25 Errors in Traversing

The effect of errors in both distance and direction is reflected in the resulting uncertainty of position. Whether it be an open figure or a closed polygon, a traverse is a chain of "directed distances," and the errors will carry through the chain. Fig. 10-9 shows the realm of uncertainty and a typical error situation. Note that it is very likely for the error to be zero at any point. The relative error of closure is not as qualitative as the relative uncertainty.

294 / MEASUREMENTS, ERRORS, AND COMPUTATIONS

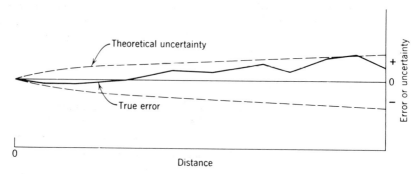

Figure 10-9 Traverse errors and uncertainties.

10-26 Errors in Triangulation

A position resulting from triangulation will have uncertainties from errors in distance and direction. The principal difference between triangulation and traversing is that the distance is not measured directly, but it is a result of previously measured distances and angles.

Considering a simple triangle (Fig. 10-10), one side is known (base line), and all three angles have been measured. Side AB will be computed to determine the position of B. The side AB will then contribute uncertainties because of the errors in its length. The length has been computed from the length of side AC and the two angles, B and C. To determine the uncertainty in this length, one must differentiate the expression

$$AB = \frac{AC \sin C}{\sin B}$$

The resultant is found by taking the square root of the sum of the squares, and the resulting error in position is found in the same manner as that in Section 10-19.

EXAMPLE. *In the triangle shown in Fig. 10-10, the uncertainty in the base length is ±0.05 foot, and the uncertainties in any one of the angles is ±10″. The direction is uncertain by ±10″. What is the uncertainty in the position B in respect to A?*

SOLUTION

1. Due to base $\overline{AC}$:

$$\sigma\overline{AB} = \frac{\sin C}{\sin B} \sigma AC = \frac{0.988 \times 0.05}{0.984} = \pm 0.05 \text{ foot}$$

2. Due to angle C:

$$\sigma\overline{AB} = \frac{AC \cos C}{\sin B} \qquad \sigma C = \frac{697 \times 0.152}{0.984} \times 0.00005 = \pm 0.005 \text{ foot}$$

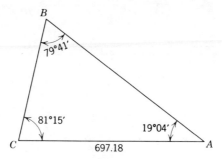

Figure 10–10 Error in triangle.

3. Due to angle B:

$$\sigma \overline{AB} = \frac{\overline{AC} \sin C \cot B}{\sin B} \sigma B$$

$$= \frac{697 \times 0.988 \times 0.182}{0.984} \times 0.00005$$

$$= \pm 0.006 \text{ foot}$$

The theoretical uncertainty in length AB is

$$\sqrt{(0.05)^2 + (0.005)^2 + (0.006)^2} = \pm 0.05 \text{ foot}$$

It should be noted that there is a choice in "distance angles"; either C or A could be used. If A had been used, the error due to an uncertainty in that angle alone would have been 4.5 times greater than the error of ± 0.005 due to C. A *distance angle* is any angle that is used to compute a distance; therefore both the angle opposite the base and the one opposite the unknown distance are distance angles. As the sine changes at a faster rate for small angles, it follows that better solutions are obtained when the larger angles (nearer 90°) are used.

10-27 Uncertainties in Areas

When areas are calculated, there is often the temptation to extend the numbers beyond a reasonable significance. As area is computed by the product of two numbers or a sum of several such products, it will have uncertainties accordingly. The area in Fig 10–11 is determined by measuring x and y and applying the expression $A = xy$. A, in this case, would be 600 feet^2.

If x (30 feet) were uncertain by ± 0.5 feet and y (20 feet) had an uncertainty of ± 0.2 foot, the theoretical uncertainty could be computed as follows:

$$\sigma A = \pm \sqrt{(x\sigma y)^2 + (y\sigma x)^2}$$

$$= \pm \sqrt{(30 \times 0.2)^2 + (20 \times 0.5)^2}$$

$$= \pm 12 \text{ feet}^2$$

296 / MEASUREMENTS, ERRORS, AND COMPUTATIONS

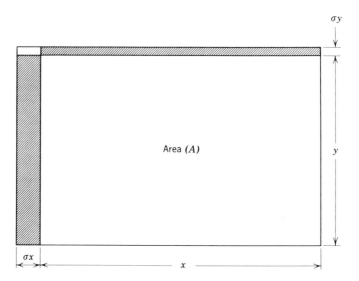

Figure 10–11 Error in area.

A little more convenient and generally satisfactory means of predicting the uncertainty in the area quantity is provided when the significant figures are respected (see Section 10-16).

Many surveyors want to determine the certainty of the area without the necessity of computations. As a rule of thumb a surveyor can determine the certainty of the area by determining the error of closure of the closed figure and then divide by 2. That is the relative area accuracy. As an example the area of a parcel is computed as 7.99 acres, and the error of closure is 1 in 12,000. The 1 in 12,000, divided by 2 gives 1 in 6000. The expected accuracy of area is then 1 in 6000 or 0.001 acre in 7.99 acres as a close approximation.

10-28 Purpose of Survey Specifications

Many organizations such as the Massachusetts Land Court, various title insurance associations, and the ACSM have drafted standards and specifications governing property surveys. Each of these has served well to bring about some uniformity in the property surveys and to improve the quality of the work. With the advances in technology and the ever-increasing value of real property, it may be necessary to reexamine the question of standards and present criteria that are workable, realistic, and will assure a minimum of title and boundary-location difficulties.

The purpose of survey standards is to specify the quality of work to be performed. Only the accepted theories on probability and errors can be of help in estimating the uncertainties in expressed positions.

Table 10-3 Table of Traverse Closures

	First-Order	Second-Order	Third-Order
Number of azimuth courses between azimuth checks not to exceed	15	25	50
Astronomical azimuth Probable error of result	0.″5	2.″0	5.″0
Azimuth closure at azimuth checkpoints not to exceed[a]	2 second $\sqrt{N}$ or 1.0 second per station[b]	10 second $\sqrt{N}$ or 3.0 second per station	30 second $\sqrt{N}$ or 8.0 second per station
Distance measurements accurate within	1 in 35,000	1 in 15,000	1 in 7500
After azimuth adjustment, closing error in position not to exceed[a]	0.66 foot $\sqrt{M}$ or 1 in 25,000[c]	1.67 feet $\sqrt{M}$ or 1 in 10,000	3.34 feet $\sqrt{M}$ or 1 in 5000

[a] The expressions for closing errors in traverse surveys are given in two forms. The expression containing the square root is designed for longer lines where higher proportional accuracy is required. The formula giving the smaller permissible closure should be used.

[b] N is the number of stations for carrying azimuth.

[c] M is the distance in miles.

The specifications suggested in the following pages are a guide for performance standards. Requiring field crews to follow rigid specifications is better assurance than is errors and omission insurance.

10-29 Adaptability of Existing Standards

Standards of accuracy have been developed by and for the governmental agencies engaged in surveying activity. The table of standards for traverses prior to 1975 is shown Table 10-3. Many states, counties, and local organizations adopted these standards and also the required performance of surveys to conform. Nongovernmental organizations have written contracts using these specifications or a modified version thereof. The new specifications are in metric units and have additional classes of survey.[3]

It must be borne in mind that these standards were devised to suit the needs of the federal agencies that were engaged in control surveys usually involving large areas. The closure discrepancy (not really an error) is but an indication of the quality of the observations and does not necessarily reflect the amount of the uncertainty in the quantities. Although such specifications when used for

[3] See *Specifications to Support Classifications, Standards of Accuracy*, and *General Specifications of Geodetic Control Surveys*, U.S. Department of Commerce, Rockville, Md., July 1975.

298 / MEASUREMENTS, ERRORS, AND COMPUTATIONS

property surveys are better than none at all, they are not suited for that purpose. As an example, consider a city lot that is 50 × 100 feet in size and 300 feet from the nearest acceptable monument (see Fig. 10-12). Suppose the property surveyor is required to conform with second-order traverse standards as defined in Table 10–3. The "closing error" in position must not exceed $1.67\sqrt{M}$ or 1 in 10,000. The traverse that encloses the perimeter of this sample city lot would be 50 + 100 + 50 + 100 or 300 feet. The traverse needed to connect the lot with the known monument would be 2 × 300 or 600 feet long. Considering the perimeter traverse first, a closure of 0.22 foot or 0.03 foot is called for; and, since it is required to take the smaller of the two, the 0.03 would be controlling. If distance and direction contribute equally to this discrepancy, the share for distance would be $0.03/\sqrt{2}$ or ±0.02 foot. This would limit the error to no more than ±0.01 per side or 1 in 10,000 for the long sides and 1 in 5000 for the shorter ones. If 1 in 8000 is used for each side, the ±0.02 can be satisfied just as well.

The azimuth closure is required to be 0°00'20" or 0°00'12"; and, since it is required to take the smaller, 0°00'12" will govern. Three seconds per station will introduce an error of ±0.0015 foot on each of the long sides and ±0.0008 foot for each of the short sides. The resulting effect of this restriction is about ±0.0024 foot or only about one-tenth of the amount that will still be small enough to satisfy the traverse closure of 0.0103 foot. If the direction error is to carry the same share as distance (±0.02 foot), an uncertainty of about 0°00'25" per angle will be adequate to realize the required closure. Permissible angle and distance errors are not in balance.

The traverse needed to locate the lot within the block is somewhat longer but would require only two sides. The limiting closure, according to these standards, would be somewhat relaxed because of the much longer peripheral distance.

From the foregoing, it is obvious that to comply with the closure specifications, it is not necessary to comply also with the distance and direction requirements. In fact, it is practically impossible to determine a distance of 50 feet on the ground within 1 to 15,000. It is equally impossible to obtain directions to ±3" for such short sights, even if this were necessary. Certainly, the location of the lot in the block is more critical for title purposes than is the relative size of the lot itself. The use of the government standards reverses this order.

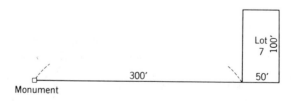

Figure 10–12

10-30 Uncertainty Expression

A statement that Mr. Jones' property corners were located within 1/10,000 is meaningless. If the surveyor had commenced at a point located 2 miles away, the error could be 2 feet; but, if the nearest original monument position was only 10 feet away, the error could only be 0.001 foot. A person erecting a building, a fence, or any improvement would only be bewildered by an accuracy ratio; but, if he were told the error could be plus or minus 0.5 inch, he would have concrete information.

A statement that a building encroaches 2 inches is positive in terms of recognized units; it is never expressed as an accuracy ratio.

Every client should be entitled to a quantitative statement of uncertainty of position. Survey standards of accuracy should likewise be expressed relative to positive units rather than as an accuracy ratio.

Error ratios are helpful to evaluate errors due to a particular element of measurements. An error of $15°$ in temperature contributes a nearly 0.01 foot error for 100 feet measured, or, as more conveniently expressed 1/10,000. This is a proper usage of error ratio; the error is in direct proportion to the distance.

When making a given set of surveying measurements, the sum total of all errors can probably be said to be some function of the distance traversed, but that function is not in direct proportion to distance as would be indicated by standards based on error ratio. Survey standards based solely on a given permissible error ratio are arbitrary.

10-31 Theoretical Uncertainty

The theoretical uncertainty (tu) is a value derived from the theory of probability and the propagation of accidental errors. If all features are taken into proper account, it will provide an indication of the quality of the position. The statement "tu = 0.10 foot" indicates that the actual quantity, as based on probability, has a 70 percent chance of being a value of anything from 0.10 foot less than to 0.10 foot more than the stated number. In a statement 327.62' ± 0.10, the 327.62 is the best quantity from the given evidence and observation. It is impossible to ascertain the true value, but it is within a tenth either way.

10-32 Specifications for the Location of Property Boundaries

Class	tu
A	±0.10
B	±0.25
C	±0.50
D	±1.00

The uncertainty of boundary location is the theoretical uncertainty determined by proper analysis. The estimate of this quantity can be determined by two different methods. If a number of redundant observations are taken (a closed-figure traverse usually has two redundants), the standard deviation may be computed by applying statistical theory. If no redundants or only a few are provided, the uncertainty can be determined by compounding the various contributing uncertainties according to the best theories of probability. The results should be about the same in the absence of blunders. Since it is impractical to take a large number of observations, or to measure the traverse many times, analysis of the contributing uncertainties is generally the best approach for surveying locations. Such a theoretical uncertainty has a likelihood of being the 70 percent error. To allow for the 100 percent uncertainty, an infinite value would result, but twice the standard deviation will be approximately 95 percent error. Figure 10-13 shows the probability curve and where these theoretical values lie.

10-33 Value of Property

There is much merit to the viewpoint that the value, cost, or potential of a property should not be a consideration of the care with which the survey is made. Who can predict the value of a tract of land a few years or a hundred years from now? Also, at times, expensive structures are erected on relatively inexpensive land. The fact that present value is low should not preclude a more precise survey.

10-34 Size of Properties for Each Class

Assuming that each property will be located with equal diligence and care, irrespective of the value of the land, the listed classes of surveys will fall into the following sizes:

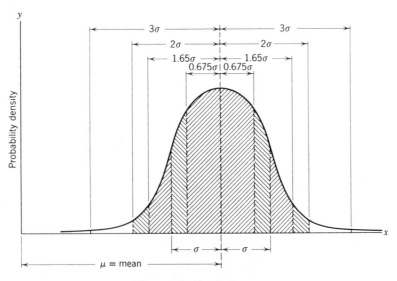

Figure 10-13 Bell curve.

Class A   Small areas wherein dense monument controls exist, as in a downtown commercial area, where lots are 50 × 100 feet.
Class B   Longest side is under 250 feet.
Class C   Longest side is from 250 feet to 2500 feet, and no side is under 100 feet, unless the periphery exceeds 500 feet.
Class D   All sides are 1000 feet or larger; lots have a periphery of 5000 feet or more.

A farm survey of 40 acres would probably be included in Class D. It would also be possible for a small rural farm tract of 1 acre to fall in Class B and a large city property to be under Class C or D.

10-35 Required Procedures

The foregoing specifications, based on uncertainties, are sufficient to define the accuracies desired. The surveyor will need to decide on the necessary procedure to attain such accuracy, and it will vary somewhat, depending on the type of equipment used. Techniques used when measuring with an electronic measuring device certainly differ from those used when measuring with a tape.

Assuming that equal care will be used for all sizes of surveys and assuming taping and angular measurements will be made with usual instruments, the following procedures will ordinarily come within the limiting uncertainty for the sizes listed in Section 10-34.

Taping

1. The length of the tape under the prevailing conditions must be known according to the U.S. Standard of length, and appropriate correction must be made. Length must be known to ±0.005 foot.
2. All lengths must be reduced to horizontal. The correction method must be such that the error due to uncertainty in observing the slope, angle, or difference of elevation will contribute not more than 65 percent of the limiting tu.
3. The temperature must be observed and proper correction made. The field temperature must be correct to the nearest 5°F for steel tapes and can be neglected for invar tapes.
4. The pull exerted on the tape must be gauged by some device such as a spring balance. Standard pull for fully supported tapes is 10 pounds. The pull for end-supported tapes must be sufficient to allow the tape to clear the ground. This pull for 100-foot tapes must be observed to the nearest 5 pounds when fully supported and to the nearest 3 pounds when the tape is end suspended. The proper corrections for pull and sag must be made in each case.
5. The weight of the tape between the end supports must be known to the nearest 0.1 pound.
6. The tape must be aligned so that it is never off line by a distance exceeding the value $\sqrt{nL/50}$, where L is the length of tape and n is the

number of tape lengths employed. On long lines, the tape should never be more than 5 feet from the true line.
7. The ends of the tape and the fractions should be set in the ground with steel taping arrows or other devices that will adequately mark the distance within ±0.005 foot.

Directions
1. Directions should be observed with sufficient precision that an uncertainty in the last place will contribute no more than 65 percent of the limiting tu. Consider, for example, the following: if the line is 1000 feet long and the required tu is ±0.25 foot, the limiting angular uncertainty in radians is

$$\frac{0.65 \times 0.25}{1000} = 0.00016, \text{ or about 33 seconds}$$

The angles in this case would need to be expressed to the nearest half minute.
2. Targets or range poles must be centered on the stations within 0.01 percent of the distance being sighted.
3. The instruments must be centered within 0.005 foot on the station occupied.
4. All angles must be checked by repeating or doubling.
5. The theoretical error due to the accumulation of direction uncertainties must not exceed 65 percent of the limiting tu.
6. The transit or theodolite must be leveled to assure horizontal angles and minimize errors in inclined sights.

Other Distance Devices When stadia, subtense, electronic, and trigonometric devices and methods are used for distance, the procedures must be such that the resulting uncertainty in the position does not exceed 65 percent of the limiting tu.

10-36 Closures

The table of tu in the specifications gives theoretical values (Section 10-32) and will be in effect only if the proper procedures have been followed and if there have been no blunders. This is affected by providing at least one redundant observation or check to compare the closure with the theoretical. Realizing that the closure has as good a chance of being zero as any other number and that it can fall anywhere in the range, plus or minus, its limiting factor is a function of the tu and the distance to the closure point. In most cases, the closure tolerance can be more than that of the tu. The following expression will serve as a guide: $LC = \text{tu}\sqrt{D_c/D_p}$, where LC is the linear closure, D_c is the total distance to the closure point, and D_p is the distance to the point.

10-37 Adjustment

All positions must be adjusted to balance out the closure discrepancy and provide a more probable value for the position of the points. The adjustment method must be chosen with reason and logic to suit the surveying conditions. Generally, a "least-squares" adjustment will be satisfactory, although it is not justified in many cases. If the uncertainty has been caused by distance and direction in equal proportions and is a function of the distance, the "compass rule" will be satisfactory. In any case, *the adjustment must not shift the position of the point more than the limiting tu.*

10-38 Monuments

The monuments must be chosen for their durability and identification. Wood should not be used in those areas where soil conditions will cause early decay. If iron pipe or pins are used, they should be long enough to hold their position against frost, heavy enough to last against rust for a reasonable length of time, and should be identifiable at a future date. Concrete monuments and stones are most satisfactory. All monuments should be set low enough so that they will not be disturbed by wheel traffic. Referencing is essential for all monuments.

10-39 Computations

The method of computations must not be such as will detract from the value of the fieldwork. All calculations should be independently checked, and sufficient significant figures must be used so that the computations will be adequate for the quantities of the survey.

REFERENCES

Adams, O.S., and Claire, C.M. *Manual of Plane-Coordinate Computations*, Special Publication No. 193, U.S. Coast and Geodetic Survey, U.S. Government Printing Office, Washington, D.C., 1948.

Bauer, S.A., "Factors in Surveying Accuracies," *Surveying and Mapping,* Vol. II, No. 2, 1951, pp. 145–148.

———, "The Use of Geodetic Control in Surveying Practice," *Surveying and Mapping*, Vol. 9, No. 3, 1949, pp. 187–191.

———, "The Value of Control Surveys to the Property Surveyor," *Surveying and Mapping*, Vol. 12, No. 4, 1952, pp. 338–342.

Bouchard, Harry, and Moffitt, Francis, *Surveying*, International Textbook Co., Scranton, Pa., 1959.

Brinker, Russell C., and Wolf, Paul R., *Elementary Surveying*, 6th ed., Harper & Row, New York, 1977.

U.S. Coast and Geodetic Survey, *Use of Coast and Geodetic Survey Data in the Surveys of Farms and Other Properties*, Serial No. 347, U.S. Government Printing Office, Washington, D.C., 1940.

Danner, Charles S., "Some Applications of Aerial Photography in the Field of Land Surveying,"

Proceedings of the Illinois Land Surveyors Conference, pp. 130–136, Division of University Extension, Urbana, 1958.

Fullerton, C.N., "The Use of Aerial Photographs in Land Descriptions," *Canadian Surveyor,* Vol. 7, No. 9, 1942, pp. 6–11.

Garder, Irvine C., "What Should Surveying Instrument Specifications Contain?," *Surveying and Mapping,* Volume 11, No. 2, 1951, pp. 132–137.

Hempel, Charles B., "Electrotape—A Surveyors Electronic Eye," *Surveying and Mapping,* Vol. 21, No. 1, 1961, pp. 85–88.

Hoke, William B., "The Use of Plane Coordinates in Property Surveys," *Surveying and Mapping,* Vol. 6, No. 1, 1946, pp. 6–8.

King, Jasper E., "Photogrammetry in Cadastral Surveying," *Photogrammetric Engineering,* Vol. 23, No. 3, 1957, pp. 493–505.

Kissam, Phillip, *Surveying for Civil Engineers,* McGraw-Hill Book Co., New York, 1956.

Mitchell, Hugh C., and Simmons, L.G., *The State Coordinate Systems,* Special Publication No. 235, Coast and Geodetic Survey, U.S. Government Printing Office, Washington, D.C., 1945.

Monroe, C.A., "The Use of Plane Coordinates in Property Surveys," *Surveying and Mapping,* Vol. 6, 1946, pp. 8–10.

Morse, E.D., "Concerning Reliable Survey Measurements," *Surveying and Mapping,* Vol. 11, 1951, pp. 14–21.

Schermerhorn, W., "Photogrammetry for Cadastral Surveys," *Photogrammetria,* Vol. 10, No. 2, 1954.

Shartle, Stanley M., "The Application of Empirical Formulas to Property Surveys," *Surveying and Mapping,* Vol. 11, No. 3, 1951, pp. 264–270.

Slessor, D.R., "Use of Photogrammetry on a Legal Survey," *Canadian Surveyor,* Vol. 14, No. 8, 1959.

Trager, Herbert F., "Photogrammetry Applied to Cadastral Surveys," *Surveying and Mapping,* Vol. 16, No. 1, 1956, pp. 29–36.

Weissenstein, Henry G., "Simple Machine Method for Area Computation," *Surveying and Mapping,* Vol. 19, No. 1, 1954, pp. 75–79.

Welch, Harold J., "Standardized Computations for Subdivisions," *Surveying and Mapping,* Vol. 17, No. 3, 1957, pp. 287–298.

CHAPTER 11

Preservation of Evidence

11-1 Scope and Purpose

As previously stated, all conveyances must, somehow, be related to known monument positions on the face of the earth. If monuments are destroyed, certainty of location of land is endangered. The object of this chapter is to discuss means of preserving evidence called for in former conveyances and to discuss duty to preserve this evidence.

In a few selected states the law delegates the authority and duty to preserve evidence to the private practitioner. This is accomplished by requiring the recording of public documents disclosing evidence discovered, disclosing witness evidence taken under oath, and disclosing new monuments and measurements perpetuating original monument positions.

Measurements to preserve monumented positions are of two types: (1) those made prior to conveyancing and called for by the conveyance and (2) those made after completion of the conveyance. Measurements made after a conveyance is made, if they are to serve to reestablish monument positions as acceptable evidence, must be made by an expert and must be recorded so that the evidence is available in the event of monument destruction.

Two important forms of evidence, State Plane Coordinates and photographs, have not been used to a large degree.

If all surveyors always located property lines in the same position, their prestige and standing would be greatly enhanced. It is not frequent that surveyors disagree about the length of a line or the angular direction between lines; surveyor's differences usually arise from varying opinions about which monument they should commence measurements from and the dignity of other monuments. If all surveyors had equal knowlege of evidence and its application to monument position, the certainty of all surveyors locating lines in the same position would be increased. The need for public records disclosing former monument positions is great.

11-2 Vanishing Evidence

Although the early surveyors set stakes, monuments, blazed trees, and recorded the facts in field notes, precious little, if any, of their original marks are left today. New roads, cultivation, cutting timber, and improvements have all taken their toll. With the advent of the machine age, the inventions of the bulldozer, the plough, and the ditch digger have contributed substantially to the loss of permanent markers.

Time eradicates everything living—along with many physical inanimate objects. Trees eventually die, and, with their death, corners and witness monuments to corners disappear. People die and with them vanishes knowledge of old corner locations. Erosion and weathering will eventually, although sometimes very slowly, erase mounds, pits, and even stones. Iron will rust and eventually fade away. Documents are lost by fire, age, or destruction.

In early times trees were the most common witness objects; today, it is quite obvious that trees fall far short of being satisfactory. In some states witness trees have almost completely disappeared. Even in states still wooded, fire and old age have caused original trees to vanish.

Evidence of physical objects cannot be expected to last forever. By learning from past experience the surveyor today, by better selection of monument material and better methods of referencing monuments, can preserve the certainty of a monument's position.

As previously pointed out, the spot occupied by the original monument at the time it was set is the controlling consideration. Hence, if its former position is known, or can be determined with a degree of certainty, the monument is merely obliterated, not lost, even though the original monument may be destroyed. After an original surveyor sets a monument the problem is one of perpetuating its position, even though the physical evidence of the monument itself may be destroyed. This chapter is devoted to means of perpetuating monument positions.

11-3 Perpetuation of Evidence

It is the duty of the original surveyor to set monuments indicating his footsteps; it is not this surveyor's duty to see that those monuments are preserved forever, but his efforts should be directed in that direction. Although the original surveyor can do much toward establishing the permanency of monumentation by selecting better monument materials, setting them in positions unlikely to be disturbed, and selecting adequate reference monuments, he obviously cannot prevent destructive forces. Once an owner enters on land, he has greater control over preservation of his monuments than does the surveyor.

The surveyor is often given by law the exclusive right to prepare and survey subdivision plats and to resurvey lands. In exchange for these exclusive rights he has obligations to the public, and one of those obligations ought to be devising some means of perpetuation of monument positions.

The United States devised one of the finest survey systems in the world. And with this system numerous monuments have been set. Perpetuation of the system was not provided for; it was delegated to the states as their responsibility. Some states enacted laws that assisted perpetuation; most did nothing. It is the disappearance of monument evidence and the lack of remonumentation that causes survey difficulty today.

11-4 Authority to Perpetuate

The authority to perpetuate monument positions and the responsibility of doing so are not the same. Many states delegate by law the privilege of allowing licensed surveyors to remonument and replace original corners, but they do not often delegate to the surveyors the responsibility of remonumentation. The usual provision of the law is that surveyors *may*, among other things,

> locate, relocate, establish, reestablish, or retrace any property line or boundary of any parcel of land or of any road, right-of-way, easement or alignment; perform any survey for the subdivision or resubdivision of any tract of land; set, reset or replace any monuments or reference points.
> —California Land Surveyors Act.

If a surveyor does remonumentation and if he keeps his acts of remonumentation as well as the evidence discovered his personal secret, with his death all is lost. Normally, what a surveyor did or found is not allowed as evidence unless accompanied by the personal testimony of the surveyor. Even though the surveyor may leave evidence in the form of writings, it is usually impossible to introduce these writings as evidence.

In the usual course of a survey, the surveyor discovers a section corner post that is deteriorated and well rotted and replaces it with a 2-inch iron pipe. Since the post is now gone, how can any other surveyor or the court prove the correctness of the pipe? Perhaps a farmer used the pipe for staking out a cow or possibly someone set a pipe where he believed the corner ought to be. The only link in the chain of evidence is the surveyor, who does not live or remember forever. Only the surveyor's properly recorded record may be everlasting. Delegating authority to remonument old original corners without the responsibility of perpetuating the chain of evidence is futile.

11-5 Responsibility of Perpetuating Evidence

If the surveyor is delegated the privilege of remonumentation of deteriorated corners, he should also be delegated the responsibility of perpetuating the evidence. In a few states the land surveyor's law has such provisions as these:

> After making a survey in conformity with the practice of land surveying, the surveyor may file with the County Recorder in the county in which the survey was made a record of such survey.
>
> Within 90 days after the establishment of points or lines the licensed land surveyor shall file with the County Recorder in the county in which the survey was made, a record of such survey relating to land boundaries or property lines, which discloses:
>
> (*a*) Material evidence, which in whole or in part does not appear on any map or record previously recorded or filed in the office of the County Recorder, County Clerk, Municipal or County Surveying Department or in the records of the Bureau of Land Management of the United States.

(b) A material discrepancy with such record.

(c) Evidence that, by reasonable analysis, might result in alternate positions of lines or points.

—California Land Surveyors' Act.

This law places the responsibility of perpetuating discovered evidence on the private practitioner. If evidence of monument positions is preserved by public records and if new monuments are set with a continuous chain of evidence from the time of the original monuments, the problems of future location of land are greatly diminished.

The filing of a public record of survey is insufficient unless there is a mandatory provision to require the disclosure of evidence found and monuments set. In conjunction with the foregoing law in California it is mandatory to show the following:

1. All monuments found, set, reset, replaced, or removed. A description of the monuments' kind, size, location, and other data related thereto must be provided.
2. Bearing or witness monuments, basis of bearing, bearing and length of lines, and scale of map.
3. Name and legal designation of tract or grant in which the survey is located and ties to adjoining tracts.
4. Memorandum of oaths.

Within the last 30 years surveyor organizations have been active in obtaining laws requiring land surveyors to preserve evidence in public offices. One such law, enacted in 1963 in Colorado, illustrates the modern trend:

38-53-101. Legislative declaration. It is hereby declared to be a public policy of this state to encourage the establishment and preservation of accurate land boundaries, including durable monuments and complete public records, and to minimize the occurrence of land boundary disputes and discrepancies.

38-53-102. Definitions. As used in this article, unless the context otherwise requires:

(1) "Accessory" means any physical evidence in the vicinity of a survey monument, the relative location of which is of public record and which is used to help perpetuate the location of the monument. Accessories shall be construed to include the accessories recorded in the original survey notes and additional reference points and dimensions furnished by subsequent land surveyors or attested to in writing by persons having personal knowledge of the original location of the monument.

(2) "Aliquot corner" means any corner in the public land survey system created by subdividing a section of land according to the rules of procedure set forth in the "Manual of Instructions for the Survey of the Public Lands of the United States."

(3) "Bench mark" means any relatively immovable point on the earth whose elevation above or below an adopted datum is known.

(4) "Board" means the state board of registration for professional engineers and land surveyors.

(5) "Control corner" means any land survey monument whose position controls

the location of the boundaries of a tract or parcel of land. The control corner may or may not be included within the perimeter of said tract or parcel.

(6) "Monument record" means a written and illustrated document describing the physical appearance of a survey monument and its accessories or of a bench mark.

(7) "Public land survey monument" means any land boundary monument established on the ground by a cadastral survey of the United States government and any mineral survey monument established by a United States mineral surveyor and made a part of the United States public land records.

38-53-103. Filing of a monument record required. Whenever a land surveyor conducts a survey which uses as a control corner any public land survey monument or any United States geological survey or United States coast and geodetic survey monument, he shall file with the board a monument record describing such monument, but said monument record need not be filed if the monument and its accessories are substantially as described in an existing monument record previously filed pursuant to this section. A surveyor shall also file a monument record whenever he establishes, reestablishes, restores, or rehabilitates any public land survey monument. Each monument record shall describe at least two accessories or reference points. Monument records must be filed within six months of the date on which the monument was used as control or was established, reestablished, restored, or rehabilitated.

38-53-104. Surveyor must rehabilitate monuments. Whenever a monument record of a public land survey corner is required to be filed under the provisions of this article, the surveyor shall, if field conditions require it, restore or rehabilitate the corner monument so as to leave it readily identifiable and reasonably durable.

38-53-105. Forms to be prescribed by board. The board shall establish and revise, whenever necessary, the forms to be used for monument records and shall prescribe the information to be presented on said forms. These forms and necessary instructions shall be furnished to all Colorado registered land surveyors without charge.

38-53-106. Monument records to be certified. No monument record shall be accepted for filing unless it is signed and sealed by the land surveyor who was in responsible charge of the work.

38-53-107. Filing permitted on any survey monument. A land surveyor may file with the board a monument record describing any land survey monument, accessory, or bench mark.

38-51-104. Violations. Any person, including the responsible official of any agency of state, county, or local government, who willfully and knowingly violates any of the provisions of this article is guilty of a misdemeanor and, upon conviction thereof, shall be punished by a fine of not less than twenty-five dollars nor more than two hundred fifty dollars. It is the responsibility of all district attorneys of this state in all cases of suspected willful and knowing violation of any of the provisions of this article to prosecute the person committing such violation. The state board of registration for professional engineers and land surveyors may revoke the registration of any land surveyor convicted under the provisions of this article. Such a person is entitled to a hearing on the revocation, pursuant to article 4 of title 24, C.R.S. 1973.

In addition to these monumentation requirements, the Colorado law also gives minimum standards for filed plats as follows:

> **38-51-102. Land survey plats.** (1) All land survey plats or maps recorded in any county of this state on and after July 1, 1967, shall include, in addition to any information required by local authorities, at least the following items:
> (a) A scale drawing of the boundaries of the land parcel;
> (b) Recorded and apparent rights-of-way and easements;
> (c) All dimensions necessary to establish the boundaries in the field;
> (d) A statement by the land surveyor that the survey was performed by him or under his direct responsibility, supervision, and checking;
> (e) A statement by the land surveyor explaining how bearings, if used, were determined;
> (f) A description of all monuments, both found and set, which mark the boundaries of the property, and a description of all control monuments used in conducting the survey;
> (g) A statement of the scale or representative fraction of the drawing, and a bar-type or graphical scale;
> (h) North arrow;
> (i) Title description or reference thereto; and
> (j) Signature and seal of the land surveyor.

On many occasions in the conduct of surveys, surveyors recover evidence of original bearing trees and monuments. At times surveyors are required to cut into trees to verify markings, yet to do this they may have to enter onto private land not directly related to the immediate project at hand.

Initially, all monuments and their accessories belonged to the government, and surveyors could do whatever was necessary. Today, there seems to be some confusion about responsibility and liability. The courts have repeatedly held that the surveyor is bound to recover all of the best-available evidence, yet the land no longer is in the public domain, and trespass charges could be likely.

In recent years a surveyor was sued for the defacing of an original GLO bearing tree. The court held that the surveyor was not liable because of the necessity to prove the authenticity of the corner, yet the question was not addressed about what if he had cut into the tree and it proved not to be the tree he was looking for?

In visiting surveyors' offices one may witness examples of bearing trees, monuments, and other "spoils" exhibited in display cases, shelves, and as doorstops. Although the question has never been addressed, there is some question about the authority to remove these items and make them personal property. If this practice of altering and removing evidence is conducted and no public recording is made of the found evidence, the authenticity of the original corner could be seriously jeopardized.

In those states where it is not mandatory by law for the surveyor to file and record public records of found and altered evidence of monument positions, it should be the duty of the state surveyor organization to seek legislation to that effect.

11-6 Oaths and Witness Evidence

In the event that an original corner monument is obliterated and there are reliable witnesses to testify about the former location of the monument, the surveyor should be authorized to swear in witnesses and record their testimony under oath. Witnesses do not live forever; it is highly desirable to perpetuate the evidence of witnesses while they are available. In California, Section 8760 of the Land Surveyors' Act states:

> Every licensed land surveyor may administer and certify oaths;
> (a) When it becomes necessary to take testimony for the identification or establishment of old, lost or obliterated corners.
> (b) When a corner or monument is found in a perishable condition, and it appears desirable that evidence concerning it be perpetuated.
> (c) When the importance of the survey makes it desirable to administer an oath to his assistants for the faithful performance of their duty.
> A record of oaths shall be preserved as part of the field notes of the survey and a memorandum of them shall be made on the record of survey filed under this article.

Although affidavits relative to land and surveys in and of themselves cannot be used as evidence, the law of Georgia provides that an affidavit concerning real property can be submitted as evidence if it is filed on record pursuant to the requirements of the law.

11-7 Identifying Marks on Monuments

A set monument is worthless if it is unidentifiable in the future. The property owner may at any time set markers along his claimed line, and he can use iron pipes or any other material similar to surveyor's monuments. A found monument without a background history of who set it and how it got there is of little value as evidence.

Without doubt the most certain method of identification of monuments set by surveyors is a legal requirement that their license number be permanently attached or marked on each monument. Not only does the marking lend authority to the monument, but also it serves as a means of checking its past history.

11-8 Recording Documents

The most certain and best means of perpetuating evidence is to record copies of original documents in a public place. This has three advantages. (1) The document itself is not apt to be destroyed; two copies exist, one in the possession of the owner and the other in the public records. (2) The general public is charged with knowledge of the document's contents. (3) The recorded document is admissible as evidence in court actions.

Deeds, easements, quitclaims, and similar documents related to land are those most commonly recorded. Maps and plats are the second most important recordings. All land must be located with reference to monuments, and one of the most important documents disclosing evidence of monuments and monument positions is the recorded plat and description.

Plats admissible as evidence include all publicly stored maps that a public official is charged with preparing. Road surveys, maps prepared for taxing purposes, and maps showing monuments or property-line locations that are prepared by the county surveyor *as a part of his official duties* are similar to recorded plats.

The law often provides for the recording of private surveys. Recording has many advantages. Evidence disclosed in the way of monuments is preserved. If a client adversely occupies lands not his own, a recorded document showing adverse possession is public notice. Recorded surveys tend to decrease arguments between surveyors; both act on the same evidence. After a surveyor's death, publicly recorded documents are admissible evidence.

The surveyor can perpetuate evidence by including descriptions of all monuments found within the writings describing newly created parcels of land such as this: "Beginning at the southwest corner of Section 10, said corner being marked by a stone 10 inches in diameter and 18 inches long; thence. . . . " But the opportunity for including such evidence is limited, since not all resurveys are accompanied by a new parcel description.

11-9 Private Survey Records

Private survey records, after the death of the surveyor, are rarely of value as evidence. Generally speaking, unless properly identified by witnesses to the act of surveying, unrecorded notes are not admissible as evidence. If the surveyor is alive and if he testifies about the correctness of the notes or plats, the facts are evidence.

Private notes are not entirely useless after the death of the surveyor; the notes enable another surveyor to duplicate measurements and to discover places to seek evidence. If the former private survey can be restored by a present surveyor, the new surveyor can testify about what he did.

11-10 The Use of Aerial and Terrestrial Photographs to Preserve Evidence

Perhaps the photograph (Fig. 11-1) itself is of more value to the property surveyor than measurements made from it. Following are some of the more important uses to which aerial (and terrestrial) photographs may be put to aid, implement, and enhance land surveys.

Identification. Many organizations photo-identify all points as soon as found. The U.S. Forest Service, BLM, the Geological Survey, and other such agencies will, by photo identification or other positive identification means, tie the location of a found corner or other boundary evidence to other monuments on the ground. Reference ties to three or more points that are easily identified on the photograph will in effect reference the corner to *all* the images on the photograph. This identification forms a permanent record for location (on the ground) even long after all the references have disappeared.

Use of Old Photographs. Old photographs show evidence of ancient lines and conditions at the exact time the photograph was exposed. After a road has been obliterated, or a fence removed, the traces of these lines may still be seen on the photograph, even though no evidence appears on the ground. Comparing old photographs with new ones will indicate some of the changes that have taken place. Thus in a 1938 aerial photograph a farm fence that was accepted as a property line between two tracts was discernible. A recent picture of the same area after a new subdivision was improved shows a road purporting to be along the old property line is clearly out of position.

Unfortunately, most of the older aerial photographs of the United States only go back to about the mid-1930s. It would have been a great advantage to have had photographic coverage of this country dating back to the time of the original subdivisions, especially for riparian lines.

Old photographs may be of more value to the property surveyor than are old maps. Ancient fence lines, hedgerows, field tiles, old buildings, and ruins of buildings all appear in their true photographic positions at the time the film was exposed. There is no danger of a surveying or drafting blunder in the picture.

Riparian Evidence. By comparing old and new photographs, the action of the water (accretion, reliction, avulsion, etc.) can be determined with some certainty. Photographs will indicate shallow areas and shoals as well as relative beach and shorelines. The location of the shoreline at the time of the original survey is essential information for determining accretion rights.[1]

Evidence Undetectible on the Ground. Pipelines and field drains may be valuable title evidence, but they can become completely undetectible on the ground. Usually, these effects, even when abandoned for many years, will be evident on aerial photographs. In one case a description called for a certain railroad right-of-way, long since abandoned and not visible on the ground because of farm cultivation over the former location. Photographs clearly showed a slight discoloration along a county road and through a field with about the same geometric shape as the old railroad. Careful measurements from the pictures enabled the surveyor to retrace the old alignment until an old stone culvert was recovered. From there on, more physical evidence was uncovered until the old right-of-way was positively located.

Infrared photography will reveal even the most subtle change in the character of the land. Law-enforcement officers were perplexed by the operation of an illegal still in the remote pine area of one of the East Coast states. Infrared photographs were taken from an airplane. Because such film is sensitive to heat, a strange, light spot among the forest trees disclosed the position of the still, well hidden under the pines.

[1] See Curtis M. Brown, *Boundary Control and Legal Principles*, 2nd Ed., John Wiley & Sons, New York, 1969, Section 10.18.

Figure 11-1 Photograph showing edge of subdivision.

Detection of Encroachments. A building wall, or corner, may appear to be over the property line on a photograph. The extent of overhang or the obstruction of natural drainage is often clearly illustrated on the pictures. Public usage can be determined from examining aerial photos, and such evidence has been admitted in court.

Identification of Lost Tracts. Sometimes tracts described by metes and bounds have insufficient title identity. If these parcels are platted to the same scale as an aerial photograph, and if this shape is tried like a piece of a jigsaw puzzle until a similar pattern on the aerial photograph is discovered, often title identity can be determined. In a township tax map in an East Coast state, many of the record descriptions were identified by this method. Overlaps, gores, and gaps were revealed.

Locations of Monuments. The search for ancient cornerstones, landmarks, and section corners can be greatly aided by a thorough study of the aerial photograph. Faint field lines can be projected, and their intersection will localize the area to search.

11-11 Daily Use of Photographs

As a daily working tool, each office should maintain a permanent file of the most recent aerial photographs in the area of practice, and a print of an aerial photograph, when available, should be filed with each survey. All corners, either located, recovered, or restored, when photo-identified become permanent evidence, and surveyors should avail themselves of this means of preserving evidence. Geodetic control points, if available, should also be photo-identified whether used in the survey or not. In the future, these photographs can be rectified into more recent photographs from which precise measurements can be made. The photograph on file will serve as a permanent record of the land at the time of the survey.

By all means, present-day photographs and old pictures should be preserved for cadastral purposes. These may do more to stave off costly boundary litigation than anything else. The process of making precise measurements from photographs, although in the infant stage, is improving at a tremendous rate. Such measurements have considerably more freedom from blunders than do usual ground surveys, and the error is of more uniform character, not a function of the length of the line as in ground surveys.

11-12. Preservation of Evidence by State Plane Coordinates

Original monuments, if lost, can be restored to their former position, provided (1) some acceptable witness remembered its former position and (2) measurements were known from other monuments.

NGS has numerous monuments set and interrelated to one another by

11-12 PRESERVATION OF EVIDENCE BY STATE PLANE COORDINATES / 317

accurate measurements. The surveyor, by tying his discovered original monuments in this network, can greatly insure the certainty of perpetuation of a monument's position.

Resurvey work and the location of properties as based on the record are interpretive in nature; once the lines called for are located, tying the monuments into a plane-coordinate network is an extra burden of cost that the property owner will object to. And where it is not a legal necessity, it is seldom done. The density of control monuments, with known coordinate positions, is a factor in voluntary usage of plane-coordinate nets. If control monuments are 10 miles apart and a surveyor has to run 3 miles in one direction and 7 in another to obtain tie-out information, the likelihood of it being done is small. But if control monuments are found every half mile, the problem of usage is reduced.

Assigning coordinate values to discovered monument evidence does not in itself mean certainty of future location; the coordinates must be correctly determined by tying into monuments not too far away. A surveyor measuring 3 miles to tie into a monument has an uncertainty of measurement of 1½ feet more or less. In reestablishing the position after it is lost, remeasuring the 3 miles will introduce uncertainty of another 1½ feet. Thus nearby monuments might be far more certain than coordinates. Coordinates have no greater value than other known measurements; it is a question of what is the best-available evidence. And coordinates carefully determined will probably be the most certain evidence.

Establishing a net of state plane-coordinate positions is a cost impossible to bear by the private surveyors; after the net is established it is not unreasonable to expect that the surveyors should use it. In some areas, such as Los Angeles City, the state plane-coordinate net is sufficiently complete that all surveys ought to be related to the system. But in other states the enabling legislation has yet to be passed.

Surveyors who utilize state plane coordinates must understand basic control surveying. Each basic control monument is subject to some error of measurement and adjustment. The land surveyor must identify the monuments from which his coordinates were calculated so that a subsequent surveyor who wants to "follow his footsteps" may duplicate not only the work but the inherent errors as well.

Without question, if a state plane-coordinate system were accurately established with high density of monument locations, and if all surveys (including resurveys) were tied into it, an ideal system to perpetuate evidence would exist. But until such time as the public appropriates sufficient funds to provide for the density of monuments needed in the net, the property surveyor will continue with local monument control.

In the matter of new land divisions the public can regulate how new land divisions shall be surveyed, and it is within the public's power to require state plane-coordinate data.

CHAPTER 12

Guarantees of Title and Location

12-1 Scope

Ownership of land consists of (1) a good written title with the right of possession or (2) possession with the right to acquire written title as a result of a lawful, unwritten conveyance. Because most laymen are incapable of determining when written title defects exist or when others have lawful title rights arising from possession, a need has arisen for experts willing to guarantee titles to be free of defects.

The professional surveyor certifies to land location and status of encroachments; he does not issue a policy of location guarantee, but he is liable for failure to exercise due care. Title companies, attorneys, courts, or the state (Torrens titles), depending on the jurisdiction, guarantee written title adequacy; and, sometimes on request, they will guarantee location. How titles are guaranteed and who guarantees written titles are the subjects discussed in this chapter. Such discussion cannot be complete without mention of possession guarantee, since lawful occupancy can defeat paper title.

Land is more than the soil one sees on examination of a parcel. Land is rights, and the rights extend from a point at the center of the earth outward, piercing the land surface and extending into the heavens. Land is not one facet; it includes many different rights. Such rights include, but are not necessarily limited to, rights of possession, ownership, timber rights, mineral rights, water rights, and aerial rights. All rights can be subsurface, surface, or aerial.

In its true perspective, ownership of land is never complete; someone else always has a right against the land (taxing authority, zoning, etc.). Ownership is more properly a right of possession and title.

A deed to land is never proof of ownership; it is evidence of ownership. Others may have rights against the land because of taxes, liens, mortgages, easements, judgments, bonds, improvement assessments, incompetency to alienate, restrictions, zoning, and many other items. Because of the numerous hazards, new landowners demand ownership guarantee or at least a statement of ownership condition, along with a guarantee that no rights exist other than those stipulated.

Exposures to written title defects are not visible and cannot be seen; exposure to possession defects can be seen by the purchaser. The demand for title guarantee has been greater than the demand for possession guarantee. In recent years, because of instances of damages resulting from possession matters, an increased demand for guarantee of both possession and title matter exists.

The basic concept of land in the United States is one of the commingling of two ancient doctrines or concepts of law—one finding its origin under the English Common Law in which title to land is paramount and the other having its origin in Roman Common Law in which possession of land is superior.

12-2 Registration of Titles versus Registration of Ownership

Recording of a deed in a public place is registration of *evidence* of landownership. Torrens registration of land includes both registration of *title* and registration of *ownership*. When an automobile is registered, the name of the owner is registered; when an automobile is sold, the state requires that the new owner must be registered within a definite period of time. The Torrens registration system is similar; new transfers of land must be accompanied by a change in registration of ownership. The difference between the two systems is that one *registers title*, whereas the other registers *ownership*.

12-3 Aids to Title and Location Guarantee

Unwritten title rights are aids to cure location defects. The statute of limitations is of material assistance in maintaining the status quo and preventing spite litigation. These aids to cure defective titles do not include the element of financial guarantee and are not a part of this chapter.

Courts, in general, do not guarantee title; courts declare by decree what are the existsing rights between litigants. Parties not involved in the action have their rights unimpaired. But in the Massachusetts Land Court the court actually guarantees ownership of both title and location; an assurance fund is set up to pay damages to those inadvertently harmed by the court's decree.

In Torrens title registration a similar situation exists, provided the registration includes both written title and location guarantee. However, in many Torrens registrations location is not a part of the registration, and location is not guaranteed.

12-4 Title and Possession Guarantees

Within the United States title guarantee, possession guarantee, and statements of title condition guarantees (abstracts) vary from state to state and can be classified here:

1. Warranty deed.
2. Abstract (guarantee of facts).
3. Abstract with attorney's opinion.
4. Written title insurance.
5. Written title and possession insurance.

6. Torrens title (including Massachusetts Land Court system).
    a. Written ownership of title guarantee.
    b. Written ownership guarantee.

12-5 Warranty Deed

The grantor of a warranty deed personally guarantees the title to be free of defects. Such a guarantee is sufficient, provided the grantor has adequate funds to pay in the event of damages. From the grantor's viewpoint he is often unwilling to issue a warranty deed and would rather pay to have others assume the risk of guaranteeing title. Death or financial ruin of the grantor leaves the grantee without guarantee; therefore the grantee is usually willing to pay for responsible assurance of title.

12-6 Patents

Sovereigns issuing patents to lands do not guarantee title. A patent is merely a quitclaim to any rights or interests possessed by the sovereign. If the land had been previously patented to others or if the patent is contrary to law, it is without force.

In any patent from a sovereign, legal opinion is needed about whether the patent was properly executed in accordance with the then-existing laws, whether the correct description of the land involved was properly described, and whether any prior patent was issued on the land described.

12-7 Chain of Title

All land titles originate from a sovereign, and the sequence of title transfers from the sovereign to the present is called a *chain of title*. Like a chain, the strength of the present title is no stronger than the weakest link. If a defective link exists, the written title may be voidable or faulty from that point on. A person guaranteeing title must, of necessity, examine all transfers back to the creation of the title.

Although all titles originated from the sovereign, it is not always necessary to search the chain back to the original patent. The sovereign, by operation of law, may declare a new start for a title through court decrees or Torrens title actions.

12-8 Abstract of Title

An abstract of title is a statement of publicly recorded facts relating to the chain of title. An abstract is not a complete statement of every detail of past title transfers; it is a summary of the essential facts necessary to pass judgment on the sufficiency of written title.

An abstract is usually completed for a period of time that varies from state to state according to the accepted title standards in effect. In Maine and New Hampshire abstract search is usually for a period of 40 years or to a warranty deed at 40 years or further back. For title insurance the search usually extends to at least 60 years. In Massachusetts the minimum search is for 60 years, and in Georgia it is 50 years. In California search usually stops at the previously issued title insurance policy.

12-9 Abstract Companies

In early years it was the custom for attorneys to search public records, extract essential information, and pass opinion on condition of title. With the increase in the volume of title transfers, the problem of searching more numerous records created a demand for those specializing in title-abstract information.

Many abstract companies have complete title records that enable them to prepare a title abstract with little outside search. Such systems are indexed by location, grantees, and grantors. The result is a more efficient operation; a few serve many and save the cost of duplicating research efforts.

12-10 Sources of Abstract Information

Public records (an abstract pertains only to public records) from which the title searcher obtains title history are scattered. More usual sources are (1) public records offices (Recorder of Deeds, Office of Recorder, Hall of Records, County Clerk, etc.); (2) courts (federal, state, and municipal); (3) county offices (surveyor, engineer, clerk, treasurer, assessor, etc.); (4) state offices (Division of Highways, Division of Water Resources, Division of Beaches and Parks, Department of Mines, etc.); (5) state and national archives, where original land patents and records are preserved; (6) quasi-public bodies (irrigation, sanitation, water, and other taxing utility districts); and (7) educational districts.

12-11 Typical Abstract of Title

The following abstract was abridged and insertions were made when it was believed the changes would not detract from its value as an example of an abstract of title.

1. CAPTION. An abstract of title in the southeast quarter of section 9, Township 22 North, Range 8 East of the Third Principal Meridian, situated in the County of Champaign, State of Illinois.
2. Beginning with Title in the United States Twp. 22 N, R 8 East of the third P. M. Plat of this title is recorded in the Book of Township Plats, p. 25 in the County Clerk's Office.... (At the beginning of the book is the following certification)...

    Office of Land Titles
    St. Louis, Missouri
    June 16, 1869

    ... The correctness of plats appearing in this book is certified...

    Albert Sigel, Recorder
    by E. H. Hesse, Deputy

3. A copy of the original township plat showing distances and areas.
4. East half of southeast quarter containing 80 acres entered March 15, 1854 by Robert Houston as appears from Book of Original Entries, County Clerk's Office, Urbana, Illinois.

5. United States
    to
Robert Houston, assignee of
    Michael Brown

PATENT
Dated January 10, 1855
Filed July 18, 1870
Book 23, p. 2
Military Warrant No. 53232

Grants east half of Southeast quarter, sec. 9, T. 22 N, R. 8 E. of the 3rd P. M. in district of lands subject to sale at Danville, Illinois, containing 80 acres, according to official plat of said land.

Signed by President Franklin Pierce by H. E. Baldwin, assistant secretary. Recorded in Vol. 268, p. 181.

6. United States
    to
Robert Houston

The west half of the southeast... was entered Mar. 15, 1854, as appears...

7. United States
    to
Robert Houston, assignee of
    Phebe Dunn

PATENT
Dated January 10, 1855
Filed July 18, 1870
Book 23, p. 4
Military Warrant #53862

(Grants the east half of the quarter similarly to the west half as indicated under item 5) Recorded in Vol. 268, p. 180.

8. Robert Houston
    to
School Commissioners for the
    Use of Inhabitants of Twp.
    22N, R. 8E.

MORTGAGE
Dated March 1, 1855
Filed March 3, 1855
Book G, Page 578 of deeds
Secures the sum of $856.00 payable in one year with interest at 10% payable semi-annually.

Conveys the south half of sec. 9, Twp. 22N, R 8E of the 3rd PM and other lands.

Acknowledged March 1, 1855 before Isaac Devore, Justice of the Peace of Champaign County, Illinois.

9. On the margin of the record of the above mortgage appears the following: Be it remembered that on this third day of March, 1856, having received... (satisfaction)... (I release)....

    Thompson Dickson, Treasurer, Twp. 22N, R. 8 E of the 3rd PM.

10. Robert Houston and
    Eliza M., his Wife
    to
Theodore L. Houston, son of
    above

WARRANTY DEED
Dated April 1, 1872
Filed May 2, 1872
Book 33, page 156
Consideration: $5,000.00

Conveys the south half of sec. 9, Twp. 22 North, Range 8 East of the third Principal Meridian.

Acknowledged April 1, 1872 before James S. Jones, Notary Public of Champaign County, Illinois.

11. Theodore L. Houston,
    having no wife
    to

MORTGAGE
Dated August 9, 1876
Filed August 10, 1876

Joshua M. Clevenger   Book 17, page 608
                               Secures loan of $3500.00 . . .
Conveys south half of sec. 9. . . .
Acknowledged August 9, 1876 before C. H. Yeomans, Notary Public, of Ford County, Illinois.

12. J. M. Clevenger   RELEASE
    to   Dated August 12, 1880
    P. C. Houston   Filed April 28, 1881
                               Book 52, p. 630
                               Consideration: $1.00

Releases all interest acquired by a mortgage dated . . . and recorded . . . to premises described therein, to-wit: south half of sec. 9 . . .
Acknowledged August 12, 1880 before George F. Beardsley, N.P.

13. Theodore L. Houston,   WARRANTY DEED
      a bachelor   Dated April 22, 1879
    to   Filed May 9, 1879
    Laura Houston and Ida   Book 56, page 459
      Houston, both unmarried Consideration: $8,000.00
Conveys south half of sec. 9 . . .
Acknowledged April 22, 1879 before James S. Jones, N.P.

14. Theodore L. Houston and   WARRANTY DEED
      Babie L., his wife   Dated August 30, 1880
    to   Filed September 10, 1880
    Laura Morden and Ida   Book 58, p. 485
      Houston   Consideration: $2000.00
Conveys south half of sec. 9 . . .
Remark: Second grantor does not acknowledge as wife.
Acknowledged August 30, 1880 before Thomas E. Hayes, Notary Public of Essex County, New Jersey.

15. Affidavit of George F.   Subscribed and sworn to
      Beardsley   Sept. 19, 1884
                               Filed September 19, 1884
                               Misc. Records Book 3, p. 364

(States that the affiant personally knows Ida Houston and Laura Houston as referred to in the deeds preceding and that Ida Houston is the same person as Ida Nelson and Laura Houston is the same person as Laura Morden.)

16. Ida Houston, unmarried   QUIT CLAIM DEED
    to   Dated February 24, 1881
    Laura Morden, wife of   Filed February 24, 1881
      W. J. Morden   Book 41, page 460
                               Consideration: $750.00

Conveys interest in the south half of sec. 9 (and other lands) . . .
Acknowledged Feb. 24, 1881 before John L. Pierce, N.P.

17. Affidavit of Geo. F. Beardsley   Subscribed and sworn
                                      February 2, 1901
                                      Filed May 7, 1901
                                      Misc. Records Book 8, p. 204

(Affiant states that Laura Morden referred to in item 16 is the same person as Laura Houston.)

324 / GUARANTEES OF TITLE AND LOCATION

    Subscribed and sworn to Feb. 2, 1901, before E. S. Clark, N.P.

18. to 24. These entries, not included, show a mortgage, a release, a trust deed, a release, a mortgage, an assignment, a release.

25. Laura H. Morden   WARRANTY DEED
    to   Dated July 28, 1906
    William J. Morden   Filed July 31, 1906
                                    Book 141, page 339
                                    Consideration: the natural love and affection she has for her son, William J. Morden and for his better maintenance and support.

Conveys the south half of sec. 9 . . . with all buildings and improvements thereon.

Acknowledged . . .

26. William J. Morden and   WARRANTY DEED
      Florence Happer Morden,   Dated June 29, 1922
      his wife   Filed July 6, 1922
      to   Book 186, page 183
    Stacy C. Mosser, as trustee   Consideration: $1.00 and other good
      under provisions of trust   and valuable consideration.
      agreement dated June 29,
      1922

Document 158283

Conveys south half of sec. 9 . . . subject to trust agreement . . .

27. Stacy C. Mosser, as trustee   TRUSTEE'S DEED
    to   Dated May 2, 1939
    William J. Morden   Filed May 11, 1939
                                    Book 246, page 227
                                    Consideration: $1.00 and other . . .

Releases all interest for consideration in south half of sec. 9 . . .

28. William J. Morden and   DEED IN TRUST
      Myrtle Irene Morden,   Dated December 8, 1952
      his wife   Filed December 27, 1952
      to   Book 471, p. 88
    First National Bank of Chi-   Cons.: $1.00 and other good. . . .
      cago under trust agreement
      dated November 28, 1952
      and known in its records as
      Trust No. 42675

Document 501663

Conveys south half of sec. 9 (and other lands).

(Grants full power and authority to improve, manage, subdivide, sell, lease for 198 years or less, grants easements etc. . . .)

29. to 30. (Terms of the trust   TRUSTEE'S DEED
      agreement)   Dated October 8, 1958

31. First National Bank of Chi-   Filed October 15, 1958
      cago, as trustee for William   Book 607, page 437
      J. Morden   Cons.: $10.00 and other . . .
        to

Irene Morden
conveys and quits claim to southeast quarter and southwest quarter of sec. 9.... (subject to taxes for 1958)
32. In matter of Hillsbury Drainage Assessment Roll... (assessment of $1585.40)
33. NOTE: Property is within limits of Hillsbury Slough Drainage District and (subject to taxes and assessments levied by the district)
34. NOTE: Property is located in Champaign County Soil Conservation District and subject to (rules, regulations, assessments, etc.)
35. Taxes for 1957 on southeast quarter of sec. 9 ... (have been paid by) ... The First National Bank of Chicago.

12-12 Effect of Abstract of Title

An abstract by a reputable abstract company is a guarantee of a statement of facts, and liability is limited to the effect of omission of publicly recorded facts. Issuance of an abstract is not a guarantee that a title is free of defects; someone else passes opinions on the sufficiency of the title, a legal question. The abstract is merely a history of ownership.

The abstractor's job is to compile a complete statement of all publicly recorded matters affecting the ownership of a particular parcel of land. Who signed a conveyance will be noted, but whether the person had the authority to sign the conveyance is not always noted. If a word is misspelled in a conveyance, the misspelled word is perpetuated and usually underlined to direct attention to it.

Completion of an abstract reflects the paper history or condition of the real property. The desire of the attorney and the abstractor is to give certainty about the "health" of the written title. Changes in the description of the property through subsequent conveyances disturb these individuals. The most certain title is that property that retains today the exact same description as in the original conveyance. The surveyor, on the other hand, is interested in describing the same parcel in relation to boundaries and monuments on the ground today. This discrepancy can be overcome by describing parcels of land relative to the boundaries and monuments on the ground and then making reference in the description to the original and subsequent source of title.

12-13 Abstract and Attorneys' Opinions

An abstract by itself is not a guarantee of ownership; someone else assumes the responsibility of proclaiming the sufficiency of the facts stated, usually attorneys. The attorney's opinion does not include possession matters, and unless there is a survey, location is not guaranteed. The opinion of the attorney is limited to *facts revealed by the abstractor* through the abstract. Attorneys' opinions do not include guarantee against fraud, forgeries, or acts of omission or commission not of record. Title policies often do.

They undertook to furnish him an abstract of what appeared from the public records affecting the title to his property, and he was authorized to rely upon their competency and fidelity in this respect. Failure to note the judgement on the sale of the land for taxes, it was held, that the party making the abstract was liable for damages to the purchaser.

—*Chase v. Heaney*, 70 Ill. 268.

12-14 Title Insurance Policy

Title insurance is a written guarantee that as of a particular moment of time no title defects, other than those stated, exist. The limit of financial liability is stated on the face of the policy and may be increased or decreased in proportion to the fee paid. Unless otherwise stated, a policy is limited to title facts of record, not matters of location or unrecorded documents.

Policies show the amount of the policy, the person or persons whose title is guaranteed, the estate (fee simple, life estate, lease, etc.) being guaranteed, and the description of the real estate covered by the policy.

Title companies are in the risk business; for a fee they will assume a financial responsibility. If a person examines any title close enough, he can almost always find something that will present an uncertainty. In the matter of obtaining Torrens titles through court quiet-title actions, it may take months or even years to clear a small defect that may never materialize. Title companies generally do not delay title transfers because of remote defects; they can and do take calculated risks. For speed and quick service the title companies are far superior; for this reason more and more people are relying on them for title matters. Most lending agencies insist on title insurance.

Matters that title policies ordinarily insure against are taxes, bonds, assessments, trust deeds, mortgages, easements (sewer, water, gas, electricity, oil lines, drainage, etc.), liens, insanity, minor children's rights, forgery, false impersonations, wives with community rights, irregularity of conveyance form, omissions, heirs, leases, attachments, judgments, restrictions, reservations, reversal of court decision, senior rights, errors in public records, errors in transcribing and interpreting the public records, attack on the title, and many other items.

Title policies do not insure against everything. No guarantee is made against bankruptcy power, eminent domain, future police power or law, ordinance or regulation limiting the use of property such as zoning ordinances, building ordinances, or other regulations. Defects arising from a person's own violation (fraud or things known by the person to whom the guarantee is issued), matters not of record, such as a mechanic's lien not recorded, possession, and survey are excluded.

Owners' policies are based on the record, and no attempt is made to investigate matters of possession. Mortgage policies, however, are subject to claims of possession and questions of survey. Ordinarily in a mortgage policy the title company sends out an inspector to determine possession, and, if doubt exists, a survey is required.

A title insurance policy is of considerable aid to the surveyor in locating land, since he then has a statement that includes senior rights of others and a list of all easements. If there is an omission or error of statements, others assume that responsibility.

12-15 Wording of Title Policies

The wording of title policies varies, depending on the title association. In general a named person, persons, and/or corporations are insured for a maximum stipulated sum of money against loss or damages sustained because of (*a*) unmarketability of the described title unless it is caused by conditions listed; (*b*) title to the land being vested otherwise than as stated; (*c*) any defects in title not listed (includes easements); (*d*) defects in previous execution of mortgages, deeds of trust, liens, and the like; and (e) priority of other claims over mortgage being insured. Among the exceptions listed are (*a*) taxes, liens, easements, encumbrances, and assessments not shown on the public records; (*b*) rights of persons in possession of land not shown on the public records; (*c*) mining claims, water rights (in the West), claim or title to water, and reservations in patents; (d) any rights, interests, claims or facts that could be ascertained by an inspection of the land or by inquiry of the person in possession or by a correct survey; (*e*) laws, government acts or regulations (zoning, restrictions, use of land, setback, occupancy, etc.).

12-16 Title and Location Insurance Policy

If a parcel of land is surveyed by a competent surveyor and encroachments are certified to, most title companies will extend the coverage to include matters of possession without extra charges. But it must be remembered that if a title company is insuring location as based on a survey, it is its prerogative to accept or reject the results of a surveyor.

12-17 Location Guarantee

Surveyors, because of professional liability, must guarantee their work to the extent required by law. Liability is not stipulated as in a title policy but is one of mistakes or omissions caused by failure to do what an ordinary skilled surveyor would do and one of subsequent reliance by the parties involved.

Many surveyors carry liability insurance to protect themselves from loss. This type of insurance is not for the benefit of the client as is a title policy; it is designed to protect the surveyor from the client.

The amount of damages that can be collected from a surveyor is in proportion to his assets and liability insurance; it is wise for those using the surveyor's services to inquire into his financial condition.

12-18 Corrective Instruments

In certain matters title companies will not guarantee title without corrective instruments. In general the courts of this country take the attitude that "any description by which the property might be identified by a competent surveyor with reasonable certainty, either with or without the aid of extrinsic evidence, is sufficient." Title examiners and lawyers judge the merits of the conveyance entirely from the instrument itself and matters of public record. Examination "with the aid of extrinsic evidence" may be sufficient to make an otherwise invalid conveyance valid. Courts may take testimony and arrive at a conclusion binding on the parties, whereas the title examiner does not. It follows then that titles that may be legally sufficient to the court may not be to the title examiner.

12-19 The Torrens Principle of Title Registration

In essence the Torrens registration requires a quiet title action for all first registrants, requires registration of title by the state, includes guarantee of sufficiency of title by the state, has optional and sometimes mandatory requirements for location guarantee, and has mandatory requirements that all matters pertaining to the land's title be recorded on the Torrens title. Historically, a system similar to the Torrens titles was practiced as early as the thirteenth century in Bohemia. The present Torrens title registrations, developed by shipping clerk Sir Robert Torrens of Australia, was adapted from the method of titling merchant ships. The system was passed into law in 1858 and Sir Robert Torrens became the first registrar of titles. In recognition of this genius, Torrens was knighted.

This system was adopted in England, parts of Canada (British Columbia, Alberta, Saskatchewan, and parts of Ontario), in Norway, the Philippines, Sweden, Denmark, Puerto Rico, France, and many British colonies. To a limited extent it has been put into practice in the United States. Twenty states have passed permissive legislation for such a land registration system:

Illinois	New York	South Carolina	Washington
California	North Carolina	Georgia	Utah
Massachusetts	Mississippi	Tennessee	Virginia
Minnesota	Ohio	North Dakota	Colorado
Oregon	Nebraska	South Dakota	Hawaii

California repealed the act in 1958 because of its disuse. Many other states seldom use Torrens registration.

12-20 Characteristics of the Torrens System

The validity and sufficiency of written title is dependent on a continuous chain of valid transfers from the first sovereign's alienation to the present. The chain of title grows and grows with time. One primary purpose of Torrens registration is the elimination of maintaining previous transfer records (chain of title). To fulfill this objective the written title must have a new starting point, and this is done by a quiet title action and court guarantee of ownership.

Essentially, initial Torrens title registrations are quiet title actions. A responsible party is appointed to examine the sufficiency of title; and, if the title is found wanting, a quiet title suit is instigated.

Those registering titles are charged with the responsibility of protecting other valid title rights; titles are not supposed to be registered if defects exist. But in the event a title is registered and others do have a valid claim, the state is financially responsible and pays damages from an assurance fund.

The state guarantees title; this creates a new starting point for the title. If others are found who do have rights to the land, they cannot claim possession or usage; they can only claim damages from the state. To compensate those collecting damages, an assurance or indemnity fund is provided, said fund being accumulated by a charge on each title transfer.

As soon as title sufficiency is satisfied, the land is registered in the name of owners, and a certificate of registration is issued to the owners. All new valid transfers or encumbrances on the title must be done by an entry in the register. Provision in the law is made that failure to register transfers or encumbrances operates only as between the parties of the transaction but not as against future register owners. Thus the new owner is assured that only items registered can interfere with his rights.

Since the Torrens system was first developed in Australia, it was impossible to obtain possession rights against a Torrens title holder. Since that time the policy has eroded, and, under some circumstances, land can be obtained by occupancy.

12-21 Advantages and Disadvantages of Torrens

Because the title and ownership are up to date each time a new conveyance appears, searches are unnecessary after the system is in force. The indemnity fund, if properly set up, will cover most claims that arise from clerical errors and omissions. The protection also covers loss from fraud and wrongful possession. After initial registration, there is some saving of time and money to the parties of the conveyance, and the administration realizes some benfit in improving tax rolls.

Since the condition of the title can be determined immediately, the delay and expense of having the title abstracted are eliminated. In some cases, the registered land can be transferred without the need of a new survey, unless further subdivision is made. Forgery is quite unlikely, if not impossible, since the certificate of title must be presented before a transfer is effected.

Even though it is theoretically an ideal system, the Torrens system has disadvantages. The cost involved in first entering the properties into registration prevents many landowners from taking advantage of Torrens registration. In many cases the system has not worked because of the poor ties of the property to the ground. In California it was a failure and was repealed; loan companies would not loan on Torrens titles because of the insufficiency of the assurance funds.

Torrens land acts prohibit the acquiring of land by adverse possession.

However, any state can at any time pass a statute law prohibiting adverse possession acquisition of land for any title; hence this feature of the law is not exclusively in the Torrens system.

The Torrens system is a monopoly by the state; the success or failure of the system is dependent on political whims. The inefficient and incompetent can be placed in charge.

12-22 Land Registry in Illinois

The first state to enact legislation to permit Torrens type of land registration was Illinois in 1897. After the disastrous Chicago fire of October 8–9, 1871, many of the land records of Cook County were lost forever, and it was difficult to complete the chain of property title. A statute approved May 1, 1897, made possible the land registration, particularly in the counties having more than 500,000 population. This was a means to quiet title. Cook County is the only one with Torrens registration in Illinois.

12-23 Massachusetts Land Court

In 1898 the Massachusetts Registration Act, a Torrens act, provided for a court of registration with a judge, an associate judge, and a recorder. In 1904 the court was renamed the "land court."

All original registration proceedings, as provided by law, were to be handled by the court of registration, sitting in Boston or elsewhere, by adjournment as public convenience demanded. After a decree of the court had been issued and forwarded to the assisant recorder in the district where the land was located, it would be transcribed into a book as an original Certificate of Title, and a duplicate certificate would be furnished the owner of the land. A certificate contained the following information:

1. The number of the certificate in that particular registry.
2. Name and address of the owner.
3. Description of the land with reference to a plan filed either with the certificate or some other certificate in the same registry. The locus was fixed in detail on the plan and generalized as a bounding rather than a running description.
4. All appurtenant rights and encumbrances as decreed by the Court were set forth.
5. On the back, all conveyances, mortgages, easements, liens, and so on, pertaining to the land were recorded.

The powers of the court were increased in 1904 and at later dates to include these:

1. Writs of entry.
2. Petitions to require action to try title to real estate.
3. Petitions to determine the validity of encumbrances.

4. Petitions to discharge old mortgages.
5. Petitions to foreclose tax titles.
6. Petitions to establish power or authority to transfer an interest in real estate.
7. Petitions to determine the boundaries of flats (along ocean).
8. Petitions to determine whether equitable restrictions are enforceable.
9. Petitions to determine county, city, town, or district boundaries.
10. Petitions to determine the validity of zoning laws, and the like.
11. All equity actions affecting land, except those relating to specific performance of contracts.

When a jury trial is demanded, the issues are framed in the land court, and the arguments on these issues are tried before a jury in the superior court. With the adjudication of the questions of fact, the work in the land court proceeds, and the ultimate decision is in no way amendable to the machinery of the superior court. A petitioner for registration in the Land Court has a constitutional right to demand a jury trial, but this right is seldom exercised.

The work of the court is, in general, divided into two classes of cases—contested and uncontested. An uncontested case may take more time than a contested case.

> The Land Registration Act necessitates an adjudication by the Court of the boundaries as well as the ownership of the premises, and the boundaries so determined remain definite and fixed.
>
> The procedure for registration of land is as follows: A petition setting forth the claim of the petitioner, accompanied by a plan of the property and a certificate signed by the Assessors of Real Estate of the locus setting forth the abutters at the last date of assessment, is filed with the Land Court. The Court appoints an Examiner, who must be an attorney qualified by the Court, to examine the records and report by filing a narrative, opinion, and abstract of, all the records that he feels are necessary to meet the requirements of law. He may be required to furnish supplemental reports if the Court feels that they are necessary. Upon completion of his report citations are prepared and served on the abutters, and all others who may be affected, by registered mail with return receipt. The land is posted by a deputy sheriff with a copy of the citation, and publication is made for three successive weeks in a newspaper that is circulated in the particular vicinity of the locus.
>
> Thus, all persons are given a chance to be heard and an opportunity to file appearances and answers. In due course the petition reaches the judge's chambers, and after determining the validity of the title by trial or otherwise, he issues an order for a decree setting forth exactly what the petitioner will receive.
>
> A plan is prepared by the Land Court Engineering Department, and a decree drawn in conformance with the order and plan so prepared. This is returned to the judge for his final initialing and, upon final payment of all fees, the order is sent out to the registry district where the assistant recorder issues a Certificate to the petitioner.
>
> The Land Registration Act does not provide for an engineer, but the Honorable

Charles Thornton Davis, Judge of the Land Court from 1898 to 1936, recognized the necessity of an engineering department to properly make adjudication of boundary lines upon the earth's surface. Judge Davis established the Engineering Department as an adjunct to the Recorder's Department. It is the duty of this department to prepare a "Manual of Instructions for the Survey of Lands and Preparing Plans for the Land Court." The department inspects the plans submitted to see they are made in accordance with the Instructions. The engineer for the Court is called upon to interpret engineering features, to conduct investigations, and to fix the boundaries that may be decreed by the judges.

The manual is issued for the information and guidance of all engineers and surveyors making surveys and plans for the Land Court. The manual may be amended from time to time, and new editions appear about every 5 years, to keep pace with changes in survey practices.

Each plan made for a petitioner to file for registration must have a statement that the survey has been made in accordance with Land Court instructions over the signature of the engineer or surveyor who has supervised the work. (The term survey applies not only to the actual field work, but also the preparation of the notes, computations and plans.)[1]

REFERENCES

Flick, Clinton, P., *Abstract and Title Practice*, 2nd ed., West Publishing Co., St. Paul, 1958.

Fitch, Logan D., *Abstracts and Titles to Real Property*, Callaghan and Co., Chicago, 1954.

---

[1] Llewellyn T. Schofield, "Massachusetts Land Court," *Surveying and Mapping*, Vol. 8, No. 4, October–December 1948, p. 182.

CHAPTER 13

Platting Laws and Original Surveys

13-1 Contents

Creation of new parcels of land by original surveys and their regulation by governmental agencies are the subjects of this chapter.

Land boundaries can be created in two manners: (1) by description in a legal conveyance that usually did not or does not require the benefit of an actual survey and (2) by running lines on the ground by survey, setting monuments, and then describing the resulting parcel. A harmonious parcel is one in which the boundaries described in the written description are superimposed on those lines located in the field by survey. The best results for the benefit of the public are obtained by passage of statutes regulating how land boundaries may be created.

"Subdivision law," as the term is commonly used, has a broader meaning than does "platting law." A *subdivision law* includes regulations for the use of land, how streets must be improved, zoning, and many other restrictions on land. A *platting law* is generally, although not always, thought of as a law regulating the size and shape of parcels, regulating how parcels shall be monumented and measured, who can do the monumenting and measuring, what accuracies must be attained, and how the land shall be platted.

In the early United States the purpose of laws regulating how land could be divided into parcels was to devise a simple means of describing land and a means of ensuring certainty of location. At the present time subdivision laws usually include regulations on land usage and requirements for improvements. At one time the public agencies would accept street dedications without improvements; now it is common practice to require paving, drainage, sewers, water, curbs, sidewalks, and other improvements prior to accepting street dedications. Subdivision laws are becoming more and more complex as more and more land-usage restrictions and improvement requirements are included as part of subdivision acts.

This chapter does not discuss land-usage regulations or requirements for improvements; the main purpose is to discuss laws pertaining to the creation of land parcels, with emphasis on monumentation and measurements. The preparation of plats is the subject of Chapter 14.

Laws regulating how the public domain could be subdivided into sections had greater impact than any other subdivision act. Most surprising, the acts did not provide for public roads. Many of the states passed laws declaring that a road of a certain width existed along each section line. In some areas, as in California, streets were never provided for, except by a government taking. In mountainous areas a requirement for roads along each section line would be meaningless; the roads could not be constructed on terrain where such construction would be impossible.

13-2 Regulation of Original Surveys

PRINCIPLE. *Original surveys to create new land parcels are regulated by legislative rules.*

Resurveys are interpretive in nature and are for the purpose of locating parcels already described by an existing document. Preparing subdivision plats and performing original surveys are creative in nature and are not interpretive. For original surveys the governing body usually specifies the size and shape of new land parcels, who may do the surveying and platting, what the land may be used for, and numerous other regulations. In some areas the law makes it illegal to convey land other than by lot and block in accordance with a recorded plat.

For a subdivision plat to be valid and have effectiveness, in general, it must be authorized by law and must be in substantial agreement with the law. Those creating subdivisions should be certain of their compliance with their enabling act.

13-3 Former Presurveying Practices

In places within the United States no laws regulated the creation of new parcels of land, and in other places the laws were inadequate. In many instances the better surveyors, without regulatory laws, properly applied sound surveying practices and carefully laid out new additions. On the other hand, those with an eye for extracting the last dollar of profit, hired the incompetent for a lesser fee and left land location certainty in a chaotic condition. To ensure uniformity and certainty of future land location, regulatory measures are required.

Each state and the Federal Government have had and do have laws regulating land divisions. The failure of these laws in the past has been due to numerous causes, the more important being a failure to legislate adequate laws, a failure to enforce the laws, and the public's destruction of monuments after they were set.

The enactment of subdivision laws by the Federal Government to dispose the public domain was an example of the best thinking of the time. One flaw was that once land passed into private hands, the federal law had no binding effect on the states. In some states the federal rules for resurveys have not been adopted, but in most, either by statute or by court approval, they have been followed.

13-4 Objectives of Platting Laws

The objective of most platting laws is to specify simple means of identifying land and simple means of insuring certainty of location after land has been identified.

Plats are used to identify land (Chapter 14); monuments and measurements are used to locate land. In recent times, land-usage laws have been tacked onto platting acts. Although some mention is made of land-usage laws, the purpose of this chapter is to discuss methods of identifying and locating land.

13-5 Certainty of Land Location

PRINCIPLE. *A monument set by the original surveyor and called for by the conveyance has no error of position.*

The courts have given those creating new land divisions a precise method of positioning land boundaries; by merely setting a monument and calling for it, land has an exact location. A parcel of land is certain in location if an indestructible, unique monument is set and called for at each corner of the land. But certainty of location is not necessarily proof of ownership, since a senior right might be interfered with or the subdivider may have had a defective title.

Once a monument and all evidence of its former position disappear, measurements are relied on to replace the monument. The uncertainty of reestablishing a monument to its former position is then dependent on the uncertainty of originally reported measurements and the uncertainty with which remeasurements can be made. Original reported measurements cannot be analyzed, checked, or rechecked after monuments are lost; measurements must have been reasonably certain prior to a monument's destruction.

The objectives of platting laws with respect of insuring future certainty of land location are to require the following:

1. Durable permanent monuments.
2. Unique monuments having certainty of identification.
3. Frequent setting of monuments.
4. Accurate measurements tying monuments together.
5. Reference to the monuments in later writings.

The responsibility of future certainty of the location of described land parcels resides entirely in the hands of the persons setting monuments, making measurements, and preparing descriptions or plats. The only time that location can be assured for future years, without corrective documents, is at the time the conveyance is being made. After a conveyance is made, it becomes a question of interpreting what has been written. It is far easier to correct an ambiguity at the time a conveyance is being made than it is to try to correct it some time later.

13-6 Features of a Platting Law

Although there are many variations of platting laws, some variations being due to climatic or use conditions, the essential features of a platting law are these:

1. Definition of when a plat must be made.
2. Approval by governing agencies.

      a. Planning.
      b. Health (sewer, water, etc.).
      c. City engineer (grades, design, etc.).
      d. Tax assessor (pay back taxes and set new valuation).
      e. City attorney (approval of form).
      f. Inspection (usually by city engineer).
      g. Final approval by the governing body and various environmental agencies.
3. Title guarantee.
4. Performance bonds.
5. Definition of who may prepare maps.
6. Dedications and street widths.
7. Monumentation.
8. Setbacks.
9. Easements.
10. Measurement data required and accuracies needed.
11. Recordation.
12. Prohibiting of lot sales prior to recording map.

13-7 Platting Laws

How many parcels may be sold prior to the necessity of filing a plat varies with the need of the locality. The California law states:

> "Subdivision" refers to any real property, improved or unimproved, or portion thereof, shown on the last preceding tax roll as a unit or as contiguous units, which are divided for the purpose of sale, whether immediate or future, by any subdivider into five or more parcels within any one-year period.
>
> It is unlawful for any person to offer to sell, to contract to sell or to sell any subdivision or any part thereof until a final map in full compliance with the provisions of this chapter and any local ordinance has been duly recorded or filed in the office of the recorder of the county in which any portion of the subdivision is located.

Because of the escape clause (five parcels per year) many cities have passed augmenting ordinances which say in effect that "a legal building site is a lot within a subdivision and building permits may only be issued for legal building sites." Because under such an ordinance a lot divided in half must go through the subdivision map act, the local name "split-lot ordinance" has developed. In many states local ordinances may not augment state laws, but in California it is specifically permitted.

In Missouri land divisions are regulated by Section 137.185 of the Missouri Revised Statutes of 1949:

In all cases where any person, company or corporation may hereafter divide any tract of land into parcels less than one-sixteenth part of a section or otherwise, in such manner that such parcels cannot be described in the usual manner of describing lands in accordance with the survey made by the general government, it shall be the duty of such person, company or corporation to cause such lands to be surveyed and a plat thereof made by the surveyor in the county where such lands are situated, which plat shall particularly describe and set forth the lots or parcels of land surveyed, as aforesaid; the lots and blocks shall be numbered in progressive numbers, and the plats shall show the number, location and quantity of land in each lot, and the description of tract of land so divided, provided, that whenever it shall appear to the county court . . . that tract or parcel of land less than one sixteenth of a section . . . have been conveyed without having been surveyed . . . the court may require . . . a survey and plat . . . at the expense of the owner.

137.195. Any person, company or corporation that may hereafter violate the provisions of section 137.185 shall upon conviction be deemed guilty of a misdemeanor.

13-8 Planning Boards

Within the last 50 or more years there has been a marked change in the attitude of people toward land usage. Formerly, it was the opinion of most that they had a right to do as they saw fit with their own land; now, in many cities, the reverse is true. Land is designated for a particular use by a governing body, and the land may not be used for other purposes inconsistent with the desires of the majority without the consent of the governing body.

The agency charged with planning for land use has been variously designated as the planning commission, planning board, or planning agency. It is this agency's privilege to recommend to the governing board a course of action for land usage. Normally, the agency has no legislative authority, and only by approval of the governing agency is its opinion given force.

If a planning board rejects a given plan, the owner has the privilege of appeal to the governing body.

13-9 Considerations of Title

There may be others who have valid claims against the title of the property that will appear in the form of senior fee titles, easements or lessor estates, or encumbrances. As the quality of title to the lots of a subdivision will be no better than the title of the parent tract, these must be investigated before the development starts. Usually, such matters are referred to an abstractor, title company, or others guaranteeing ownership. At times this responsibility does fall on the surveyor. If the surveyor undertakes this responsibility, he may assume all liabilities resulting from his actions.

Easements are particularly troublesome when subdividing land. Because an easement is abandoned does not imply that it is released and that the title has reverted. The surveyor, attorney, or title examiner must examine the records of all possible utilities to determine if they have service running across the tract. *An unrecorded easement may still be valid.*

In some jurisdictions *abandonment* is the formal proceeding whereby the easement holder gives up his rights forever. Other jurisdictions use the word *discontinuance* to mean the same thing, although many times the two are not the same. The main point to remember, or to make clear perhaps, is that mere nonuse does not constitute abandonment.

> Abandonment of an easement necessarily implies nonuser, but nonuser does not create abandonment, however long continued.
> —*Adams v. Hodgkins*, 84 A 530, 109 Me. 361.

The plat resulting from the property survey should show all record utilities whether there is evidence of them on the ground or not and all features on the ground that have the slightest hint of being an easement on the fee title. *All easements on the parent title must be a condition binding on the lots created therefrom.*

> A reserved right in a conveyance which is not in its very nature a mere personal and temporary right will always be held to be an easement running with the land, absent some controlling provision to the contrary. Once an easement has become appurtenant to a dominant estate, a conveyance of that estate carries with it the easement belonging to it, whether mentioned in the deed or not.
> —*Burcky v. Knowles*, N.H. Supreme Court, Mar. 3, 1980.

When easements have been discovered on the tract, it must be ascertained if these may exist without greatly hindering the planned subdivision or if there is a possibility of having them relocated to a position that will allow the land development to proceed. In some cases, it is possible to obtain a release from the lessees or easement holders, but often this is impossible because of the restrictive language of the parent easement. If the easement is to remain in the subdivision tract, the restrictions in the easement must be respected.

In some areas there may be oil and mineral exceptions. Such mineral owners are, in fact, interested parties to the tract, and they must sign releases before their rights are extinguished. They may have a senior right of entry to mine and drill and to otherwise use the land for mining. In many states, these rights must be extinguished before land can be subdivided.

Covenents and restrictions "run with the land," and the mere act of subdividing does not destroy their effect. A subdivision developed on a tract that was conveyed years back, "provided no motor vehicles enter there upon," caused considerable difficulty before streets could be opened. Restrictions and conditions that have been created in a will can only be overcome by court action, usually by eminent domain procedure.

If covenants are existing, it must be determined if they are in conflict with the subdivision and if they can be modified.

A fee title to the tract may be clouded because of a prior unsettled lien or tax judgment against the owner or former owner. *An encumbrance against the parent tract is against the whole, even though only a small portion of the land is subdivided.* Therefore an owner of a quarter-acre lot that has been carved from a hundred-acre tract having an encumbrance may find that he is in danger of losing his title, or at least some money, when the party who holds a lien or judgment presses claim.

13-10 Title Guarantee

To protect the new purchaser of land, it is advisable to require that the landowner have some financially responsible body guarantee that if the title is found wanting they will have financial compensation. The statute law may require the following:

> The subdivider shall present to the recorder evidence that, upon the date of recording, as shown by public records, the parties consenting to the recordation of the map are all the parties having a record title interest in the land subdivided whose signatures are required by the provisions of this act, otherwise the map shall not be recorded.

In those states where title insurance is common practice, the following notes are endorsed on the face of the map. (This certificate includes dedications, see Section 13-16.)

> We hereby certify that we are the only owners of, or are interested in, the land embraced within the subdivision to be known as Speer Tract, and we hereby consent to the preparation and recordation of this map consisting of two sheets and described in the caption thereof. I hereby dedicate to public use Midway Drive, the easements for sewer, water, drainage and public utilities, and that portion marked "Reserved for Future Street," together with any and all abutters' rights of access in and to Rosecrans Street and that portion marked "Reserved for Future Street" adjacent and contiguous to Lot L, as shown on this map within this subdivision.
>
> Signed and notarized.
>
> ... Title Insurance Company, a corporation, hereby certifies that according to the Official Records of the County of ..., state of ... on the ... day of ... 19—, at ... o'clock, A.M., Mr. So and So were the owners and the only persons interested in and whose consent was necessary to pass a clear title to the land embraced within the subdivision to be known as Speer Tract as shown on this map, consisting of two sheets and particularly described in the caption thereof, except the City of ..., a municipal corporation, owner of easements which cannot ripen into a fee.
>
> In witness whereof, ... Title Insurance Company has caused this instrument to be executed under its corporate name and seal by its proper officers, thereunto duly authorized, the day and year in this certificate first above written.
>
> Signed and notarized.

When the law requires such certification, the initial step in preparing a subdivision is a boundary survey presented to the title-guaranteeing agency for its approval. Such a survey must show monuments found and set, along with existing possession.

13-11 Boundary Survey

The location of the boundary, in former years, was the only place the surveyor incurred liability for errors. An interior monument set, even though not in its correct position, is, after it is used, correct. In recent years the tendency has been for a developer to produce house and lot. With delayed staking, the surveyor sets temporary interior stakes for house location, and if these are wrong, problems ensue.

Boundary monuments of a subdivision, along a line adjacent to another owner, do not always have the dignity of no error of position. Subdividing land does not automatically erase the rights of the adjoiner; if part of the adjoiner's land is taken, the adjoiner may assert his rights up until the time imposed by the statute of limitations.

In the past, errors have occurred along subdivision boundaries with resultant gaps or overlaps. Within a platted subdivision, wherein one line represents the boundary between adjacent lots, it is not possible to have gaps or overlaps. But along boundaries, even though the adjoining property is shown as abutting and only one line is drawn, the new subdivision plat does not change the former status of the adjoiner's rights. In resurvey procedures, lots adjoining boundary lines of the parent tract represent a great hazard to the surveyor.

13-12 Requirements for Monumentation

All platting laws should require complete lot monumentation. The requirements should extend beyond that of merely setting lot corner stakes; they should specify the use of identifiable monuments (bounds) at frequent intervals in indestructible locations.

Monuments required may be of two types: (1) those that are control monuments to be used to relocate lost lot corners, and (2) those that mark every lot corner. Control monuments are placed where they are least likely to be destroyed and where they can be conveniently used. Ordinarily, they are located within a street right-of-way. Property monuments have a high propensity for destruction within a subdivision. They are often treated as temporary stakes, and they are often no more than that.

The most important feature of a monumentation law is the requirement for permanent, indestructible, identifiable monuments, located at frequent intervals, sufficient to enable relocation of any lost lot monument, within specified accuracies. Control monuments are usually located at each street intersection, at each point of change of direction of street lines, and at the beginning and end of each curve; they are buried below the frost line or below the street grade line and are situated where the installation of new underground utilities will not destroy them. From the standpoint of ease of computing, monuments set at the intersection of street centerlines are best, but they present traffic hazards.

Better platting laws require that every lot must be monumented prior to the sale of any one lot. In those areas where full improvements are required, the law often requires delayed staking; that is, lot monuments are placed after the curbs, sidewalks, underground utilities, and paving are installed. Some regulations call for lead and discs (with license numbers) in the sidewalks, on an offset, after the installation of sidewalks. Such regulations are of little benefit where sidewalks creep because of sandy soil. In frost-free areas, specifications for lot monuments often include $\frac{1}{2}$-inch pipes driven 18 inches in the ground or even redwood stakes. But in frost areas, iron pins driven below the normal frost line are required.

13-13 Density of Monuments

The distance between control monuments, as required by law, should be determined by accuracy considerations and individual needs. Since all original monuments set have no error of position, it is desirable to have frequent monument control. As has been so frequently stated, all measurements have some error, and any time it is necessary to measure to relocate a property corner some error will exist. Measurement errors are a function of distance, although not a direct relationship; and it follows that the closer together or more densely located are monuments, the less uncertainty will exist in relocating property.

In residential areas, where buildings must be set 4 or more feet from the sidelines of property, an error of 2 inches in relocating property lines would not be excessive. Permanent monuments located 600 feet apart can be tied together with measurement uncertainties of plus or minus 0.06 foot or less; in the event of destruction of a permanent monument, it could be replaced from other permanent monuments with an error of plus or minus 0.12 foot. It is logical to require permanent monumentation not further than 600 feet apart in residential areas and preferably at each block intersection. This would insure certainty of lot location within 2 inches, plus or minus, even with the destruction of one permanent monument.

In commercial areas, where buildings are permitted to be constructed exactly on the property line, uncertainties of relocation of original positions must be reduced. An error of $\frac{1}{2}$-inch in relocation of positions may be significant. Permanent control monuments, not more than 300 feet apart, are needed.

Monument density needs depend on the need for future certainty of parcel locations. Since future land use is not always foreseeable, it is better to err on the safe side and require more monumentation than is needed.

The following monumentation law for subdivisions is applicable in Colorado under Chapter 38, Article 51:

> **38-51-101. Monumentation of land surveys.** (1) On and after July 1, 1967, the external boundaries of all subdivisions shall, prior to the recording of any plat thereof, be monumented on the ground by reasonably permanent monuments solidly embedded in the ground. Affixed securely to the top of each such monument established on and after July 1, 1967, shall be the Colorado registration number of the land-surveyor responsible for the establishment of said monument. These monuments shall be set not more than fourteen hundred feet apart along any straight

boundary line, at all angle points, and at the beginning, end, and points of change of direction or change of radius of any curved boundaries.

(2) Before a sales contract for any lot, tract, or parcel within a subdivision, recorded on and after July 1, 1967, is executed, all boundaries of the block within which said lot, tract, or parcel is located shall be marked with monuments similar to those required in subsection (1) of this section. Wherever any block is bounded by streets, the monuments may be set on the centerlines of said streets or on offset lines therefrom as designated on the recorded plat. In addition, the corners of any lot, tract, or parcel sold separately shall be marked as described in subsection (4) of this section within one year of the effective date of the sales contract. If any structure is to be built on the lot, tract, or parcel before the corners have been marked as provided in this section, the seller of said lot, tract, or parcel shall provide for the services of a Colorado registered land surveyor to establish on the ground such control lines as may be necessary to assure the proper location of the structure.

(3) It is the responsibility of the land surveyor who prepares the original subdivision plat to provide external boundary monuments as required in subsection (1) of this section. It is the responsibility of the seller of the lot, tract, or parcel to provide for the services of a land surveyor to establish block monumentation and lot markers as required pursuant to subsection (2) of this section. However, if any complete block is sold as a unit, it shall become the responsibility of the subsequent seller of any separate lot, tract, or parcel within such block to provide for the services of a Colorado registered land surveyor to establish lot markers as required pursuant to subsection (2) of this section.

(4) The corners of lots, tracts, or other parcels of land, and any line points or reference points shall, when established on the ground by a land survey on and after July 1, 1967, be marked by reasonably permanent markers solidly embedded in the ground. Affixed securely to the top of each such marker shall be the Colorado registration number of the land surveyor responsible for the establishment of said marker.

(5) In the event that points designated in subsections (1), (2), and (4) of this section fall on solid bed rock, or on concrete or stone roadways, curbs, gutters, or walks, a durable metal disk or cap shall be securely anchored in the rock or concrete and stamped with the survey point and the Colorado registration number of the land surveyor responsible for the establishment of the monument or marker.

(6) In the event that monuments or markers required by subsections (1), (2), and (4) of this section cannot practically be set because of steep terrain, water, marsh, or existing structures, or if they would be lost as a result of proposed street, road, or other construction, one or more reference monuments shall be set. Affixed to the monument, in addition to the surveyor's registration number, shall be the letters "RM" or "WC." Such reference monuments shall be set as close as practical to the true corner and shall meet the same physical standards that would be required for the true corner were it set. If only one reference monument is used, it must be set on the actual boundary line or a prolongation thereof. Otherwise, at least two reference monuments shall be set.

13-14 The Use of Coordinates in Subdivisions

State Plane Coordinates, properly used, are a modern means of insuring certainty of monument restoration in the event of monument destruction. The legislature can regulate how new land divisions may be made, and it is within the scope of law to require compulsory ties to a state-authorized plane-coordinate net. The ultimate public advantages far outweighs the complaint of increased initial costs. The system, unless it has been in effect for some time offers no advantages to the immediate subdivider; it increases the subdivider's cost. But the government, in addition to requiring water, sewer improvements, and usage regulations for land, should, as an obligation to the public, see that, whenever parcels are created, the parcels shall be of full measure and shall be certain of future identity. The expenses necessary to tie to State Plane Coordinates are justifiable costs.

City or county mapping, major improvement projects, and correlation of a city's activities require a knowledge of distances between subdivided areas. The most accurate and certain method of providing this information is to require that all new parcels of land shall be located on the same basis of bearing and on the same grid system. It is only indifference to a necessary public need and failure of legislative bodies to provide necessary legislation that has prevented universal adoption of State Plane Coordinates in the United States.

State Plane Coordinates can never supplant local monument control; as long as a called-for monument exists, measurements are secondary evidence. State Plane Coordinates are an *aid to preserve evidence of monument positions*; they are not a means of changing a time-honored system of monument control.

The advantage of a call for a State Plane Coordinate system is that it increases the number (density) of monuments referred to in a deed. Instead of a newly created parcel having only measurement relationships to monuments in a local area, the new parcel is related to thousands of monuments in the entire net, if measurements are made to a State Plane Coordinate system.

To date, little or no land litigation has settled the question of what the courts think the importance of coordinates is; it is conjecture to say what the courts would or would not do. A coordinate system is just as strong as the monuments that mark it on the ground. Without doubt, all coordinate values assigned to a given control monument are in error to some extent. The monument itself is fixed in position, but measurements cannot be exact. If a surveyor measures to a known coordinate point (monument), the tie distance and direction are used to compute the new monument's coordinates. It is the measurement value itself that is the evidence, not the coordinate; it is a tie-out to a physical monument on the ground.

The value of tie-out distances to monuments marking a coordinate net is in proportion to the accuracy of the observed measurements. Poorly made measurements to determine a position in a coordinate net only confuses the record and creates a certainty of change in position when relocating a lost corner. In addition to requiring plane-coordinate position determination, the law must specify the accuracy of the measurement methods. To be of real value, errors of position of a plane-coordinate system, created by measurement errors larger than 1 foot in 2 miles, could not be tolerated.

The use of State Plane Coordinates has been recommended by the following organizations:

ASCE
American Bar Association
ACSM
NGS.

In view of the fact that original surveys are becoming more obscure each day and property is more valuable, it is imperative that better means be employed to supplement and complement the present systems of evidence determination. Of the advantages of the system of State Plane Coordinates are the following:

1. Establishes geometric identity.
2. Indicates relative position.
3. Permanently establishes lines.
4. Facilitates the relocation of land.
5. Coordinates may be scaled from government maps.
6. Self-referencing for perpetuation.
7. All surveys are tied to a common datum.
8. Proportioning becomes easier mathematically.
9. Often eliminates the need for random lines.
10. Other surveys can be used to advantage.
11. Provides a simple means of description.
12. Prevents the accumulation of error due to the shape of the earth.
13. Permits the ready check on large blunders.
14. Permits easy computations of area in "cutoff" tracts.
15. Easy adaptability to modern computing methods.
16. Provides an excellent filing system.
17. Permits easy plotting on maps and plats.

In view of the fact that after years of existence State Plane Coordinates are not widespread, there must be some disadvantages, among which are these:

1. Many surveyors, attorneys, and title men do not understand the grid projections.
2. Extra cost is involved in additional surveys.
3. Areas do not have sufficient control.
4. Many existing surveys are not of sufficient quality to contribute positions to the system.
5. Positions are dependent on control points that in themselves are subject to destruction.

13-15 Certification of Survey

In addition to monumentation a certificate is needed assuring that the monuments have been accurately set in accordance with law. Most states require such a certificate to be certified to by the surveyor, although in some states the civil engineer may do so. Such a certificate often takes the following form:

>State of . . .
>County of . . .
> I, . . . , hereby certify that I am a Licensed Land Surveyor of the state of . . . ; that the survey of the subdivision was made by me or under my direction between . . . and . . . and that the survey is true and complete as shown, that all stakes, monuments, and marks set, together with those found, are of the character and occupy the positions shown thereon, and are sufficient to enable the survey to be retraced.
>   Signed . . . .
>Date . . . .

In many instances monuments set will be destroyed by construction of improvements. Because of this, many areas have provided delayed staking provisions whereby, after improvements are installed, final stakes must be set. such a certificate may take this form:

>State of . . .   ss
>County of . . .
> I, . . . , hereby certify that I am a Licensed Land Surveyor of the state of . . . ; that the survey of this subdivision was made by me or under my direction between . . . and . . . and that the survey is true and complete as shown. A two (2) inch pipe twenty-four (24) inches in length will be set at each boundary corner except as shown, and I will set three-quarter ($\frac{3}{4}$) inch pipes eighteen (18) inches in length at all lot corners and points of curves along dedicated streets, unless otherwise noted on this map, and each monument will be stamped LS 2554. Said monuments will be set within thirty (30) days after the completion of the required improvements and their acceptance by the city, and such monuments are, or will be sufficient to enable the survey to be retraced and will occupy the positions shown thereon.
>   Signed . . . .
>Date . . . .

13-16 Dedications

Dedications in accordance with statutory law must comply with the letter of the law. There must be an offer to dedicate and an acceptance of dedication. Although the form may vary from state to state, one form used is as shown in Section 13-10. Acceptance of dedication may be as follows:

>State of . . .   ss
>County of . . .
> I, . . . , city clerk of the city of . . . , State of . . . , hereby certify that the Council of said city has approved this map of Speer Tract, consisting of two sheets, and described in the caption thereof, and has accepted on behalf of the public Midway Drive, together with any and all abutters rights of access in and to Rosecrans Street and that portion marked "Reserved for Future Street" adjacent and contiguous to Lot L, and the easements for sewer, water, drainage, and public utilities, all as

shown on this map within this subdivision, and hereby rejects that portion marked "Reserved for Future Street" shown on this map within this subdivision.

In witness thereof, said council has caused these presents to be executed by the city clerk and attested by its seal this . . . day of . . . 19—

Cities sometimes desire to have a space reserved for future streets but do not wish to accept the street until such time as the street is improved in its entirety. Provisions in state laws can be made than "any offer of dedication shall remain open and subject to future acceptance by the city." In the foregoing form where this law applied, the city rejected the area designated as "Reserved for Future Street" and can at any future date accept the offer of dedication. The only way to remove such an offer to dedicate and the rejection of such is for the city to complete a legal vacation proceeding in the same manner streets are vacated.

13–17 Setbacks

For the purpose of uniformity and to provide front-yard planting area, modern zoning laws require minimum setbacks for buildings along front and rear yards. Often such setback requirements are shown on the subdivision map as in Fig. 14-2. A blanket requirement that all lots have a given setback is not practical, especially on hillside developments. By mutual agreement between the subdivider and the planning body, setbacks can be adjusted to fit topography. On lots rapidly sloping away from a street, excessive setbacks would prevent sewer connections and poor driveways to garages.

13-18 Recording

The certainty of ownership of land sold by lot and block by reference to a map is entirely dependent on the preservation of the map in a place that allows inspection of the map yet insures that the map cannot be altered. Recordation of a map in a public place and appointment of a public employee to be responsible to see that no map is altered is the best solution.

> No final map of a subdivision shall be accepted by the county recorder for record unless there has been a compliance with all provisions of this chapter and of any local ordinance. The recorder may have not more than 10 days to examine the final map before accepting or refusing it for recordation.
>
> The approval in accordance with the provisions of this chapter by the appropriate governing body or bodies, and the recordation of the final map shall automatically and finally determine the validity of the map under the terms and provisions of this chapter and local ordinances.
>
> —California Map Act.

Under these provisions the burden of proof of compliance is placed on the recorder. After the map is filed, it is valid. This is as it ought to be, although there are states without this provision.

13-19 Examination by City Engineer

Prior to recordation of a map, the city engineer—or if in the county, the county engineer—must express approval about compliance with improvements and all other things required by law.

The certificate of the city engineer may take this form:

> I, ..., city engineer of the city of ... hereby certify that I have examined the annexed map of this subdivision to be known as Speer Tract, consisting of two sheets and described in the caption thereof and have found that the design is substantially the same as it appeared on the tentative map and any approval alteration thereof: that all the provisions of the Subdivision Map Act and of any local ordinances of said city, applicable at the time of the approval of the tentative map have been complied with, and I am satisfied that said map is technically correct, I hereby approve and recommend said map.
>
> City Engineer................

Date..........

13-20 Presurveys without Recording

In some states the law requires the surveyor to record a plat whenever he makes an original survey to divide land; in other states the owner may record the plat delivered, at his option; and in other states no plat may be recorded, since no office is delegated to accept plats for recording. Recording a plat is similar, in effect, to recording a deed; it is recorded evidence that may be used in court.

Many surveys made for the purpose of dividing land are never recorded. It is unfortunate, because, after the surveyor dies, the evidence of location will be lost.

13-21 Summary

This chapter is not a complete discussion of how subdivisions are made, since part of the discussion is in Chapter 14, "Survey Plats." Subdivision laws are for the purpose of regulating how and when land may be divided into smaller parcels. It is within the scope of law to regulate the use of land, how new divisions of land may be created, and who is responsible for monumenting and measuring new parcels. Each state has its own regulations, and the laws are variable from place to place. The subject matter presented is general and will not fit a specific area, but it does give general concepts. Whenever a surveyor accepts the responsibility of undertaking work within a subdivision, he is charged with knowing the full and complete requirements set forth in the statutes and ordinances of his local area of practice.

CHAPTER 14

Survey Plats

14-1 Introduction

The use of drawings to identify land predates written descriptions. Man relies on drawings to supplement words and phrases, and many conveyances depend on a plat to depict the intent of contracting parties. In this chapter the features of a survey plat, as well as the effect it may have on boundary location, are discussed, but drafting techniques and cartographic expression are not included.

14-2 Definition of a Survey Plat

A *survey plat* is a surveyor's diagram showing land boundaries and/or a subdivision of land. A *map*, as contrasted with a plat, graphically represents to a scale the physical features of an area and may show some general land boundaries, especially political boundaries.

A plat is more restrictive in scope than is a map, and it has added features. As used by surveyors, a *plat* is a plan showing property lines and interrelationship of property lines with dimensional data on lines; it does not normally express relief. A map rarely has dimensional data; the quantities can be determined by scaling. On a plat, dimensional data may eliminate the necessity of scaling, and the value of the plat is not dependent on the accuracy with which points are plotted.

14-3 Types of Plats

Plats logically fall into three classes: (1) those made to represent the results of original surveys, (2) those made to represent the results of a recent property survey as based on a recorded conveyance or plat, and (3) those made to represent a compilation of record property-boundary data (assessor plats, abstractor plats, tax plats, etc.).

Original plats are called for by a conveyance and are those showing the results of a property survey made prior to conveyancing. They are often given a name as "Pine Hollow Subdivision" or "Cyprus View Subdivision." Original plats are generally regulated by and filed with a governing body, but they may also be attached to a deed and called for by the deed.

Those plats made to represent the results of a resurvey or the results of any boundary survey as based on the record do not have a special name and are usually called a "survey plat." Sometimes the survey plat does show newly created parcels, but no conveyance calls for the plat. These plats are not a consideration of, nor are they called for by, a deed's writings. They show information pertaining to property-line locations; they perpetuate a record of found evidence; and they give new tie-outs to found evidence.

Original plats act as written instructions about how future surveyors or owners are to identify a parcel of land; a resurvey plat, or any plat representing the results of a survey based on the record, explain how and where the surveyor identified the land called for in a conveyance. The first gives instructions to be followed; the latter explains how instructions were followed. Nowadays, an original survey is almost always preceded by a resurvey or a survey as based on the record; the boundary of an original plat, or at least part of it, almost always represents the results of a resurvey or a survey as based on the record.

Compilation plats, such as assessor plats, legislative plats, and abstractor plats, are plats showing a compilation of boundary data from numerous record sources; they do not represent the results of a recent survey. Technically, they are not survey plats, but surveyors commonly use them. A compilation plat is a representation of property boundaries and interrelations of property boundaries as defined by the record. When the record dimensions are established on the ground, they may be found in error; a compilation plat is not necessarily an accurate representation of location but may be accurate with respect to record information.

14-4 Purposes to Be Accomplished by Survey Plats

PRINCIPLE. *The essential purposes to be accomplished by a survey plat are* (1) *to represent the correct size and shape of a property to a scale*, (2) *to define by dimensions the correct size and shape of a parcel of land*, (3) *to specify locative points (physical monuments, including cultural features), and* (4) *to show title identity (record monuments).*

In addition to the purposes listed, plats also show data that lend authority to the plat such as date, surveyors's name, certificate of accuracy, and the client's name.

Compilation plats generally lack expression of locative points and correct dimensional data as would be revealed by a recent survey.

14-5 Features of Plats

Cartographic expressions, such as symbols, north arrow, dimension arrows, and like expressions, are generally identical for all types of plats; but cadastral features, such as legal descriptions, monuments found, and setbacks, are variable. Since the purpose is not to discuss the techniques of plat making, much of the cartographic expression is only briefly mentioned.

14-6 Drawings

The size, shape, and kind of drawing needed to record the results of a survey are so variable, depending on the state, need, and purpose, that no standard exists.

350 / SURVEY PLATS

In some areas the law requires recordation of plats, such as a record of survey in California, and it specifies the details.

In court trials the resultant print needs to be large enough that a juror or judge may see it at considerable distance. For ordinary lot surveys a small drawing is usually adequate (see Fig. 14-1). At times it is necessary to make one large-scale plat and several detailed sheets.

Drawings should be reproducible. Many types of tracing paper, cloth, and film are available, although no standard exists.

A plat should tell a complete story; it should show sufficient information to allow any other surveyor to understand how the survey was made and why the survey was correct. It also should show complete information on encroachments to enable any attorney or others to evaluate properly the effect of continued possession.

No set rule exists about how much area needs to be shown. If a lot within a

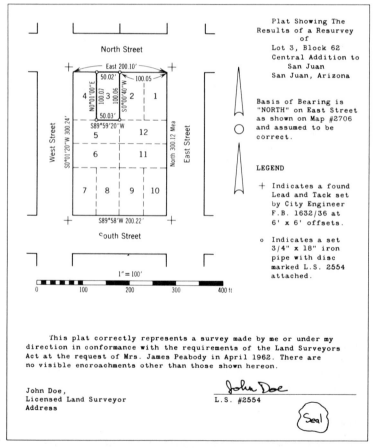

Figure 14–1 Plat showing the results of a resurvey of a lot within a subdivision.

block is being surveyed, often it is only necessary to show the dimensions and size of the block. But if a parcel is abutting a land-grant line, it may be necessary to show the land-grant boundary monuments found several miles apart. If a portion of a section is being measured off, complete proof of the survey, as a minimum, calls for four quarter-corner locations and at least one section corner.

14-7 Title of Plat

The eye should instantly see a title that identifies the survey; this means the title should be prominent, short, and clear. Ordinarily, the words "Plat of a Survey of" precede the general description to follow. Appropriate titles are "Plat of a Survey of Rancho Santa Fe," "Plat of a Survey of Lot A, Block 12, Horton's Addition," or "Plat of a Survey of a Portion of S.12, T.13 S., R.2 W. as described in Book 1213, page 21, Official Records."

The purpose of a title is to direct attention to a general area or neighborhood. It does not detail the survey, and it should not be long.

14-8 Symbols

Symbols quickly convey a meaning to a viewer. Crosses along a line on a plat indicate a fence; at a glance, the viewer instantly knows the position of the fence relative to other objects and lines, and he knows the length, extent, and direction of the fence.

Symbols, to be effective, should be consistent and should suggest the object being symbolized. Trees are suggested by the shape of the symbol; waterlines look like waves coming in along a shore. A good procedure to follow is to adopt the conventional symbols established by NGS or other well recognized mapping organizations.

The object of a plat is to reduce the number of written words to a minimum and present the results of a survey graphically to scale. Profuse writings on a plat only detract from its value, but sufficient writings must be included to avoid ambiguities. Symbols decrease the number of necessary words.

14-9 Scale

The relation between the plat dimensions and those on the ground can be expressed by (1) a bar scale, (2) a ratio, or (3) a representative fraction. The bar scale is most useful because its value is not changed when the plat is reproduced by either an enlargement or reduction process. Along with the bar scale, a ratio scale is indicated as 1 inch = 100 feet. Architects' scales ($\frac{1}{16}$ inch = 1 foot) are not convenient to use on survey plats.

A representative fraction, such as 1:1200 (meaning 1 inch = 100 feet), is commonly used on government topographic maps but is not used on survey plats.

14-10 Indicating the Direction of North

Every map, plat, or sketch should bear an orientation arrow, and if a grid system is used, grid north is indicated. Ordinarily, the north arrow refers to astronomic north, but under some circumstances the magnetic meridian is indicated. If the magnetic meridian is used, the plat should bear a date, have a statement that the magnetic meridian is used, and show the relationship between magnetic and astronomic meridian (declination). Often, alongside the north arrow, the scale is indicated.

When convenient, the drawing should be oriented with north at the top of the sheet. If this is not practical, a larger-than-usual north arrow should be included on the plat. The north arrow need not be elaborate. The ancients were sometimes carried away with embellishments on their compass rose, but plats made today can be best served with a simple, efficient arrow (Fig. 14-1). If other than astronomic north is used, it should be noted at the bottom of the arrow or as a special note indicating the basis of bearings.

14-11 Basis of Bearings

Every plat showing the results of a survey should carry on its face a note explaining the basis of bearings. Without doubt, the most frequent cause of ambiguity is the result of failure to specify a datum meridian. For every survey by a surveyor, the surveyor (1) must assume a meridian, or (2) must establish a meridian by astronomical observation, or (3) must establish a meridian from geodetic control, or (4) must establish a meridian by a magnetic observation. By some type of note on the plat, the surveyor must explain what he has done. Examples of notes found on plats are as here:

> Basis of bearings is a Polaris observation as made Jan. 12, 1962 at corner No. 1.
> Basis of bearings is a sun observation made July 2, 1961 at corner No. 3.
> Basis of bearings is N 0°12'00" E along the center line of 6th Street as assumed from the city engineer's records.
> Basis of bearings is N 7°21'30" W as shown on Map 2132 for the easterly line of Lot 13 between monuments indicated.
> A magnetic observation was made to determine the bearing of the east line of the property platted; all other bearings were determined by calculation from deflected angles as measured by a one-minute transit.
> Basis of bearings is grid north as based upon ... State Plane Coordinate System (the drawing will show the stations occupied).

All platting laws should require the surveyor to specify his basis of bearings; and, where a suitable grid control exists, the basis of bearings should be specified as grid north. If a magnetic basis of bearings is permitted, it is particularly important that the date of magnetic observation be given. Magnetic north changes with time.

14-12 Elevation Datum

Elevation is rarely the subject of property-line measurements, but occasionally a deed is defined to run along the mean high-tide line or along some other water-contour line (for dams or lakes). For original plats, mean sea-level datum is preferred and should be defined on the plat. The use of colloquial expressions such as "U.S.G.S. Datum" or "Government Datum" is not definite and should be avoided.

For any survey as based on the record, the datum is usually defined by the record, and, of course, that datum must be used to determine property location. But on the plat, the datum as used should be explained, and it may be advisable to relate the datum to other data.

14-13 Dimensional Data

One of the essential features of a survey plat is the dimensioning of the size and shape of the land represented by a drawing. Too many dimensions detract from a clear picture, yet too few dimensions can create ambiguity. Dimensions, as symbolized on plats, are reduced to their simplest expression; dimension lines are omitted when not needed for clarity.

PRINCIPLE. *Unless otherwise proved by scaling, distances given on a plat are from the nearest point on each side of the dimension as written.*

Often dimension lines are unnecessary and do not contribute to a plat; in fact, too many lines will detract from the clarity of a plat. Plats are drawn to a scale; most distance data, uncertain in the mind of the viewer, can be made certain by scaling. Occasionally, dimension lines are necessary to prevent ambiguity in indicating limits of measurement—in which event, dimension lines as shown in Fig. 14-1 are used. Arrows of the type used on house plans are not used.

Most subdivisions, as developed today, are rarely rectangular with a gridiron pattern. Curved street lines and nonuniform bearings require dimensions on every lot line and, in addition radial bearings (see Fig. 14-2). Good platting laws require complete dimensional data on every line.

Distances are expressed consistently, usually to the nearest 0.01 foot, except where absolute values are implied. A street shown as "50 feet" will be construed consistent with other units as 50.00 feet.

Most subdivision acts require each lot to mathematically close; for this reason angles are often shown to an accuracy greater than that which can be measured. Angles shown to the nearest second for lines 50 feet long cannot be measured that accurately, but they comply with the law. Sectionalized land subdivisions (U.S. Government) have all dimensions written on the map as measured; seldom does a section close mathematically. As long as closures come within permitted errors, the survey is accepted, and dimensions are written on the map as shown in the field book.

354 / SURVEY PLATS

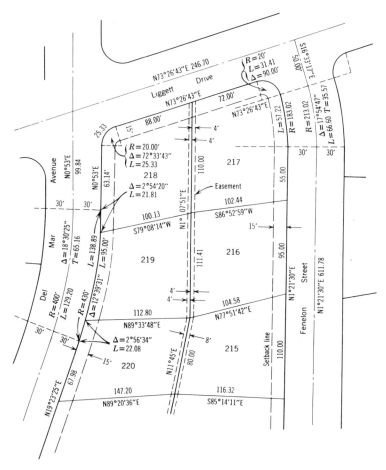

Figure 14-2 Plat showing lot dimensioning for a portion of a subdivision. Often platting acts require additional curve data such as the chord and tangent. If the title sheet does not specify what the easements are for, it must be stated on the face of the plat.

Formerly, the use of ditto marks on plats was permitted to allow expression of the same dimension; presently, because of ambiguities caused by the ditto marks, most laws prohibit their use.

Other dimensional data, such as curves (Section 15-44), are discussed elsewhere.

14-14 Monuments

All land must be located by one or more locative calls; all plats should disclose locative points, that is, the monuments used by the surveyor.

Monuments as disclosed on plats are of two essential types: (1) those visible to the eye such as stones, iron pipes, rivers, and lakes and (2) those often invisible to the eye such as an adjoiner property called for, the sideline of a street, or a lot in a particular block. The latter are record monuments and are discussed under "Title Identity," Section 14-16.

Every monument is either found or set, and most monuments indicated should be so noted. A few monuments, such as trees, marshland, fences, and structures are known to be found without the necessity of saying so. In addition, sometimes it becomes necessary to say what was not found. If a deed calls for a 1-inch iron pipe and it is not found, it becomes important to say so. In other words, whenever a monument is called for in a conveyance and it does not exist, such fact should be stated on the map.

A found monument has a past history, and as much past history as is known should be included. A note of "Fd. orig. mon. as per Map 2102" or "found city engineer's 2 " × 2" stake per field book 56, pg. 32" or "found old barbed-wire fence known to have existed more than 20 years" or "gradient line of Red River" or "mean high-tide line" or "centerline of existing concrete paving" or "Eucalyptus tree" or "hole in ice caused by warm spring" or "found stone marked with three notches south and two notches west" all convey the idea of a found existing monument. If the person who originally set the found monument can be identified, he should be mentioned.

All monuments set should be identifiable in the future by some unique means such as the surveyor's license number firmly attached or stamped on all set monuments.

14-15 Cultural Improvements

Cultural improvements, such as cultivation, barns, houses, streets, paving, sidewalks, storm drains, sewers, power lines, telephone lines, and any encroachments, may have a controlling consideration on property ownership and are usually shown on survey plats. In litigation involving unwritten rights these features are essential data for the court. Ordinarily, the position of any of these improvements is shown relative to some boundary line as defined by a written conveyance. Conventional symbols, as used on topographic maps, represent cultural features.

14-16 Title Identity

Title identity is established by identifying the position of "record monuments" relative to the property being surveyed. A call of "lot 10, block 3, Barn's Addition" is a call for a record monument. It can happen that lot 10 may not be marked by a single physical monument; yet Lot 10 is a record monument. A call of "to the east line of section 12" is a call for a record monument.

Record monuments on the ground are identified by locative points (physical monuments) or measurements from locative points. Both record and locative monuments are essential to identify property, and both are necessary features of a survey plat.

Nowadays, irrespective of whether a plat is for the purpose of showing the results of an original survey, a resurvey, or a survey of a property as based on the record, it is always necessary to show title identity. In former years it was possible to survey and describe a parcel of land that was not near any other parcel of land, but that condition no longer exists. For the purpose of obtaining a clear title, it is now necessary to show the relationship of the parcel being surveyed to that of other parcels of record.

This does not mean that title identity must be established for junior landholders. Ordinarily, only those with senior or equal rights (senior parcels, lot lines, section lines, original lines marked and surveyed, etc.) are identified on plats. At times junior title holders are shown, but it should be made clear that they are junior in standing.

PLATS OF SURVEY RESULTS

14-17   Effect of Plat Showing Survey Results

A survey plat properly certified (1) serves as evidence of the limits of title rights of the client and (2) serves as evidence of the surveyor-client contract.

A survey plat, showing the results of a survey, is merely evidence for the benefit of both the client and the surveyor. The client can use the plat to show to neighbors or buyers, to prove area for taxation, as evidence in court, and as a basis for an American Title Association policy. The surveyor can use the plat for ready reference when performing other surveys in the area, and he can clearly state the limits of his responsibility. The plat serves as a conclusion to a contract and should clearly define what has been done and what is certified in exchange for the fee paid.

14-18   Plats of Title Surveys for Title Associations

Various title associations have established minimum standards for acceptable title surveys wherein location as well as title matters are insured. The proposed minimum standards for land surveys made for title insurance purposes in Florida are the following:

> A land survey to be acceptable for title insurance purposes must be a full and complete survey showing every detail affecting title, and shall be certified to the insuring agency as meeting the following minimum standard requirements approved by the Florida Society of Professional Land Surveyors [F.S.P.L.S.] and the Florida Land Title Association [F.L.T.A.]. The certificate shall read as follows:
>
> "I hereby certify that the survey represented hereon meets the minumum standard requirements approved and adopted by the F.S.P.L.S. and the F.L.T.A."
>
> All measurements made in the field must be in accordance with the United States Standard and made with a transit and steel tape, or other modern devices proven equal or superior. All measurements shall refer to the horizontal plane. All computed distances and bearings must be supported by careful and accurate preliminary measurements made as required above. Wherever possible, the accu-

racy of the field work thus performed shall be substantiated by the computations of a closed traverse. The relative error of closure permissible shall be no greater than the following:

Locality	Max. Error of Closure
Commercial and high risk areas	1 foot in 10,000 feet
Suburban	1 foot in 7,500 feet
Rural areas	1 foot in 5,000 feet

The plat drawn of that survey must bear the name, address, certificate number, and signature of the Land Surveyor in responsible charge. The date of the survey must be shown.

A reference to all bearings shown must be clearly stated, i.e., "Bearings shown refer to True North," or "Bearings shown refer to Grid North as established for the Eastern Peninsular portion of Florida by the U.S.C. & G.S.," or "Bearings shown refer to Assumed North based on a bearing of S 5°-30'-00" W used for the centerline of County Road No. 100," or "Bearings shown refer to the Deed Call of N 22° E for the easterly line of Lot No. 4," etc. References to Magnetic North should be avoided except in those cases where a comparison is necessitated by a deed call. Where bearings are recited in the deed description, or on an original plat of the lands being surveyed, a comparison of the deed or plat bearings with the bearings used shall be shown on all courses. In all cases the bearings used shall be referenced to some well established line.

A "North Arrow" shall be prominently shown and, whenever possible, placed in the upper right hand corner of the plat.

The caption of the prepared plat must be in complete agreement with the record title. Whenever possible the correct caption will be furnished by the Title Company concerned along with a record description to which the survey must refer. In all cases the survey must make reference to and identify the source of information used in making that survey, such as: the recorded deed description, or other conveyance; a recorded or unrecorded plat; or other claim of rights.

Where inconsistencies are found, such as; overlapping descriptions, hiatuses, excess or deficiency, erroneously located boundary lines and monuments, or where any doubt as to the location on the ground of survey lines or property rights exists, the nature of the difficulty should be shown and the range of possible differences indicated upon the plat.

All outer boundaries and any interior subdivision lines lying within the lands surveyed must be shown upon the plat with all the data necessary to determine the correctness of the survey by both mathematical computations and by plotting.

All angles must be either given directly or indicated by the bearings shown. Where lines are curved, the significant elements of the curve, such as; the radius, the arc length, and the bearing and distance of the chord, must be shown. Centerline data for an adjoining curved right-of-way should also show sufficient information to permit the computing or plotting of the curve in its entirety, such as: the Central Angle (Δ), the Tangent (T), and the Radius (R). The degree of curve shall be shown only in those cases where it is recited in the deed description or shown upon a controlling plat.

In all areas where lots and blocks are established, the distance to the nearest intersecting street or right-of-way must be shown. Distances to intersecting streets or rights-of-way in both directions must be shown if either of the distances vary from the original plat. Surveys of parcels described by metes and bounds within a large tract of land should show the relationship of that parcel to at least one of the exterior lines of that tract, preferably by the bearing and distance along an established right-of-way.

If the survey is of all, or any part, of a lot, or lots being a part of a recorded subdivision, all lot numbers, including those of adjoining lots, and the block number must be shown upon the plat.

All legal public and private rights-of-way shown on record plats adjoining or crossing the lands surveyed shall be located and shown upon the plat. Where ties are made to an intersecting street or other right-of-way pertinent data concerning the right-of-way shall also be given. Changes in rights-of-way should be noted and the date of and the authority for the change shown. Where a proposed change in a right-of-way is known, such as; the knowledge that appropriations have been made by the authorities to acquire additional right-of-way for the widening of a public thoroughfare, that proposed change should be shown and noted as such upon the plat. The Land Surveyor, however, is not to be held responsible for the accumulation of such information. If streets abutting the lands surveyed are not physically opened, a note to this effect should be shown.

The character of all evidence of possession along the boundary lines, whether fence, wall, buildings, monuments, or otherwise, must be shown and the location thereof carefully given in relation to reference or record description lines.

All encroachments such as; eaves, cornices, doors, blinds, fire escapes, bay windows, windows that open out, flue pipes, stoops, steps, trim, etc., by or on adjoining poperties or on abutting streets must be indicated with the extent of such encroachment clearly shown. Openings, such as; windows, doors, etc., in walls of premises or adjoining premises adjacent to the boundary lines must be noted. Whenever possible, foundations near the boundary lines should be located and shown upon the plat. In all cases, where the locating of foundations is prevented by pavements or other obstructions, the failure to determine the location shall be noted upon the plat. If building on premises has no independent wall, but uses any wall of adjoining premises, this condition must be shown and explained. This also applies when conditions are reversed.

All physical evidence of easements or rights-of-way created by roads, surface drains, telephone or telegraph lines, electric or other utility lines on or across the lands surveyed must be located and noted upon the plat. Surface indications or markers of underground easements should also be shown. If location of easements of record, other then those on record plats, is required, this information must be furnished to the Land Surveyor.

The character, construction and location of all buildings on the lands surveyed must be shown and referenced to the boundaries. On large tracts, buildings and structures remote to the boundary lines may be shown in approximate scaled positions and noted thus upon the plat. If interior improvements are not located, a note to this effect such as; "interior improvements, if any, not located," must be shown. The same rules shall apply with respect to interior cross fences. Joint or common driveways must be indicated. Independent driveways must be shown completely to the property line.

Cemeteries and burial grounds located within the premises must be properly located and shown upon the plat.

All monuments, stakes and marks found or placed must be adequately described and data given to show the actual location upon the ground in relation to boundary lines.

Redates of "foundation" or "under construction" surveys must take into consideration all of the requirements contained herein. Special attention should be given to any indication of changes and to any encroachments that may have arisen since the date of the original survey, including, but not limited to, possible encroachments of driveways, aprons at the corners of irregular lots on curved streets.

If survey order indicates F.H.A. [Federal Housing Administration] or V.A. [Veterans Administration] insurance of the title for which the survey is to be used, compliance with the appropriate requirements must be made. City, County, or State requirements must be complied with where applicable.

Most of these standards, as in the foregoing, fail to use an effective method of specifying accuracy. The error of closure, as noted in Chapter 10, is never an indication of accuracy; it merely indicates a possible absence of blunders.[1]

14-19 Surveyor's Certificates

On every survey plat the surveyor should certify about what he has done. The form used varies, depending on the purpose it is to accomplish. In populated areas, usually the client's purpose of having a property surveyed is to determine the status of encroachments as in this certificate:

Certificate of Survey

This is to certify that the hereinbefore plat correctly portrays a survey made under my direction of Lot 12, Block 3 of Grantville, and that the monuments shown thereon are to the best of my knowledge and belief wholly in accordance with Map thereof No. 4390 filed in the office of the County Recorder of . . . County, State of . . . , and that there are encroachments thereon as shown above, but no others.

John Doe
Licensed Land Surveyor No. 2554
Sealed.

In connection with title insurance surveys, the Pennsylvania Title Association requires:

To Whom It May Concern:
This is to certify that this plat and the survey on which it is based were made in accordance with Mininum Standards for Title Surveys adopted by the Pennsylvania Title Association on May 20, 1961, and that this plat correctly sets forth all data obtained in the said survey.

Signed by the surveyor. License No. . . . .

The title company limits its policy to persons named on the face of the policy, yet it asks surveyors to guarantee location for present as well as future owners

---

[1] See "Minimum Standard Detail Requirements for Land Title Surveys," *Title News*, Vol. 41, No. 5, 1962.

by using "to whom it may concern." The better procedure for the surveyor is to state specifically for whom he has made his survey, such as John Doe and A.B.C. Title Company.

14-20 Contents of Plats

A plat should be complete in itself and should present sufficient evidence of monuments (record and locative) and measurements so that any other surveyor can clearly, without ambiguity, find the locative points and follow the reasonings of the surveyor. A plat does not show the client's land alone; it shows all ties necessary to prove the correctness of location. If it is necessary to measure from a mile away to correctly locate a property, that tie, as measured, is shown.

On most plats it is necessary to show the following:

1. Record monuments called for (title identity), including abutting streets and easements.
2. Found physical monuments that locate the record monuments and a complete description of these monuments.
3. Proof of correctness of the found monuments (history).
4. Notation of monuments called for but not found.
5. Basis of bearings.
6. Expression of measurements on all lines.
    a. Direction.
    b. Distance.
    c. Coordinates (not always given).
    d. Curve data (central angle, radius, length, chord, tangent, etc.).
7. All monuments set and their descriptions (including a description of monuments replaced).
8. Oaths or witness evidence.
9. Date of survey.
10. Client's name.
11. Surveyor's certificate (signature, seal, statement of accuracy, and guarantee of location in accordance with a particular description furnished).
12. Easements located in accordance with descriptions furnished.
13. Encroachments and possession on the title lines as described in the title furnished.

ORIGINAL PLATS

14-21 Purpose

Original surveys have two parts: (1) a boundary survey that is a resurvey or a survey as based on the record and (2) newly created parcels of land. Original plats of original surveys show the results of a resurvey or a survey based on the record and show how new parcels of land were created. On an original survey plat the exterior boundary must show all the information previously discussed; in addition, the plat also shows new divisions of land and the dimensions and locative points that determine their location.

14-22 Laws Regulating Platting

The creation of new parcels of land is usually, though not always, regulated by law. When the law does regulate how land must be divided, the surveyor must obey the law; in many areas, how plats shall be made and their scale and size, the amount of information disclosed, and numerous other items are all specified by platting acts.

The regulations contolling the contents of plats often include the following:

The final plat shall conform to all of the following provisions:

(a) It shall be a map legibly drawn, printed, or reproduced by a process guaranteeing a permanent record in black on tracing cloth or polyester base film, including affidavits, certificates and aknowledgments, except that such certificates, affidavits and acknowledgments may be legibly stamped or printed upon the map with opaque ink when recommended by the county recorder and authorized by the local governing body by ordinance. If ink is used on polyester base film, the ink surface shall be coated with a suitable substance to assure permanent legibility.

(b) The size of each sheet shall be 18 by 26 inches. A marginal line shall be drawn completely around each sheet, leaving an entirely blank margin of one inch. The scale of the map shall be large enough to show all details clearly and enough sheets shall be used to accomplish this end. The particular number of the sheet and the total number of sheets comprising the map shall be stated on each of the sheets, and its relation to each adjoining sheet shall be clearly shown.

(c) It shall show all survey and mathematical information and data necessary to locate all monuments and to locate and retrace any and all interior and exterior boundary lines appearing thereon, including bearings and distances of straight lines, and radii and arc length of chord bearings and length for all curves, and such information as may be necessary to determine the location of the centers of each curve.

(d) Each lot shall be numbered and each block may be numbered or lettered. Each street shall be named.

(e) The exterior boundary of the land included within the subdivision shall be indicated by colored border. The map shall show the definite location of the subdivision, and particularly its relation to surrounding surveys.

(f) It shall also satisfy any additional survey and map requirements of the local ordinance.

—Calif. Subdivision Map Act.

14-23 Compilation Plats

To the surveyor the most useful of the compilation plats is the arbitrary plat prepared by an abstractor or title-insuring agency. It is a compilation of all land descriptions within an area showing controlling measurements and other data.

Not all deed dimensions are shown; those that are a controlling consideration are always emphasized to show their importance. In a deed reading "N 20°10' E, 128.02' to the westerly line of that land described in . . . ," the notation on the platted line would be "N 20°10' E, 128.02' ±." The abstractor has shown that the call of 128.02' is more or less in character, since the distance must go to the record monument called for.

Abstractor plats vary from office to office, and the contents of the plats depend on local customs or local abbreviations. Each state has its own recording system or method of numbering plats, and the method of disclosing recording data varies. The following notations and their meaning are sometimes used:

TK 10,921 (Tract Map 10,921).

B1721/231 OR (Book 1721, page 231, Official Records).

B122/31 D (Book 122, page 31, Deeds).

SC52127 (Superior Court Case 52,127).

Using these notations, an abstractor's plat could be as shown in Fig. 14-3. In this figure each parcel is given an arbitrary number; the plat is called an "arbitrary plat" or just an "arb plat." On the plat arb Lot 1 is clearly a 200 × 200 feet parallelogram as measured along each street. Arb Lot 2 is 200 × 200

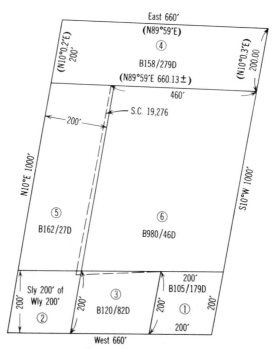

Figure 14–3 Abstractor's plat.

feet (westerly 200 feet of the southerly 200 feet of Lot 10) as measured at right angles to each street. Arb Lot 3 is a remainder between 1 and 2 and is formed by a straight line connecting the corners of 1 and 2 as shown. Lot 4 was created next and presents no difficulties. Lot 5 was sold as the westerly 200 feet, minus 2, 3, and 4. When Lot 6 was sold it overlapped 5 and 3. The deed author failed to recognize the 200 feet as being measured at 90° to the street. The overlap on 5 was settled by Superior Court Case No. 19276. The overlap on 3 still exists, although 3 is senior and is entitled to the disputed area. This arbitrary plat clearly shows the conditions existing and who has senior rights as defined by written titles.

14-24 Summary

Original survey plats and resurvey plats are similar in cartographic expression. Original survey plats are called for by a conveyance, and the contents of the plat are considered as part of the writings. A resurvey plat serves as evidence of what the surveyor did in following and relating evidence to the ground.

On an original plat the object is to explain how to identify and locate a newly created parcel of land. On a resurvey plat the object is to present sufficient evidence to prove that the surveyor correctly identified and located a previously described parcel.

An original plat usually shows the results of a resurvey of a boundary.

The object of all survey plats is to show sufficient information to enable any competent surveyor to identify with certainty a given parcel of land by the locative points (physical and record monuments), by the measurements, and by the scaled drawing shown.

CHAPTER 15

Writing Descriptions

15-1 Contents

The law in general requires that conveyances should (1) be in writing, (2) identify grantors and grantees, (3) identify the interests being conveyed, (4) express an intent to convey the interest identified, and (5) identify the location of the land in which the interests are being conveyed. This chapter pertains to item 5.

A *land description* identifies by writings the physical location of the land interest conveyed. Whether that interest is a fee title, life estate, easement, royalty, or other enjoyment is not the subject matter of this chapter.

Words used in land descriptions have a presumed order of importance attached to them; that is, a call for a monument has more force than a recited distance. Topic headings are arranged to follow this presumed order of importance.

15-2 Graphic and Written Descriptions

The purpose of a land description is to identify a particular area of land. There are many, many ways to identify an area; hence land descriptions take many forms.

There are two fundamental methods of identifying a land area: (1) by a call for a graphic representation (plats) and (2) by written words. All graphic descriptions include some written words and many symbols. For purposes of explanation, this discussion of description writing is divided into graphic problems (including symbols and words commonly used as part of a graphic presentation) and wholly written descriptions.

Many descriptions combine both graphic representation and written words. Descriptions logically fall into three classes: (1) those calling for a graphic description (lot and block on a given map), (2) those describing an area by written words (metes and bounds), and (3) those using combinations of 1 and 2.

Areas are enclosed by a series of lines, and in metes and bounds descriptions these lines start at a point, have a direction of travel, do not cross, and return to the point of beginning. Essentially, knowing how to describe an area is knowing how to describe a sequence of lines.

15-3 Subdivision Descriptions

Subdivision descriptions identify land by reference to a complete parcel of land represented on a plat. The most common examples of such descriptions are (1)

Lot 2, Block 27, Hillside Estates, according to Map Number 2761 filed ... and (2) Section 12, T 3 S, R 2 E, Third Principal Meridian.

The term *subdivision descriptions*, as used herein, refers to entire parcels described by reference to a plat. Descriptions describing a part of a parcel, such as "the westerly 100 feet of Lot 7," are herein considered as intermediate types.

Reference to a plat or map in a land description incorporates into the description all the data or facts written on the plat; knowledge of how to prepare subdivision descriptions is a knowledge of how to incorporate the data revealed on specific plats. Subdivision descriptions are "reference descriptions"; they call for a plat that identifies the land.

Some subdivision descriptions have a call for both a plat and a metes and bounds description. These often lead to confusion, since one may conflict with the other. The U.S. government in conveying sectionalized lands sold the land by reference (Sec. 10, T 3 S, R 2 E, SBM), and the reference automatically included the field notes. The field notes are a written perimeter description of each section, having all the similarities of metes and bounds descriptions. Many states, when alienating their lands, also required a survey, a plat, and field notes.

15-4 Metes and Bounds Descriptions

The term "metes and bounds description" has a general but not a specific meaning. The word *metes* means to measure or to assign by measure, and the word *bounds* means the boundary of the land. These words are similar and probably represent the same thought. As used in deeds, the word *bound* is sometimes used in a restricted sense when referring to a monument as a "stone bound."

> Bounds means the legal, imaginary line by which different parcels are divided.
> —*Walton* v. *Tifft*, 14 Barb. 216.

> Metes and bounds mean the boundary lines or limits of a tract.
> —*Moore* v. *Walsh*, 37 R.I. 436.

As commonly used by surveyors, the *metes and bounds description* means complete perimeter descriptions wherein each course is described in sequence and the entire description has a direction of travel around the area described. The distinguishing feature of this type of description, herein called *true metes and bounds description*, is that each course identified must be described one after another in the same direction of travel that would occur if a person walked around the entire perimeter. Either of two directions can be used, clockwise or counterclockwise, but once a direction is selected it must be consistent for the remainder of the description. Technically, all written descriptions can be metes and bounds descriptions, even those calling for a lot and block by referral. But surveyors and the public do not, by common usage, classify lot and block descriptions as metes and bounds descriptions.

Rare indeed is a metes and bounds description written without a "reference" call. A description reading, "Beginning at the southwest corner of section 10;

thence, . . . " has a reference call for the township plat and the field notes. Metes and bounds descriptions often state, "Along the easterly property line of . . . "; this is a bounds call and includes any plat called for by the adjoiner deed. "According to the map of Hillside Acres" is a reference call for a plat and survey.

Within a metes and bounds description, taking the term in its broadest sense, lines defining the limits of the area can be classified as being described (1) by measurements and (2) by monuments (bounds). A line described by measurement must have a starting point, a distance, and a direction. The term "monument" is used here in its broadest meaning.

Specific names have evolved to describe certain types of metes and bounds descriptions such as proportional conveyances, bounds conveyances, linear conveyances, strip conveyances, area falling on one side of a line and true metes and bounds conveyances. Some names, such as "bounds descriptions," have become so commonplace that title men refer to them as a distinct type.

15-5 Bounds Descriptions

A bounds description calls for an adjoiner or monument on each side of the described area; it is a referral description.

A description reading "bounded on the north by the Ohio River; bounded on the south by Jones; bounded on the west by Rose Creek, and bounded on the east by lot 12, Acre Park Subdivision" is bounded by two natural objects (Ohio River and Rose Creek) and two legal monuments (Jones and Lot 12). The lack of necessity of direction of travel around a bounds description distinguishes it from true metes and bounds descriptions.

Many of the descriptions of the New England area are of the bounds type. A large number just name the abutter or a road, and it becomes necessary to do a title search back to the point where a metes and bounds or some other type of description can be found.

Bounds descriptions are best used to describe a remainder (a junior title holder). Bounds descriptions calling for an adjoiner junior parcel should never be used. Let the junior parcel call for the senior, but do not let the senior call for the junior. New England scriveners did not understand this.

15-6 Proportional Conveyances

Proportional conveyances are a fractional part of a whole area. Ordinarily, unless a statute exists to the contrary, proportional conveyances are a proportion of the total area. In sectionalized lands a federal statute defines the method of surveying the SW 1/4; hence it is not a proportion of area.

Statutory proportional conveyances are those described by fractional parts (SE 1/4 of SE 1/4, Section 2) and are not mathematically proportional parts of the area. Fractional parts conveyed under common law are always proportional parts of the area; the only exception is the sectionalized lands where statute defines otherwise.

A proportional conveyance can be a fractional part of either a subdivision description or a metes and bounds description. Ordinarily, the term is applied only to whole lots in a subdivision.

15-7 Area Described as Being on One Side of a Described Line

Areas can be described as being all of a certain described parcel lying on a given side of a line, said line being described by monuments or measurements. Examples are "Lot 12, except the easterly 20 feet," "Lot 12 lying westerly of the easterly 20 feet," "all of Smith's land lying southerly of the following described line," and "all of section 10 lying northerly of Alvarado Creek."

15-8 Strip Conveyances

By describing a line and by including a given number of feet on each side or one side of the line, an area is described. The strip conveyance is a special form of the description given in Section 15-7.

15-9 Area Conveyances

Area conveyances (north 5 acres of Lot 5) are used to convey a specific quantity of land. To be effective one dimension of the area must be omitted, so that area becomes the determining quantity to calculate the omitted dimension.

Generally, area conveyances are written without the benefit of a recent survey. The "north 10 acres of Lot 12" conveys exactly 10 acres. If there were a recent survey, a better description could be written. This description is uncertain, since one dimension is omitted (direction of the southerly line of the north 10 acres). An infinite number of areas is possible. Even though the courts have, at times, given definiteness to such descriptions, writing an area description without defining the direction of the dividing line is creating an ambiguity.

15-10 Linear Conveyances

Such descriptions as "the easterly 50 feet of lot 16" are linear conveyances. The wording of the description is short, concise, and exact, provided everyone understands that the "easterly 50 feet" means "the easterly 50 feet is measured so as to give the person owning the 50 feet the maximum amount of land." This description is a combination of a "subdivision description" and a linear distance and is just as certain of location as are the lots on the plat called for.

Linear conveyances can be used with metes and bounds descriptions as "the westerly 50 feet of the following described property. . . ."

15-11 Indispensable Parts of a Description

PRINCIPLE. *A description is legally sufficient when it relates how to identify a particular area of ground relative to monuments on the face of the earth.*

Descriptions never completely describe a particular area of land; they merely relate how to find it. As yet humankind has not invented a system whereby land can be identified without reference to monuments. Even longitude and latitude are relative to the earth's poles and the Prime Meridian. Every description must somehow *identify an area relative to fixed monuments.* Although some descriptions seem to be written without a call for a monument, the call, in actuality, is always there. A call of "Lot 1, Block D, according to Map 2201" calls for all the monuments originally set by the surveyor when surveying Map 2201. A call for an adjoiner also is a call for any monuments necessary to locate the adjoiner. Measurements of bearing and distance are always started relative to a monument on the ground. Bearing itself may be relative to a monument (north or south pole).

In legal literature and elsewhere it is often stated that a conveyance must identify a *unique* area, meaning one, sole, lone, or single area. A conveyance cannot be equally applicable to two areas; it must be unique in this respect. The interest being conveyed can be attached to land being used for some other interest. The right to use an underground location or the sky rights to erect a building above a certain elevation may all attach to the same description; it is the location that needs to be unique, not the interest.

Again, a description may overlap another described area; the description as a whole is not always unique with respect to its entirety. A conveyance that overlaps another is not always void or even, at times, voidable; hence described areas with respect to their entirety are not always wholly unique.

15-12 Objectives when Describing Land

PRINCIPLE. *A land description to be legally sufficient in the matter of locatability must identify a particular locatable area or areas to which the interest conveyed attaches. It is desirable that a land description (1) should contain title identity, (2) should not interfere with the senior rights of others, (3) should be so written that either at the present or at a future date it can be readily located by a competent surveyor, (4) should not contain words capable of alternate interpretations, (5) should contain measurement data sufficient to describe a geometric area that closes mathematically, and (6) should be based on a recent survey.*

15-13 Sufficiency of Descriptions

The purpose of a description is to identify a particular area, not to fully describe it; a description is sufficient if it identifies the land. What is legally sufficient and what is desirable are two entirely different things.

Competent title authors strive to make a description complete within itself; that is, the description recites all senior claimants, describes all monuments, describes the size and shape by measurements, and agrees with existing possession on the ground (a fact that requires a recent survey).

The courts try to declare a conveyance valid rather than void. Only in cases where the area described by the writings is equally applicable to more than one location or where the deed fails to identify an area does the court reluctantly declare a deed void.

How land may be described and what is sufficient are properly regulated by law, and, if the law prohibits a certain form of description, it may not be used. The right to refer to plats or maps (subdivision descriptions), without including a perimeter description, is excluded in patenting federal mining claims and in describing certain political boundaries. In some areas building-site conveyances by metes and bounds descriptions are illegal.

15-14 Title Identity

Title identity is the relationship between a particular description and its adjoiners. Certaintly of location can be attained without certainty of title identity as in the following description body:

> Commencing at a 40 inch blazed pine tree Zone 3 ——— State Plane Coordinates of $x =$ ——— and $y =$ ———; thence east 300 feet to a 38" blazed pine tree; thence north 301.00 feet to a 60" oak tree; thence west 300 feet to a stone mound; thence south 301.00 feet to the point of commencement.

This description is certain of location but is completely unidentifiable relative to adjoiner titles. Whether the land overlaps another's rights is not stated, nor can it be determined without a survey. If this description describes land not owned by the grantor, the conveyance is without force. Title-guaranteeing agencies generally will not issue title insurance on the basis of such titles; it is not a marketable title.

When composing a new conveyance the most desirable description from the title-guaranteeing agencies' viewpoint is one calling for adjoiners without reciting measurement information other than that necessary to complete a perimeter such as shown here:

> Beginning at the southwest corner of Thomas Brown's land as recorded in . . . ; thence along the southerly line of said Brown's land to the southeasterly corner thereof; thence in a straight line to the northeasterly corner of Doris Muller's land as recorded in . . . ; thence westerly along the northerly line of said Doris Muller's land to Third Street; thence along Third Street to the point of beginning.
>
> Bounded on the north by the land of Thomas Brown as recorded in . . . ; bounded on the west by Third Street; bounded on the south by the land of Doris Muller as recorded in . . . ; bounded on the east by a straight line connecting the southeasterly corner of said Thomas Brown's land and the northeasterly corner of said Doris Muller's land.

In this type of description exposure to damages resulting from discrepancies between monument locations and measurement data is eliminated; size is not stated. But such a description is or should be unsatisfactory to the new owner; it does not disclose on its face the quantity of land purchased. To be satisfactory to a new owner or to a surveyor, a description should be dimensionally complete.

Although title identity is not necessarily essential for validity, it is a part of every good description. To the title insurer the most important feature of a land description is its stated relationship to adjoiners, whether it is readily locatable is of little moment. To the surveyor the most important feature of a land description is its simplicity of location on the ground and its dimensional completeness; title identity is a bother. It is this difference of viewpoint that causes opinions to clash. Both are essential features of good descriptions.

15-15 Senior Rights

Whether a conveyance interferes with the rights of an adjoiner or not does not necessarily alter the validity of the conveyance; but, in instances, it may operate to make a conveyance partially void. If an author inadvertently included portions of an adjoiner's lands, the land inadvertently included is not conveyed, since a person cannot sell another's land. Such inclusion does not operate to void the remainder of the conveyance, since the grantee can claim the land remaining after the overlap is excluded. Because a person may be held liable for selling land not his, it is *desirable* in composing descriptions to exclude lands of others. From a title insurer's point of view, it is a necessity that others' lands are excluded; the title insurer is liable for such inclusions, and by recent court opinions, in some states, the surveyor may also be liable.

15-16 Effect of Loss of Evidence on Location

PRINCIPLE. *If a written conveyance is valid, recorded, and locatable when executed, no later act or event will operate to make the conveyance nonlocatable.*

The fact that the conveyance can be located at the time it was made is all that is necessary; later destruction of monuments, along with the death of all witnesses who knew where the monuments were and resulting confusion over location cannot operate to make a description nonlocatable or void. The land must be located; it is a matter of evidence and law that determines where the land belongs. Descriptions, when composed, should contain sufficient words to insure certainty and ease of future location. Unfortunately, since this is not a legal necessity, many (especially title people) compose descriptions without considerations for future location.

Unrecorded deeds are valid as between the parties but generally not as against the rights of innocent parties. Usually, if a party sells land twice, the first recording is recognized. If the second buyer records first, and he has no knowledge of the first sale, he can usually obtain ownership.

Surveyors want descriptions dimensionally complete; that is, related accurate bearings and distances or some other accurate equivalent dimension should be recited for all lines. Monuments called for should be permanent or certain of future location.

15-17 Ambiguity

English words often have multiple meanings and, if improperly selected, may cause ambiguity. Words in one part of a description may conflict with words in another part. Ambiguities or conflicts in general do not operate to void a conveyance; they merely require legal interpretation and sometimes court action to clarify the meaning of the description. Thus, in the event that "North" is written instead of "South," the deed is not necessarily void, since errors can be corrected if sufficient writings remain to identify the land. Conflicts or ambiguities are eliminated from the writings by competent scriveners.

15-18 Mathematical Correctness

In many existing conveyances the area described is not mathematically correct and sometimes is not geometrically correct. This does not in itself invalidate a conveyance, since identity is all that is required. If a surveyor can locate the land described and measure and determine its area, the conveyance is sufficient. When there is a call for monuments (including record monuments) it is almost impossible to assure mathematical and geometric closures without an accurate survey. Unfortunately, mathematical and dimensional completeness of a description is not a legal necessity; it is merely a desirable feature of a good description.

15-19 Description Based upon a Survey

Unwritten conveyances or prolonged possession may transfer title to lands without writings. Possession is one of the many necessary elements of ownership. The purpose of a deed is to convey the ownership of land; without knowledge of a possession, relative to the described lines, certainty of transfer of ownership is not always possible.

Occasions do arise wherein a description can be written without benefit of a survey, but it usually takes expert knowledge to determine when a survey is not necessary. If a property is well monumented by definite monuments, it may be a simple matter to describe a rectangular parcel.

Nowadays, especially in urban areas, the demand for title policies that guarantee both title and possession is greatly increasing the necessity for surveys. In many areas, new divisions of land are prohibited without surveys.

15-20 Desirable Qualities of a Scrivener

Competent scriveners should have a knowledge of (1) trigonometry, geometry, and all elementary mathematics, (2) the legal meaning of all words and phrases used in land descriptions, (3) the location of known controlling monuments in the area of the description, (4) the science of measurements and calculations to prove the correctness of measurements, (5) senior rights of adjoiners, and (6) the order of importance of conflicting elements.

Unfortunately, the activities of scriveners are seldom regulated by law, and many scriveners are incompetent to perform their duties. Those better qualified are surveyors, title-company employees, abstractors, and attorneys. Attorneys specializing in title work are eminently qualified about the legal meaning of words and phrases used in land descriptions, and they know how to use proper conveyance forms. But they are often weak in mathematics, and the science and techniques of measurements and lack knowledge of possession on the ground.

In some states, wherein title companies or abstractors keep complete title records, that is, sequence of conveyances, the knowledge of employees with regard to seniority of conveyances and legal meaning of words and phrases qualifies them to compose descriptions that do not need the benefit of survey. Unfortunately such land descriptions, although legal, are often difficult or almost impossible to locate on the ground owing to the uncertainty of monument locations.

The skilled surveyor is eminently qualified in the techniques of measurements, mathematics, and monument locations and is usually qualified in the legal interpretation of words and phrases, but he is often weak in the knowledge of the rights of adjoiners. In some states, such as in the sectionalized land areas, the necessity of knowledge of seniority of conveyances is minimized, whereas in other states, such as Texas, the necessity of seniority knowledge is imperative.

Better land descriptions can be made with the benefit of survey. Others occupying the area described may have occupancy rights, or monuments may be in positions other than that indicated by the record.

Subdivision descriptions of land (lot and block by map) are a surveying approach to a description problem. Only the qualified surveyor should prepare such maps, although the preparation of the land description (Lot 1, Block C of Map 1842) can be done by many.

Perhaps, the best land descriptions of the metes and bounds type are prepared by the team work of the surveyor and those engaged in title-guarantee work. The surveyor locates the lines, describes the lines, and certifies as to possession; and the title-guaranteeing agency, abstractor, or attorney seeks out possible adjoiner rights that might overlap the property described and corrects legal defects.

15-21 Changing Description Wordings

Once a description is written and a conveyance is made using that description, no matter how poorly it was made, it is rarely advisable to change it, except by agreement deed between the interested adjoiners or by court order. The object in changing a poor description is generally for the purpose of including additional evidence that makes future location of the property more certain. But who is to say what evidence may be added without changing the original intent of the description? A deed reading "commencing at a white oak; thence N 10° E a distance of 200.00 feet; thence S 80° E a distance of 300.00 feet; thence, . . . " could be more locatable by changing the form to "commencing at a white oak located N 16° E, 300.00 feet from the southwest corner of Lot 10; thence N 10° E along a rail fence 200.00 feet to an iron pipe and disc stamped LS 2554; thence S 80°20' W along a rail fence 300.00 feet to a found iron axle;

thence...." But no title person would insure or accept such changes without a written agreement between the adjoiners. The new description changes the basis of bearings (along a rail fence), adds a monument (iron axle) not intended in the original deed, and also changes the deflection angle. A new monument added to a description often changes the intent of the description.

Some items may be added to descriptions without altering the original intent and may improve the description. A description reading "commencing at a white oak; thence N 10° E a distance of 200.00 feet to an old Ford axle; thence S 80° E a distance of 300.00 feet to an old Ford axle; thence..." is greatly improved by writing as "commencing at a white oak found to be located N 16° E, 300.00 feet from the southwest corner of Lot 10; thence N 14°20′E astronomic north (N 10° E magnetic north), as determined by surveyor Nolan, for a distance of 200.00 feet to an old found Ford axle; thence S 75°40′ E (S 80° W magnetic north), a distance of 300.51 feet (record 300.00 feet) to an old found Ford axle; thence...." Since monuments control over bearing and distance, correcting the errors of measurements does not alter the intent to go to the monuments. But even in harmless instances title people object to changes and often will not insure changes in wording.

As stated, the object of changing a deed is usually to add evidence that makes location more certain. But such evidence can be perpetuated by other means. The best way to correct errors and discrepancies in a land description is to file a public record of survey disclosing evidence found and new points set (see Section 11-5). The surveyor discovering an error in the deed wording and filing a public record puts all future surveyors and title people on notice of the facts. The title-insuring agency, abstractor, or court can judge on the merit of the evidence. Evidence of this generation is often lost to the next generation; hence it is advantageous to have surveyors give public notice of discovered evidence and to give new tie-outs to found old evidence. This method of preserving evidence is superior to arbitrary attempts of one person to change a deed's wording.

15-22 Technique of Writing

The art of writing descriptions is not simple; carelessly used words—or omission of words—can alter the entire meaning of a conveyance. Too many words may cause conflicting statements in a deed, whereas too few words may cause ambiguity and uncertainty. The best deed authors use a minimum of terms that give a clear intent without error, conflict, or ambiguity. The selection of the proper words to use comes from knowledge, experience, and practice. It is not the verbose writer filling many foolscap pages who wins acclaim in writing descriptions; the writer who is applauded is the one who condenses but omits nothing essential, who creates no conflicts and is clear.

15-23 Parts of a Description

Descriptions are divided into four parts: caption, body, qualifying clauses (including reservations), and augmenting clauses. Although in writing a particular description the four parts may be intermingled to make it difficult to distinguish one part from another, it is better practice, when feasible, to keep the four parts distinctly separated. In general, the distinctions between the four parts are as follows: the caption recites a general area or locality and directs attention to a general vicinity; the body pinpoints a particular area in the given locality described in the caption; the qualifying clause takes back part of that given by the body or by the caption; the augmenting clause gives a right of usage of land outside that conveyed (usually easements). If the description author develops the habit in his writings of following this order just cited, he will make fewer mistakes.

An example of parts of a description follows:

> (Caption) All that portion of one quarter section 30 of Rancho de la Nacion according to Map thereof No. 166, surveyed by Morell and filed in the office of the Recorder of San Diego County, California, being more particularly described as follows:
> (Body) One acre of land forming an equal sided parallelogram in the southeast corner of said quarter section 30 of said map.
> (Qualifying clause) Excepting therefrom the easterly 100.00 feet.
> (Augmenting clause) And granting an easement for road purposes over the westerly 25 feet of the above excepted easterly 100 feet.

This description is not dimensionally complete, since the angle in the southeast corner of quarter section 30 is not recited; also no monuments are described.

15-24 Caption

A logical arrangement in writing descriptions is first to recite the general area or locality of the land so that attention is directed to a vicinity. The caption or introductory part of a description serves this purpose. A checklist of the items that often, but not always, appear in the captions is as follows:

1. State.
2. County (parish).
3. City.
4. Subdivision.
    a. Map name and number.
    b. Meridian, township, range, and section.
    c. Land-grant name.
    d. Court map.
    e. Any other identifiable map.
5. Recorded conveyance of which instant description is a part.
6. Place where record map or recorded conveyance is filed.

The wording of the caption may take many forms, either one of the following examples being sufficient.

> All that portion of Section 10, Township 15 South, Range 2 East, San Bernardino Meridian according to the United States Government plat filed January 12, 1885, located in .... county, state of ...., more particularly described as follows...
>
> All that portion of Lot 12, Block 15 according to tract Map No. 16,213 recorded in the office of the County Recorder of Los Angeles County, California, more particularly described as follows...

15-25 Body of Description

The body of a description identifies a particular land area within the locality designated by the caption.

The body, taken together with the caption, must identify a certain area; otherwise, the conveyance will be void. In addition the body ought to contain complete dimensional information that renders the intent of the deed more certain and clear.

Monuments called for in a conveyance locate the land. Monuments are subject to destruction. If a monument is to be replaced after it is destroyed, its measured position must be known before it is destroyed. Often land areas can be described without mention of measurements (calling for monuments alone). Such a description, although legally valid, is unsatisfactory, since no measurements indicate how to replace monuments if lost. The surveyor in describing land is vitally interested in the ease of future locatability; he includes writings designed to serve as means of identification in the event of monument destruction.

Courts, when interpreting the meaning of conflicting elements within a description, presume an order of importance. Explanations of how to write the body of descriptions follow this order of importance.

15-26 Senior Rights and Calls for Adjoiners

PRINCIPLE. *Junior descriptions should call for the senior description as an adjoiner or should be written with identical calls as that of the senior adjoiner.*

In the order of importance of conflicting deed elements, except where unwritten titles exist, senior rights rank first. If a junior title is being written that fact should be noted by inserting a call for the adjoiner.

Inserting a call for a senior claimant serves the purpose of title identity; it states the relationship between the instant property and the neighbor.

Senior deeds should never call for the junior claimant. If Jones' deed reads "bounded on the south by Brown" and Brown's deed reads "bounded on the north by Jones," confusion results. One or the other is senior, yet the fact is not disclosed; it appears that both are senior.

In Fig. 15-1, Black is senior, and Smith is junior. A description of the new

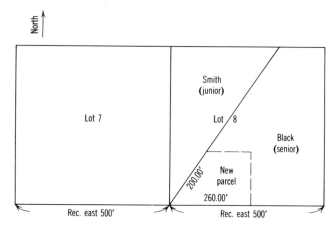

Figure 15-1

parcel being described and reading "beginning at the southwest corner of Lot 8; thence N 35° E, 200 feet along the line of Smith as recorded in . . . etc." is wrong in that it infers that Smith has the senior right.

The necessity of calling for an adjoiner can be avoided if the calls are identical. In Fig. 15-1, if Black's description read "beginning at the southeast corner of Lot 8, thence west 500.00 feet; thence N 35° E, 665.25 feet; etc." and Smith's deed, at the time it was written, commenced at the same point and called for both "west 500 feet" and "N 35° E, 665.25 feet," the two parcels would join, and the necessity of calling for Black as an adjoiner would be eliminated. But Smith's deed should not read "beginning at the Southwest corner of Lot 8; thence N 35° E, 665.25 feet; etc." A surplus or deficiency in Lot 8 would cause a gap or overlap. If identical calls are not used, a statement of "beginning at the most Westerly corner of (Black's land); thence N 35° E along (Black's land); etc." is legally satisfactory but not dimensionally complete on the face of the deed itself.

In addition to serving as a means of title identity, a call for an adjoiner serves to prevent a gap or overlap; it makes certain the last-described parcel coincides with the first-described parcel. Here, again, in describing a new parcel from a part of a senior claimant's land, the description should not call for the junior land, even though the two are intended to be coincident along the common line. Let the junior deed call for the senior.

When the adjoiner is senior, the call for the adjoiner usually takes the form "N 16° E, 1227.32 feet more or less to the southerly line of that land conveyed to Thelma Brown and recorded in . . . etc." But if the adjoiner is a junior claimant and a new call is being made that will reach the perimeter of the instant property, the call takes the form "N 16° E, 1227.32 feet to the northerly line of this parcel being divided, etc." No call is made for the junior claimant.

A call for a lot line within a subdivision is a call for a record monument, but it is never a call for a senior claimant. Lot lines are common to both lots, and both lots have equal claims to the line. A call of "N 10° E, 212.12 feet to the north line of Section 12" is considered as calling for a record monument, and the call is presumed to have force over measurements. A call of "to the north line of Section 12 and along the north line of Section 12" is for the purpose of preventing gaps or overlaps, not to disclose seniority, since none exists. It is a useful technique, if properly used.

It has been said by William C. Wattles, a well-known title authority, that a remainder description should have sufficient title identity to be able to locate the land if all dimensions were omitted. Bounds descriptions are often used for remainders.

15-27 Intent of a Description

In the order of importance of conflicting deed elements, the intent of the parties to a conveyance, as expressed by the writings, is the paramount consideration of the court, senior rights excepted. In composing a description, the objective of the deed author is to describe exactly and correctly by writings the intent of the grantor. At a later date all the verbal evidence in the world will not change the intent as expressed by the written words. Also, if there are ambiguous terms, the terms will be construed most strongly against the grantor.

PRINCIPLE. *Before attempting to describe land the scrivener first carefully determines what the intent of the parties is, then he selects words, keeping in mind the presumed order of importance of description elements, that will precisely describe that intent.*

15-28 Call for a Survey

PRINCIPLE. *If the intent is to describe land as following a particular survey, that fact must be stated.*

A call for a survey can be direct as "according to the survey made by Jones," or indirect as "according to Map No. 1272 of Lakeview Terrace," wherein the Lakeview Terrace map states "surveyed by Jones." A description reading "T 10 S, R 2 E, Sec. 12, SBM" automatically calls for a survey—that survey caused to be made by the government when Section 12 was created.

If a survey is made by a surveyor prior to writing a description, and if the owner wants that survey to be a consideration of the conveyance, he must call for that survey in the description; otherwise, it will not be a consideration of the writings. If a public record is filed of the survey called for, all the evidence disclosed on the plat is incorporated into the deed.

Frequently, the platting laws of a state require a survey for all subdivisions. If the law does require a survey for a subdivision, the presumption is that a survey was made, and any call for that subdivision also calls for the presumed survey. It is thus possible to call for a survey by reference to another document without expressly saying that a survey is called for.

In calling for a private, unrecorded survey the effect of the survey on the interpretation of the conveyance is often dependent on the description wording as "beginning at the southwest corner of Lot 10 as marked by a 2-inch iron pipe with brass disc stamped LS 2554; thence "N 10°05′ E, 200.00 feet to a 2-inch iron pipe with disc stamped LS 2554; thence "N 89°22′ E, 301.54 feet to a spike driven into an oak tree; thence ... according to the survey made by Brown." The calls for monuments, bearings, and distances describe the survey and, on recording, make such a matter of public record. Merely reciting bearings and distances without calls for monuments set or discovered deprives a called-for survey of a great amount of force. What the surveyor did must then be proved by testimony, and, after the surveyor is gone, this is difficult.

Incorporating the data given on a publicly recorded survey into a description is done by merely calling for the survey; public records of surveys are admissible as evidence. Since private, unrecorded surveys are not admissible without testimony of the surveyor, it is better to incorporate the calls for monuments in the description.

15-29 Call for Monuments

PRINCIPLE. *Other than senior rights and contrary expressed intentions, a call for a monument is presumed to prevail, and if the intent is to limit the boundaries to particular existing monuments, it must be so stated.*

Uncalled-for monuments are not a consideration of a description and will have no effect. Recital of bearing and distance that presumably go to a particular monument is not absolute assurance that the monument will be a part of the boundary; the monument must be called for. Any error of measurement, transposition of figures, or miscalculation can defeat an intent to go to an uncalled-for monument.

A call for a monument, without reciting the particular part of the monument, can cause ambiguity. If there is a possibility of confusion, the better procedure is to spell out the exact intent. Calls for natural monuments have been a continuous source of litigation because of failure to define limits of the monument. Such wording as "along a river," "along the shore," and "by the stream" are indefinite and should be explained even though various court decisions have specified how to interpret this wording. Instead of saying "along the Susquehanna river," say whatever the limits of private ownership are. "Along a river" can mean "along the average high-water mark," "along the average low water mark," "along the thread of the stream," "along the main channel of the stream," or "along the gradient boundary," depending on the circumstances. To clarify this the exact meaning should be stated.

"Along a highway," "in a highway," "by a highway," and like expressions are usually interpreted to go to the centerline of the highway, that is, if the grantor owns that far. The exact intent should be stated as "along the centerline of the highway," "along the sideline of the highway," or "together with any rights the grantor has in the highway." Although ambiguous statements do not make a conveyance void or even voidable, it is better practice to clarify the terms at the time both parties are consummating a sale.

A call for a tree, a stake, a rock mound, or like object is construed to the central part of the object, and if a different intent is wanted, it must be clearly stated.

> Where a deed conveyed land east of a line beginning four rods east of a tree, the distance is to be measured from the center of the tree.
> —*Coombs* v. *West*, 99 A 445; see 2 ALR 1424.

15-30 Lines

Lines constitute the perimeter of all described land areas. They may be (1) straight, (2) circular curved, (3) spiral curved, (4) curved to fit the intersection of the earth's surface with a given elevation (waterlines and contours), (5) gradient (along rivers), (6) of geometric relationship (parallel with, etc.), (7) geodetic (follow latitudes and longitudes), and (8) irregular (along a creek bottom, along a fence, etc.).

Excepting those continuous lines described as following the perimeter of a geometric area (circle, square, etc.), a line description forming one side of an area must describe a definite "starting point" and describe a definite "terminus."

Lines are assumed to be straight, except where otherwise qualified. These lines are irregular:

> Southerly along the ocean.
>
> Easterly along Boulder Creek.
>
> In a general northerly direction along the westerly boundary line of that land conveyed to James Case in . . .
>
> Northerly and easterly parallel with and 300 feet distant from the westerly and northerly side line of the ——— Railroad right-of-way. . . .

15-31 Straight Lines

For a local area, a straight line forming one side of an area is the shortest measurable distance between its starting point and its terminus, unless otherwise stated in the writings; it is a horizontal line. Technically, all horizontal lines are curved lines following the curvature of the earth's surface, but the effect of curvature in local areas is not discernible by plane land-survey methods.

> In the absence of something in a deed to cause a deviation, a boundary line between two points is presumed to be a straight line.
> —*Leigh* v. *La Pierre*, 312 A 2d 699.

A straight line along the perimeter of an area can have either its start, terminus, direction, or all three, defined (1) by monuments, (2) by direction and distance from a given point, (3) by coordinates as based on a defined grid system, (4) by latitude and longitude as defined by the earth's poles, and (5) by geometric relationship (parallel with, etc.).

15-32 Straight Lines Defined by Monuments

All land descriptions must somehow be related to monuments. Technically, all directions in land descriptions are defined relative to the direction of monuments, and those monuments can be the polar axis and the Prime Meridian, or any two suitably defined points or physical objects. The length of a line can be limited by either monuments or distance.

By defining any two fixed monuments, one as the start and the other as the terminus of a line, sufficient information is given to identify one side of an area, and the recital of distance and direction is not legally essential. Because monuments are so frequently destroyed and because distance and direction are aids to replace lost monuments, it is always desirable that measurements be quoted, provided, of course, that the distance and direction quoted are based on correct measurements. Quotations of incorrect measurements create ambiguities, confusion, and invite litigation. Quotations of approximate distance and direction do serve the useful purpose of defining a general area in which to search for a monument. But, as so often happens, after a monument is destroyed and witness evidence is not available about its original location, the approximate measurements become absolute, a fact not intended.

In past years the most fruitful source of land litigation has been the discrepancies between lines defined by both monuments and measurements. One of the principal functions of the surveyor is to insure harmony between monument location and measurements. The reason surveyors so often compose descriptions is that they are capable of determining monument-measurement relationships.

Record monuments (call for adjoiners) can be used to define the direction, distance, termination, or start of a line. "Along Jones' easterly line" defines direction, and "to the northeasterly corner of Jones' land" describes the termination of a line. The recital of record monuments is for the purpose of preventing gaps and overlaps and is a very desirable feature of a good description. But the practice of reciting the adjoiner without mention of measurements, although sufficient, is undesirable in that owners do not know the quantity of their land.

15-33 Straight Lines Defined by Dimension from a Point

A straight line forming the side of an area can be defined by a direction, and a distance. The starting point can be either a monument or the termination of the preceding line.

Which is more important, distance or direction, varies from state to state. For metes and bounds descriptions, Texas, Kentucky, Florida, New Jersey, and the Federal Government all say direction is more important. Generally speaking, unless a monument is called for, both are necessary to determine a bound.

15-34 Basis of Bearings for Direction

PRINCIPLE. *Directions by bearings are relative to whatever the writings define as the basis of bearings; astronomic north is merely one of the many ways to define directions and is only presumed to be the basis of bearings in the absence of other evidence.*

Direction definition in writings is probably the greatest source of misunderstandings. Too many people assume that directions, as defined by bearings, are always relative to astronomic north—an assumption far from the truth.

Direction is whatever the deed defines it as being; the astronomic definition of direction is merely one of many definitions. Direction can be defined as relative to physical monuments, and, since this method entails less expense and time, most surveyors when describing new parcels resort to this method.

Surveyor James Rogers monumented (see Fig. 15-2) a new parcel in the northwesterly corner of Lot 13 as based on the found, original corners of the lot. On the original map the westerly line was described as north 600 feet, and the northerly line was described as west 600 feet. The measured angle at the corner, as based on the original monuments, was 89°58'. The new description as written reads:

Commencing at the northwesterly corner of said Lot 13; thence East along the northerly line of said Lot 13 a distance of some 200 feet; thence S 0°02' E 200 feet,

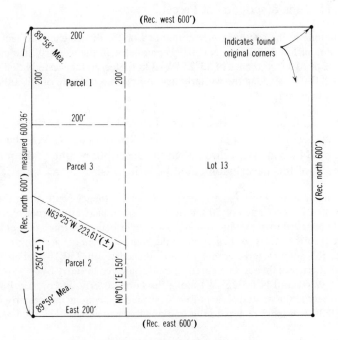

Figure 15–2 Lot 13.

thence west 200 feet to the westerly line of said Lot 13; thence N 0°02' W along the westerly line of said Lot 13 a distance of 200 feet to the point of commencement.

Translating this literally, the description author is saying, "This description commences at the northwesterly corner of Lot 13, wherever it may be, and travels along the northerly line of Lot 13 in the same direction traveled by the original surveyor. While no true determination of the direction of this line was made; if, on actual determination, it is found that this line is not astronomic east, for the purpose of this description, irrespective of the true direction of the northerly line, the direction of the second line is determined by the angle formed by the mathematical difference between east and S 0°02' E. The basis of bearings is the line as run by the original surveyor, not the astronomic east as purported to have been run by the original surveyor. If the original surveyor erred, that error is perpetuated."

If the description had been written "commencing at the northwesterly corner of said Lot 13; thence N 89°58' E, astronomic (record east), along the northerly line of said Lot 13 at a distance of 200 feet; thence . . . ," the deed author is saying in effect, "The true astronomic direction of the north line of Lot 13 was determined and the original surveyor was found to be in error, but, irrespective of this fact, the basis of bearings for this description is N 89°58' E along the northerly line of Lot 13 as originally monumented, and other lines will be relative to this basis." In the event of destruction of the monument at the northeasterly corner of Lot 13, the "N 89°58' E, astronomic," is evidence to aid in reestablishing the corner.

In Fig. 15-2 the description of Parcel 2 reads:

> Beginning at the southwesterly corner of Lot 13; thence east 200 feet along the south line of Lot 13; thence N 0°01' E parallel with the westerly line of Lot 13 a distance of 150 feet; thence N 63°25' W, 223.61 feet to the westerly line of Lot 13; thence S 0°01' W along the westerly line of Lot 13 a distance of 250.00 feet to the point of beginning.

The problem is to write a description of Parcel 3. Either north, west, east, N 0°02' W, or N 0°01' E can be used as the basis of bearing. Whichever one is used, it should be consistent throughout the description. One of the many ways that the body of this description could be written is:

> Beginning at the northwesterly corner of Lot 13; thence south along the westerly line of Lot 13 a distance of 200.00 feet to the southeasterly corner of that certain parcel conveyed to Smith by ———, said corner being the true point of beginning; thence S 89°58' E (record east) along the southerly line of said Smith's land, 200 feet to the southeasterly corner thereof; thence south 250.18 feet to the northeasterly corner of that certain parcel of land conveyed to Barry by ———, thence N 63°26' W (record N 63°25' W) along the northeasterly line of said Barry's land 223.61 feet to the westerly line of Lot 13; thence north along the westerly line of Lot 13 150.30 feet more or less to the true point of beginning.

This description as written is based on an assumed fact: the westerly line of

Lot 13 is in fact north as originally called for by the original surveyor of Lot 13. The fact that the line is later proved to be other than astronomic north will not alter the intent to make all lines relative to the assumed north called for.

Basis of bearings may be relative to (1) astronomic north, (2) grid (State Plane Coordinates) north, (3) local grid north, (4) any two defined monuments, (5) magnetic north, (6) adjoiner's recited bearing on a given line of the adjoiner's description, (7) a previous survey, (8) bearings quoted on a map, or (9) arbitrary.

One of the essential functions of the scrivener is to clearly indicate the basis of bearings used. When no basis of bearings is stated or implied, the general presumption of the state will usually be invoked as in the following New Hampshire case:

> The courses in a deed are to be run according to the magnetic meridian, unless something appears to show that a different mode is intended.
> —*Wells* v. *Jackson Iron Mfg. Co.*, 47 N.H. 235.

15-35 Changing the Basis of Bearings

In land descriptions it is assumed that all bearings within the same description are on one basis. Writing new descriptions by assembling the data contained in adjoiner descriptions may result in lines being described by unrelated bearings. Orcutt, Brown, and Jones owned land that was described respectively by astronomic north, grid north, and magnetic north. A description of the remainder reads:

> Beginning at the southeasterly corner of Orcutt's land as described in . . . thence along Orcutt's easterly line, N 12°02′ E 200.00 feet to the southwesterly corner of Brown's land as described in . . . thence N 89°01′ E along said Brown's land 129.82 feet to the northwesterly corner of Jones' land as described in . . . thence S 0°03′ E along said Jones' land 200.00 feet; thence . . .

In this particular location the grid and astronomic north varied by 0°31′ (east of north), and the angle as figured by the difference between N 12°02′ E and N 89°01′ E is 0°31′ in error. The difference between N 89°01′ E and S 0°03′ E is several degrees incorrect because of the magnetic declination.

Lines described both by monuments (physical and record) and bearings always have a potential difference in basis of bearings, and, unless the lines are related by accurate angular measurements of a recent survey, ambiguities may result.

15-36 Azimuth

Direction by azimuth is seldom used and is not recommended for descriptions. Azimuth is a horizontal, clockwise angle recorded from a fixed direction such as astronomic "south." Azimuths are not divided into quadrants but vary from 0 to 360°. Either north or south may be used as an origin of azimuth. In any description using azimuth, the direction of the origin line must be specified.

15-37 Deflection Angles.

Almost all property surveys today are performed by turning angles with a transit or theodolite. Descriptions can be written without converting the angles to bearings. Directional calls for the angles turned must be stated because four possible directions exist for a given angle. Normally, the angle is described from the prolongation of a line as follows:

> Thence 20°01' to the right from the prolongation of the last described course, 200.00 feet;
> thence northeasterly 624.26 feet along a line deflected northerly 16°02' from the prolongation of the preceding course;
> thence southerly 221.02 feet along a line deflected 168°22' from the last described course.

In the last instance the angle described is not from the prolongation of a line. The form "thence 20°01' to the right 200.02 feet" is used, although not clearly. The "to the right," as commonly used by surveyors, means "to the right from the prolongation of the previous course." The longer explanation is better and clearer.

15-38 Straight Lines Defined by Coordinates

Points on the earth's surface can be defined relative to a coordinate system, and straight lines can be defined as connecting two coordinate points. Any coordinate system used in a description must have (1) a defined point of coordinate origin and (2) a direction definition.

A deed call for latitude and longitude is a call for a system defined by three monuments, the North Pole, the South Pole, and the Prime Meridian. Utilizing the most modern equipment and exercising the greatest of care will not locate a point by latitude and longitude closer than about 10 feet, and this position is subject to the deflection of the plumb line, thus introducing even greater error. Because of this, latitude and longitude have seldom been used to define private land rights, but the system has often been used to distinguish political boundaries such as between the United States and Canada and between some of the states.

The USC and GS, now NGS, has established numerous monuments with precise latitude and longitude values assigned, but such values have uncertainty about the true geodetic position, even though their relative agreement is good.

A call of "latitude _____ and longitude _____ based on the control established by the United States Coast and Geodetic Survey" incorporates the USC and GS monuments into the deed and makes location more certain.

For the purposes of land surveying State Plane Coordinate values are far superior to geodetic coordinates.

> All that portion of Section 10, T35N, R2E, S.B.M., County of _____, State of

_____, according to the U.S. Government Survey of _____, and more particularly described as follows:

Commencing at the southeasterly corner of said Section 10 as marked by a stone mound having grid coordinates X _____, Y _____, of Zone 7 of _____ State Coordinate System; thence westerly along the southerly line of said Section 10, S 89°58′ W 200.00 feet to a point having grid coordinates of X _____, Y _____ of said Zone 7; thence northerly to grid point X _____, Y _____ of said Zone 7; thence easterly to grid point X _____, Y _____. . . .

15-39 Directional Calls

All perimeter descriptions contain definitions of how to locate a sequence of lines and, except for bounds descriptions, have a direction of travel along the lines described. In land descriptions it is never wise, even in obvious circumstances, to omit directional calls. Directional calls are not specific; they are in a general direction such as "northerly" or "southeasterly." The suffix "ly" usually designates a directional call.

> Beginning at the southerly quarter corner of Section 10, T12S, R6E, _____ Meridian; thence *easterly* along the *southerly* line of Section 10 a distance of 200 feet to the beginning of a 600 foot radius tangent curve to the *right* from the prolongation of the last described course; thence *easterly, southerly,* and *westerly* through a central angle of 180°00′; thence *westerly* parallel with the *southerly* line of said Section 10 a distance of 200 feet; thence *northerly* to the point of beginning. (emphasis added)

In the last call, "northerly to the point of beginning," the *northerly* is not essential, since "to the point of beginning" defines direction, but it is advisable to insert the directional call even when not essential.

A call of "along Red River" has two directions. If the river runs southerly, the call must be either "northerly along Red River" or "southerly along Red River" or "up the Red River" or "down the Red River."

The call "north," "east," "west," or "south" may be either directional or a specific bearing. If the wording is "thence east 200.00 feet to a 2″ iron pipe," the call of east is directional in a generally easterly direction to a 2″ iron pipe. In the absence of a qualifying term as "thence east 200.00 feet," the east is specific and is relative to the defined basis of bearing.

Any bearing call, or for that matter, any distance call, that is controlled by a call for a monument is reduced to more or less measurement status. In a call of "N 10°02′ E, 200.00 feet to a 36-ince oak tree," the N 10°02′ E is more properly directional and the 200.00 feet is plus or minus to assist in the discovery of the oak tree rather than either being specific calls.

Directional calls in themselves are indefinite as "southerly 300 feet." Directional calls are used with other locative calls that together make a certain boundary as "thence to the beginning of a tangent curve concave northerly."

386 / WRITING DESCRIPTIONS

The tangent curve could be either northerly or southerly; the directional call of concave northerly makes the direction of the curve certain.

15-40 Lines Defined by Geometric Relationships

Usage of the expression "parallel with" another line invites caution in certain situations. A line "parallel with a street" as in Fig. 15-3 introduces ambiguity at curve returns. The proper form is "parallel with Z Street and its prolongation." The expression "100 feet northerly from and parallel with the bank of Cedar Creek" is undesirable, as shown in Fig. 15-4. If the creek bank is irregular, the so-called parallel line may not have a similar shape to that of the bank. The upland limits of ownership of Fig. 15-4 are determined as of the day of the conveyance, and erosion or accretions cannot alter them. After erosion occurs, uncertainty of the property line may occur. The usage of "parallel with a particular waterline" is not recommended.

The form "the westerly 50 feet of Lot 10" is essentially a call for a line "parallel with and 50 feet easterly from the westerly line of Lot 10" and is a variant of "parallel with." Again, parallel calls with a corner cutoff radius or waterlines are to be avoided.

15-41 Area

If a description is complete, definite, and certain within itself, a statement that the parcel contains a definite number of acres does not alter the certain description. Area calls have force only when a perimeter description is dimensionally incomplete or the specific description is ambiguous and the area call makes one interpretation more probable.

Example 1. All that portion of Lot 3 of Rancho El Cajon according to the partition map thereof filed in the office of ...., more particularly described as

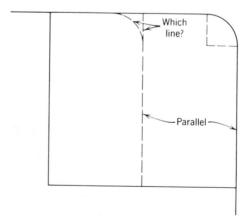

Figure 15-3 Parallel lines and corner cutoff.

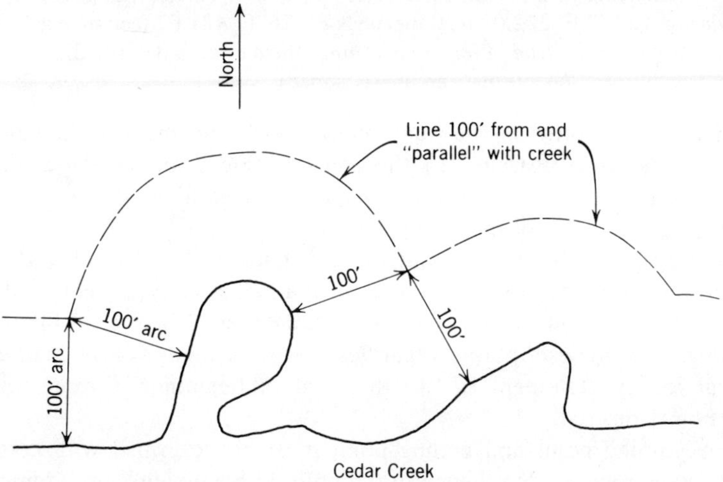

Figure 15-4 Lines parallel with a creek.

follows: The westerly 160 acres of said Lot 3, the easterly line of said 160 acres to be parallel with the westerly line of said Lot 3.

Example 2. The easterly 5 acres of the E ½ of the SE ¼ of the SE ¼ of the SE ¼ of Section 12, T 10 S, R2E, MDM.

Example 3. ... thence in a straight line to the westerly line of said Lot 13 in such a direction as to include exactly 12 acres; thence to the point of beginning.

In each of these examples, the acreage specified makes the perimeter complete. In the second example the direction of the westerly line of the easterly 5 acres is not defined and should be. This type of description has been the subject of litigation, and a yet a certain rule for direction determination in all cases has not evolved.

Also in the second example, if one-half were meant instead of 5 acres that should have been stated. Merely because the original area was 10 acres does not prove that 5 acres is one-half of the land to be found by a modern resurvey. Very rarely does a government section contain its original intended acreage, and it is probable that 5 acres is not one-half of the record 10 acres.

"Ten acres of land in Rancho Cuyamaca" is not locatable, and the conveyance is void. "Ten acres in the northwest corner of Lot G of Rancho Cuyamaca" is 10 acres as a parallelogram and is certain.

15-42 Beginning Point and Ending Point

All perimeter descriptions, excepting bounds descriptions, have a point of beginning. Various words can be used such as "commencing at," "beginning at," or even "starting at." If there is a sequence of calls prior to reaching the area being described, the form used is this:

388 / WRITING DESCRIPTIONS

> Commencing at the southeasterly corner of Lot 10 according to Map... etc.; thence S 12°10' E, 222.01 feet; thence N 87°26' E, 421.07 feet; thence S 10°22' E, 100.00 feet *to the true point of beginning;* thence (area described)...; thence *to the true point of beginning.* (emphasis added)

Sometimes the form "to the true point of commencement of the land herein described" is used. Rarely "to the true starting point of the land herein described" is used. Another form used is "Beginning at...; thence...; thence... to the point of beginning."

All closed-perimeter descriptions must go back to the point of beginning. In many instances a described perimeter, if mathematically perfect, will form a closed figure without reciting the last course as "thence to the point of beginning." But in descriptions, regardless of the perfection of the mathematics, a precautionary statement of "to the pont of beginning" or an equivalent expression is inserted.

The beginning point and ending point must be described with compatible words: "Beginning at...; thence to the point of beginning"; or "Commencing at...; thence to the point of commencement"; or "To the true point of beginning...; thence to the true point of beginning"; or "Starting at...; thence to the starting point."

Descriptions do not *commence at* and then end by going *to the point of beginning.*

15-43 Distance of Described Line from Starting Point

If a principle were stated with respect to the certainty of location of a described area it would include this, "The uncertainty of location of the land area described is a function of the distance from the nearest known fixed monument." A description reading "Beginning at the southwest corner of Rancho El Cajon, thence N 28°02' E by astronomic bearing 10,282.00 feet to the true point of beginning; thence north 100 feet; thence east 50 feet; thence south 100 feet; thence west 100 feet to the true point of beginning" and surveyed by two surveyors working independently and with better than average care would not be located at the same point within a radius of 1 foot and probably 2 feet (see Chapter 10). This means that the staking of the description could vary in position, even by competent, careful surveyors, by as much as 4 feet and very likely at least 2 feet. Such calls, unless dictated by necessity created by previous conveyances, should be strictly avoided.

The easiest way to eliminate long calls is the insertion of calls for nearby permanently fixed monuments at the time of conveyancing. If after the words "true point of beginning" in the foregoing description there had been inserted the words "from which a lead and disc stamped LS 2554, set in an 8 × 12 foot boulder, bears N 5°02' E, 101.03 feet," the future certainty of identification of the intended location is increased. Of course, such calls can only be determined by those competent to measure; hence all the more reason why surveyors should be the ones to write land descriptions. Also, to be effective, such calls must be inserted prior to conveyancing.

15-44 Circular Curved Lines

Mathematically defined curves used in legal descriptions are, with minor exceptions, based on circular curves. Curved lines defined by monuments or elevations (along the mean high-tide line, contour lines, etc.) are irregular and can at best be described by approximate measurement information.

To define the location of a circular curve, the following must be specified:

1. Location of the starting point of the curve.
2. Location of the center of the circle of which the curve is a part or sufficient information so that the center is locatable.
3. Location of the terminus of the curve.

No set rule exists about how a circular curve shall be described, since many forms are available to the scrivener. The definition of the starting point of a curve can be the termination of the preceding line, a monument, a coordinate, or other suitable method that fixes a definite point on the earth's surface.

Two points, the location for the center of a circle and the location of one spot on the perimeter of a circle, will define a circle. Curves as used in descriptions are rarely complete circles; hence the start and end of the curve must also be definite.

Most circular curves as used in descriptions are tangent curves; that is, the radial line at the beginning of the curve is at right angles to the preceding straight line; or, if the preceding line is a circular curved line and the next curve is circular and tangent, the radial lines of both curves at the point of contact coincide.

A statement "thence east, 100.00 feet to the beginning of a tangent curve" does not locate the direction of the curve, since the curve's center can be either north or south of the beginning point of the curve. The direction of the curve can be defined by "thence east 100.00 feet to the beginning of a tangent curve concave northwesterly" or "thence east 100.00 feet to the beginning of a tangent curve whose center bears north."

Compound curves or reverse curves are tangent at the point of change of curvature. (See Fig. 15-5.) The usual methods of describing these curves are "to a point of compound curvature; thence along a curve having a radius of 200 feet . . . "; "to a point of reverse curvature; thence along a curve having a radius of 300.00 feet. . . . "

In defining tangent curves the following elements must be given: (1) two or three dimensions, (2) direction of curvature, (3) direction of travel (see Fig. 15-6). An example is "to the beginning of a tangent curve of 300.00 foot radius concave northwesterly; thence northeasterly through a central angle of 10°20' a distance of 54.01 feet. . . . "

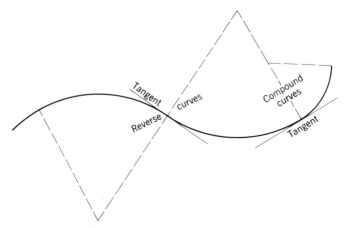

Figure 15-5 Reverse and compound curves.

Two mathematical elements of a curve are all that is needed, although three are usually quoted. Radius, central angle (delta), and curve length are more often used than are chord, middle ordinate, tangent, degree of curvature, or external distance.

A nontangent curve is normally defined by a radius bearing as

> thence N 89°00′ E to the beginning of a 300.00 foot radius nontangent curve whose center bears N 10° E, thence northeasterly along said curve through a central angle of 3°01′ a distance of _____ feet; thence . . .

The N 10° E is not at right angles to N 89°00′ E; therefore the curve is not

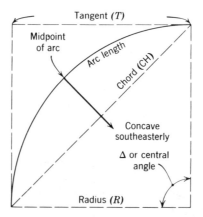

Figure 15-6 Direction of curvature.

tangent. Since most curves are tangent, it is good form to direct attention to the fact that the curve is nontangent.

The radius of a curve can be defined indirectly by degree of curvature. Caution must be used when calling for a given degree of curvature, since the degree of curvature may be defined by a number of relationships. Adjoining railroads and railroad rights-of-way the chord definition has often been used. To avoid confusion, the radius should always be stated.

15-45 Spiral Curve

A spiral curve commences with an infinite radius (straight line) and uniformly decreases its radius until the curvature of the circular curve is attained. Spirals are more properly used to design fixed works than property lines. Usage of spirals to describe the sidelines of easements is not recommended.

Many definitions of spirals are being and have been used, and in describing a spiral the definition must be specified. Reference to spiral number 7 of such and such railroad should be avoided; railroad records are not public records.

Calls of parallel with the centerline of a railroad as it exists are calls for all the spirals and adjustments of curvature. Since tracks have a habit of creeping toward the outside of a curve because of the centrifugal force of the train, certainty of future location is not assured by such calls.

15-46 Irregular Curves

Many types of undefined irregular curves result from calls for monuments such as lakes, rivers, elevation, arroyos, and like calls. Where water floods areas, as behind dams, land boundaries are often defined as all that existing below a certain elevation. In attempting to map land and determine acreage, the surveyor often describes a meander line composed of a sequence of straight lines. Elevation lines (waterlines) are rarely straight, and any attempt of the surveyor to describe a curved line by a series of straight lines is merely an approximation. This, of course, has resulted in the courts declaring meander lines as not having controlling force.

Calling for "along a monument" without dimensional information to locate the monument is unsatisfactory to the tax assessor and the new owner; the amount of land conveyed is uncertain. Specifying measurements without calling for the monument is unsatisfactory, since the intent is to go along the monument. A call of along a monument and meander information (although only approximate) is preferred.

When meander information is not given, difficulty may ensue in describing the line leaving the monument. A common form used is " . . . to the Red River; thence westerly along the Red River to a point which is S 10° E from the true point of beginning; thence N 10° W to the true point of beginning."

It is certainly never advisable to make the call for a distance along a river the controlling consideration, since erosion or accretions may change the direction of a river. A conveyance reading "thence west 200.00 feet to the Red River;

thence northerly along the river 200.00 feet; thence east 200.00 feet; thence to the true point of beginning" is indefinite at future dates. The 200-foot measurement is determined as of the date of deed; but, unless it is marked off immediately, the location of the northerly line may be made uncertain by the whims of a flood.

15-47 Forms Used to Indicate Superior Call

Occasions sometimes arise whereby the deed author wants to indicate the superiority of one course over another. This is done by emphasizing the more important line as

> thence N 89°22′ E, 321.02 feet to a point in the westerly line of the land conveyed to Dr. Maguire, said point being exactly 100.00 feet from the southwesterly corner of said Dr. Maguire's land; thence S 0°38′ E, 100.00 feet to said Dr. Maguire's southwesterly corner;
> thence N °15′ W 200.03 feet more or less to a point that bears N 89°00′ E, 200 feet from the true point of beginning; thence S89°00′ W 200.00 feet to the true point of beginning.

Each of these forms conveys the intent to make the second course controlling over the first. Repetition of statement produces this purpose.

15-48 Strip Conveyances and Stationing

Along highways, railroads, or other strip conveyances, stations are often used. Each station is 100 feet along the defined centerline of the strip; stations are numbered from some arbitrary starting point.

Stations are usually in reference to a map filed in a public agency's office, and the zero point is determined by the map referred to. The notation 24 + 16.32 means that the point is located 2400 feet, plus 16.32 feet, from the zero point and is located on the stationed line (usually centerline or construction line). A notation of right 24 feet from station 24 + 16.32 means that the point is 24 feet to the right from the direction of increasing stations, and the 24 feet are at right angles or radial to the line at station 24 + 16.32.

> Beginning at Station 0 + 16.32 on the centerline of Road Survey number 1632 as filed in the office of the County Surveyor of .... County, State of ....;
> thence N 16°02′ W, 50.00 feet to a point on the sideline of said road survey, said point being the true point of beginning ...

Strip conveyances are often described in their entirety and easements obtained for any portion crossing a specific property. The form used is as follows:

> All that portion of land lying 25 feet on each side of the following described centerline:
> Beginning at the southwesterly corner of Lot 21 of Blane's Subdivision according to Map 2167 filed in the office of the .... of .... County, State of ....;

thence N 10°01′ W along the westerly line of said Lot a distance of 21.03 feet to the centerline of the line being described; thence N 87° E, 1,222.40 feet to the easterly line of said Lot 21. The side lines of the described strip are to be lengthened or shortened to terminate in the side lines of said Lot 21.

The proper wording is 25 feet on "each side," not 25 feet on "either" side. Either side is indefinite, since it may be on one side or both sides.

Insertion of the words "a distance of" in the above description as "along the westerly line of said lot *a distance of* 21.03 feet" is necessary, since the wording "along the westerly line of said lot 21.03 feet" is ambiguous. Lot 21 and Lot 21.03 feet can be confused.

15-49 Abbreviations in Descriptions

Certain abbreviations in land descriptions, especially on plats, are in common usage and well understood. A notation of S 23, T12S, R2E, SBM is clear, certain, and identifies Section 23 as existing in only one spot. But the description "W on R NE½, Sec. 8, Tp. 5 north, range 4 east" means nothing.[1] The "R" could mean road, ridge, or river.

In the description "sections 22 and 28, Tp. 79, R 13, Poweshiek County, Iowa," the court commented:

> It appears to us that there is no uncertainty or indefiniteness in these contractions of words. They are in almost universal use in this State in describing lands, and everybody understands that they mean "township" and "range." It is true the contraction does not state whether the range is east or west, but that was wholly unnecessary, as the land was described as in Poweshiek County, and the courts in this state take judicial notice that all of the land in that County is in range west.
> —*Ottumwa, CF & St. P. RR* v. *Williams*, 71 Iowa 164.

Abbreviations, numbers, or characters in common use and understood by most people may be used in legal descriptions. But if doubt exists, it is better to spell the words out. Plats frequently use abbreviations because of the greater ease of interpretation (see Chapter 14). The following abbreviations are commonly used in written descriptions:

N	North	V	Vara
E	East	ft	Feet
S	South or section	in.	Inches
W	West	delta	Central angle
Sec	Section (preferred to S)	Ely	Easterly
T	Township	NWly	Northwesterly
R	Range or radius	°	Degree

---

[1]*Sims* v. *Rolfe*, 177 Ark. 52.

Ac	Acres	Minute or feet
ch	Chain	Second or inches

In general, it is better to spell out rather than abbreviate in written descriptions. On maps it is desirable to abbreviate, since this enables the viewer to more quickly and accurately evaluate the map. The object on a plat is to present a graphic picture by symbols in the clearest and most certain shorthand method. Spelled-out explanations only clutter up a map and make it difficult to understand.

Very often the certainty of meaning of the abbreviation is dependent on its association with other symbols, numbers, or lines. The letter "R" can be "range" or "radius." When used as R3E or R = 200', no uncertainty exists about the meaning.

By law, abbreviations may be used on tax descriptions, and descriptions of land sold at a tax sale are often abbreviated more than is permissible in a conveyance. For regular conveyancing, tax abbreviations are not advisable, since uncertainty will probably result.

15-50 Easements

An appurtenant easement—that is, an easement necessary for the enjoyment of the land—passes automatically whether recited or not. Irrespective of this fact, it is always good practice to include a description of the easement as "Lot 12, Block 40, Horton's Addition . . . together with an easement for road and utility purposes over the easterly 20 feet of Lot 4."

The best way to describe appurtenant easements is to separate the fee conveyance and the easement into two distinct parts.

The intent of a fee or easement conveyance may be reversed by a description "Lot 12, reserving therefrom the westerly 20 feet for road purposes." Ordinarily, the grantor reserves the fee subject to a 20-foot easement. If it is intended to convey the fee and reserve an easement, the form "Lot 12, reserving therefrom an easement for road purposes over the westerly 20 feet" is proper.

15-51 Exceptions in Descriptions

Exceptions in descriptions are particularly prone to ambiguities. A description reading "Lots 10 and 11, except the south 20 feet thereof" may mean an exception for both lots or an exception for Lot 11.

Double exceptions are doubly apt to be used in error. "Lot 13 except the east 12 feet, except the south 10 feet" may mean "Lot 13 except the east 12 feet and also except the south 10 feet of all of Lot 13," or it may mean "Lot 13 except the east 12 feet, except that the south 10 feet of the east 12 feet is not conveyed." This description is greatly improved by writing "Lot 13, except the east 12 feet of Lot 13 and also except the south 10 feet of all of Lot 13."

Exceptions with respect to roads and rights-of-way may inadvertently be erroneously worded as for example "Lot 13 except M Street as opened" can be construed to mean retention of fee title to M Street. Is this a fee or easement exception intended? A description reading "Lot 13, excepting an easement over the westerly 20 feet of Lot 13 as granted to . . . " is clear.

15-52 Whole Descriptions

A whole description is made up of words, and the possible number of words that can be used in a description is infinitely variable. It would be impossible to compile a complete list of all description variations.

A whole description only identifies a particular area. It can be simple, short, and to the point; or it can be long, complex, and difficult to understand. Which one is used is the choice of the scrivener. Compare these two descriptions:

(1) Lot 1, Block 4 Highland Addition to Colton, Map 1304, Riverside County, California.

(2) All that portion of land located in Highland Addition to Colton according to Map 1304 as recorded in the office of the Recorder, Riverside County, California and more particularly described as follows: Beginning at the southeasterly corner of said Lot 1 as shown on said Map 1304; thence along the southerly line of Lot 1, also being the northerly line of Hope Avenue to the southwesterly corner of said Lot 1; thence northerly along the westerly line of said Lot 1 to the northwesterly corner of said Lot 1; thence easterly along the northerly line of said Lot 1 to the northeasterly corner of said Lot 1; thence southerly along the easterly line of said Lot 1 to the point of beginning, also including reversionary rights in Hope Avenue.

Description 2 says nothing more than is contained in description 1. The desirable feature of a whole description is brevity without loss of clarity or identity (either now or in the future).

Land can never be completely described. A million volumes could be written describing the exact size, shape, and contents of every grain making up a small area of land. The deed author need not feel it his duty to do such writing, but at the same time a description cannot be so short as to cause ambiguity.

How to write whole descriptions, brief, concise, exact, and without error is the sum total of a large area of knowledge; it includes mathematics, sciences, and common customs. It cannot be completely discussed in limited space; neither is it desirable to do so.

15-53 Whole Descriptions by Referral

The most complete descriptions requiring a minimum of words, yet conveying a maximum of information, are reference calls for a given map. A map is a shorthand notation devised by people to convey, by symbols, a large amount of information without the necessity of reading many words. If a map is complete, it will in itself completely describe a parcel of land. The only knowledge needed by the deed author is how to call for the map as: "Lot 12, Block C, West Highland Addition, according to Map 1604 as recorded in the office of the Recorder, County of ...., State of ...."

396 / WRITING DESCRIPTIONS

Not all states have or always have had platting laws. In the past there were many parcels described as follows: "Lot 12, Block C, West Highland Addition to the City of ....according to the attached plat."

The strength of this type description resides entirely in the plat; and, if the plat is poor, not showing monuments found or set, the description is poor.

Reference to sectionalized lands by township plat is similar in form. Because of the large number of these descriptions used, abbreviations acceptable to everyone have evolved. Undoubtedly, sectionalized land descriptions give more information with fewer words than any other possible description.

Sec. 2, T 2 S, R 4 E, MDM.

Not only does this describe a unique area, it also calls for all the original surveyor's field notes and monuments.

15-54 True Metes and Bounds Descriptions

The infinite variations in an approach to writing a metes and bounds description can be illustrated in Fig. 15-7. In describing Parcel "A" by a perimeter description at least 12 points of beginning can be used; a good scrivener would soon eliminate all but two.

As of the time the description is being prepared, the location of the intersection of Frog Creek with either the road or the railroad would be relatively certain; but because of erosion or accretion, it would be highly undesirable to cause a land description to be dependent on a changing point.

An exact quantitative perimeter description of this property cannot be written without a survey locating the railroad, the creek, and all of Lot 10. The owner of Lot 10 sold the following parcels:

> To John J. Jones the northerly 20 acres of Lot 10 lying westerly of the National and Otay Railroad.
> To L. M. Clark the westerly 600 feet of Lot 10, excepting therefrom that portion sold to John J. Jones.
> To the National and Otay Railroad an easement described in book..., page... of Official Records.

The description to Jones is defective; the direction of the southerly line is not given. More serious is the problem, "Does the northerly 20 acres mean the northerly 20 acres of all of Lot 10 or does it mean 20 acres entirely westerly of the railroad?" The fence indicates that 20 acres are occupied, and it will be presumed that Jones did acquire 20 acres. Also, since this leaves the remainder as a lesser amount, the new buyer of Parcel "A" will not be buying future litigation. The northerly 20 acres implies that the north and south lines of the 20 acres were probably intended to be parallel. Even though the railroad has an easement, it will be assumed that the 20 acres is exclusive of the railroad right-of-way (advantage is given to grantee).

After a survey the description could be written as follows:

15-54 TRUE METES AND BOUNDS DESCRIPTIONS / 397

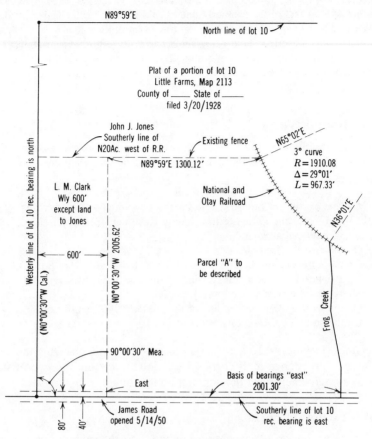

Figure 15-7 Lot 10, little farms.

All that portion of Lot 10, Little Farms, according to the Map thereof number 2113 as recorded in the Office of the County Recorder, County of ...., State of ...., more particularly described as follows:

Beginning at the southwesterly corner of Lot 10 as shown on said Map 2113; thence east 600.00 feet along the south line of Lot 10 to the southeasterly corner of the westerly 600 feet of said Lot 10; said corner also being the true point of beginning; thence N 0°00'30" W, parallel with and 600 feet easterly from the westerly line of said Lot 10, a distance of 2005.62 feet to the southerly line of the northerly 20 acres of Lot 10 lying westerly of National and Otay Railroad; thence N 89°59' E along the southerly line of said 20 acres and its extension, 1300.12 feet to a point of intersection with the centerline of the National and Otay Railroad right-of-way as described in Book . . . , Page . . . of Official Records of the County of . . . , said point of intersection being on a 1,910.08 foot radius (3°) curve whose center bears N 65°02' E; thence southerly along said curve and along the centerline of said railroad, through a central angle of 29°01' a distance of 967.33 feet to the centerline of Frog Creek; thence southerly along the centerline of Frog Creek the following course; S 4°01′ E, 300.00 feet; S 2°59′ W, 261.50 feet; S 10°31′ E, 469.20 feet; south 245.54 feet to the southerly line of said Lot 10; thence leaving Frog Creek, West 2001.30 feet more or less to the true point of beginning.

The above described land being subject to a 40-foot easement conveyed to the National and Otay Railroad and a 40-foot easement conveyed to the County for James Road, recorded in Book 1232, Page 23 of Official Records of the County of ....

15-55 Bounds Form

If the land just mentioned was not surveyed and the purchaser was willing to accept a parcel of land without knowing its size, the following bounds form could be used.

> All that portion of Lot 10, Little Farms, according to the Map thereof number 2113 as filed in the Office of the County Recorder, County of ...., State of ...., bounded as follows:
> On the east by Frog Creek.
> On the northeast by the centerline of the National and Otay Railroad.
> On the north by the southerly line of the northerly 20 acres of Lot 10 lying westerly of the National and Otay Railroad.
> On the west by the easterly line of the westerly 600 feet of Lot 10.
> On the south by the southerly line of Lot 10.

The ambiguity of the 20 acres is left just as it was found. Only the courts or the owners can change it. If corrective action is wanted, the two owners can come to an agreement, or litigation can be instituted.

15-56 Exception Form

Writing a description by exception is similar to the bounds form; instead of saying "bounded by the following," the form "excepting the following" is used, as here:

> Lot 10, Little Farms, according to the Map thereof number 2113 as filed in the Office of the County Recorder, County of ...., State of ...., excepting the following described parcels:
>
> (1) All land lying east of Frog Creek.
> (2) All land lying northeasterly of the centerline of the National and Otay Railroad.
> (3) The northerly 20 acres of Lot 10 lying westerly of the National and Otay Railroad.
> (4) The westerly 600 feet of Lot 10.

The bounds form and the exception form are complete and identify the land, but neither gives quantitative data. The advantage of the metes and bounds form given in Section 15-54 is that it gives definiteness to the meaning of the 20-acre parcel, it gives quantities by which any other surveyor can retrace what the boundaries are. If the surveyor erred in his measurements, the monument calls are superior.

15-57 Monument Calls

The metes and bounds description in Section 15-54 has one obvious defect: it does not recite the monuments set by the surveyor. At each point of change of direction the description should state what was found or set; this would enable retracement with greater certainty.

15-58 The Use of Coordinates in Descriptions

The metes and bounds description in Section 15-54 with the addition of monument calls does not completely insure future locations. Monuments can be destroyed and can be replaced, provided their position is known prior to destruction.

Ordinarily, when a conflict in calls exists, distance and direction will be superior to coordinates. This principle holds because coordinates, a product of distance and direction, are more subject to error. Nevertheless, the recitation of State Plane Coordinates in the description and on the plat provides additional evidence helpful to those interested in the property's location.

In metes and bounds descriptions, grid north may serve as a reference for bearings, provided it is properly identified in the body of the description and provided it is recognized by law in the jurisdiction. The corners can be identified by their x and y values (see sample description that follows).

It has been said that a call for coordinates is a call for all points of the system. An example of a body of a description by metes and bounds using State Plane Coordinates follows:

> Beginning at a drill hole in a stone bound which is set in the corner of a stone wall on the north line of Farm Road at the southwest corner of land of Peter I. Prince and at the southeast corner of land hereby conveyed, the coordinates of which monument referred to the Illinois State Coordinate System, West Zone, Are $x = 617,603.29$, $y = 1,316,042.17$.
>
> Thence, on an azimuth of 81°39'30" a distance of 123.39 feet along the northerly line of Farm Road to an iron pin at the southwest corner of the tract hereby conveyed . . .
>
> Zero azimuth is grid south in the Illinois State Coordinate System, West Zone. . . .

When State Plane Coordinates are used in describing parcels within the U.S. Rectangular System, it will be apparent how the section or part of the section was divided. An example of such a description follows:

> A tract of land lying in Jackson County, State of Alabama, on the left side of the Tennessee River, in the South half (S½) of the Northwest Quarter (NW¼) of Section Three (Sec 3), Township Six South (T6S), Range Five East (R5E), and more particularly described as follows:

Beginning at a fence corner at the southwest corner of the Northwest Quarter (NW¼) of Section Three (3) (Coordinates N 1,470,588; E 416,239), said corner being north six degrees twenty-four minutes west (N 6°24' W) twenty-six hundred (2600) feet from the southwest corner of Section Three (3) (N 1,468,004; E 416,529), and a corner to the land of T. E. Morgan; thence to Morgan's line, the west line of Section Three (3), and a fence line, north five degrees thirty-three minutes west (N 5°33' W) thirteen hundred and four (1304) feet to a fence corner (N 1,471,886); E 416,113). . . .

The coordinates referred to in the above description are for the Alabama Mercator (East) Coordinate System as established by the U.S. Coast and Geodetic Survey, 1934. The Central Meridian for this coordinate system is longitude eighty-five degrees fifty minutes no seconds (85°50'00").

When State Plane Coordinates are required on plats of subdivisions, these values serve to perpetuate the entire subdivision. Such ties are now required by ordinances in some areas.

15-59 Checklist for Descriptions

Because of the variable nature of land descriptions, no checklist can be thorough and complete. To be of value a checklist should be brief, and brevity itself causes omissions. The following list will serve as a reminder for the more important considerations included in descriptions.

1. General descriptions (caption with title identity).
    a. State.
    b. City, county, or parish.
    c. Subdivision: name, number, date, and place of recording.
    d. Court plat: case number, date, title, or other information necessary for identity.
    e. Recorded document: book and page, where filed, date, title, and other identify information.
    f. Township plat: date of recording (where there is more than one plat), range, township, section, portion of section.
    g. Land grant: name, date, court case number, and other identity information.
2. Seniority of deeds and record monuments: a call for all adjoining senior or superior deeds or use.
    a. The exact wording of the adjoiner deed when describing the line along a senior deed.
    b. A call for property lines of equal rights (lot lines, section lines, etc.).
    c. Proper identity of senior deed or property lines of equal rights.
3. Call for a survey.

a. Indirect: a call for a plat that calls for a survey (identify the plat).

b. Direct call for a survey: identify where plat and field notes are recorded, date, and surveyor.

4. Point of commencement.
   a. Certainty for present and future identity.
   b. Compatibility with previous deeds (is it the same point as used in previous deeds, and is the call the same?).
5. Point of beginning.
   a. Correct measurements from point of commencement.
   b. Certainty of identity: set monument, found monument, tie-out measurements from other monuments—all identified and described.
6. Call for physical monuments found or set.
   a. Described for present and future identity: size, shape, material, marks on monuments (blaze, cross, license number, etc.), peculiar means of identifying.
   b. Reference ties from other monuments by bearing and distance or other measurements.
7. Call for natural monuments.
   a. General identity: Atlantic Ocean, Columbia River, Grand Canyon.
   b. Locative position on monument: thread of stream, mean high-tide line, along the top of the bank, centerline of Keeney Street, and so on.
8. Directions.
   a. Definition of basis of bearings or basis of direction (astronomic north, magnetic north, assumed, grid, etc.).
   b. All bearings on same basis.
   c. Directional calls along a monument (northerly along Pine Creek, southerly along Lake Superior; easterly along the centerline of Cabrillo Road).
   d. Directional calls along geometric or irregular lines (southerly along a circular curve of . . . ; westerly parallel with Milar Road; southerly, westerly, and northerly along the 330-foot elevation line; etc.)
   e. Bearings: correct quadrant.
   f. Azimuth: definition of reference meridian (south or north).
   g. Grid: definition of reference meridian.
9. Distance: consistent units and definition of units where necessary.
10. Curves.

402 / WRITING DESCRIPTIONS

   a. Circular: radius, central angle, chord, arc length, radial direction, direction of concave side, compatibility of parts where more than two curve elements are given.
   b. Spiral: complete definition of basis of spiral.
   c. Elevation: definition of elevation datum.
   d. Irregular: define line (along centerline of Rose Creek, along the average low water mark).
11. Coordinates.
   a. Origin or coordinates.
   b. Basis of direction.
12. Area: more or less except where it is controlling with one dimension omitted—gross or net area?
13. Mathematics.
   a. All bearings and distances on same basis, and figure should close mathematically.
   b. Consistency of parts (length of arc, central angle, and radius, etc.).
   c. All lines and parts defined by consistent dimensions.
14. "Of" conveyances.

   a. Westerly 50' of lot 2 is 50' at 90°.
   b. Westerly 1/2—define direction of dividing line.
   c. Northerly 10 acres—define dividing line.
   d. Sectionalized lands W½ of NW¼ and W 80 Ac of NW¼ are not the same.
   e. In all except sectionalized lands, a fractional part is a fraction of the area.
15. Strip conveyances.
   a. Extend and shorten terminal lines.
   b. Each side of centerline not "either."
   c. Define station and origin.
16. Intent: Does the description properly express the intent of the seller? Did he intend to sell by a survey, monuments, dimensions, area, or by what? Does the entire description express this intent?
17. Final steps: date, sign, seal, deliver to client.

15-60 Summary

If a competent expert can locate land from the identification given in a land description, the description is sufficient. But legally sufficient descriptions are not always satisfactory descriptions. Good descriptions, in addition to identifying land, describe the geometric shape and size of the land identified, recite title identity, give sufficient locative calls to insure future location, are based on a recent survey, are clear, unambiguous, precise, brief, and certain. Proficiency in description writing is the result of a combined knowledge of mathematics and particular knowledge of the legal meaning of description words and phrases.

REFERENCES

Brown, Curtis M., *Boundary Control and Legal Principles*, 2nd ed., John Wiley & Sons, New York, 1969.

Patton, Rufford G., and Patton, Carroll G., *Patton on Titles*, West Publishing Co., St. Paul, 1938.

Thompson, George W., *Thompson, on Real Property*, Bobbs-Merrill Co., Indianapolis,

Wattles, Gurdon H., *Writing Legal Descriptions*, Gurdon H. Wattles Publications, Orange, Calif., 1976.

Wattles, William C., *Land Survey Descriptions*, Title Insurance and Trust Company, Los Angeles, 1956.

CHAPTER 16

The Surveyor in Court

16-1 Introduction

Land surveyors can, in the course of their professional practice, expect to appear in court either as a litigant or as an expert witness. Litigation over land ownership usually involves measurements or evidence, and surveyors are experts in gathering, interpreting, and evaluating both.

In some states, the court, on its own motion, is privileged to call on the surveyor as an amicus curiae to act as an expert in the determination of property lines. Attorneys may ask surveyors to act as expert investigators or witnesses in boundary disputes.

Court procedure and witness behavior, particularly that relevant to the surveyor, are discussed in this chapter.

Courts are not creative institutions and do not legislate or pass laws; they merely interpret what legislatures have lawfully decreed as proper or legal.

16-2 Initial Court Effort

In his first appearance before the court, the surveyor will find that he will be guided by the attorney and the judge and will not be permitted to say what he thinks he should be allowed to say. Unless he has thoroughly discussed the case with his attorney, he will be disappointed by the questions asked by his attorney. Further, on cross-examination he will not be prepared for the rude treatment, the twisted half-truths tossed back at him, and the attempts to discredit abilities and qualifications by the opposing attorney. If the case is lost, he will be indignant and will think that the judge was biased.

Court cases are won or lost on the merit of the evidence presented and on the foundation of law involved. The surveyor's only function in court is to present, discuss, and explain evidence; the surveyor cannot present law. In presenting evidence he will be permitted to answer only questions put to him; hence he and the attorney must have preconsultation to make certain the right questions will be asked. They become a team, with each member of equal importance.

No surveyor can predict the outcome of litigation. He may have concluded that a certain monument is an original, but the judge will decide that the evidence of a witness, whom the surveyor had good cause not to trust, fixes the corner location. The surveyor may conclude that a certain principle of law applies, yet the judge, along with the aid of a persuasive lawyer, will conclude differently.

Surveyors cannot and do not decide litigation; they are not experts on law, at least in the eyes of the court.

16-3 Duties of a Surveyor in Court

In litigation involving land boundaries, the surveyor may be either a lay or expert witness. The surveyor's major function is as a witness describing facts within his perception or as an expert witness expressing opinions within his special field. Depending on his capabilities and his relationship with the attorney, he may also be asked to accompany the attorney during the litigation and serve as an adviser.

In the practice of surveying, the client usually asks the surveyor to give him a solution to a particular boundary problem. The surveyor works until he finds a solution; he is acting as though he were the sole judge of the quality and weight of evidence and of the conclusions presented. But in court the surveyor presents the facts, and the court decides on the relevancy, admissibility, and quality of evidence and the conclusions produced therefrom. Thus in court the surveyor must change his mode of thinking and must always remember that it is not he who is solving the problem. He is merely a witness presenting the facts to the judge or the jury for determination.

> The maps which plaintiff introduced did not give the data necessary to measure the rights of parties on any theory. They gave no means of getting at just where the thread of the stream was supposed to be, or the shore conformation, except in a limited range, and were in other respects deficient.
>
> We have had occasion, in several instances, to point out that a surveyor cannot be allowed, under any circumstances, to fix private rights or lines by any theory of his own. Before a surveyor's evidence can be received at all, it must be connected with a starting point or other places called for by the grants under which the parties claim. His duty is neither more or less than to measure geometrically in accordance with those data, and his science goes no further. It is not his business to decide the questions of law, or to pass upon facts that belong to the tribunal dealing with the decision of facts. His testimony, as a man of science, is never receivable except in connection with the data from which he surveys, and if he runs lines they are of no value unless the data are established from which they are run, and those must be distinctly proven, or there is nothing to enable anyone to judge what is the proper result.
>
> —*Jones* v. *Lee*, 77 Mich. 43.

16-4 Conduct on the Stand

The surveyor's conduct on the witness stand should always reflect an attitude of a reporter of facts. At no time should he give the feeling of being an advocate; that is the position of the attorney. Once a surveyor places himself in the position of an advocate he will lose his effectiveness of treating facts and evidence impartially. Effective witnesses are those who answer questions fully, factually, correctly, and honestly; they do not try to duck or evade cross-examination questions that place the opposition in a favorable light. Harmful facts should be presented just as straightforwardly and with as much dignity as the facts relied on to make the case. To convey the idea of attempted concealment of facts is a fatal mistake and could lead to discrediting of all previous and later testimony.

16-5 Object of Litigation

Generally speaking, each side feels that he is right in most litigations; each hires attorneys and experts to assist him in proving his case. The object is to win. The object of the surveyor, as an expert, is to assist the client in winning provided, and only provided, that he is absolutely honest. The surveyor who is given an exclusive franchise to perform property-line surveys is also given the responsibility of absolute honesty.

16-6 Surveyor-Attorney Relationship

The attorney is the advocate; he is the final interpreter of the facts and the final judge of how the facts will be presented in court. He is in command.

The surveyor is in on the planning; and, often in land litigation, he is the major defense or attack. He is not there to win for the glory of winning; he is there because he thinks his side is right, or he is to present certain facts unbiasedly. If his thoughts are otherwise, he should inform the attorney at the earliest moment and withdraw, if necessary.

16-7 Court Trials

All the procedures and methods used in court trials, although interesting, are not of vital importance to the surveyor. Since the surveyor is a witness, it is important that he know how to conduct himself properly as a lay or expert witness. And it is important that he know what fees, if any, that he is entitled to.

16-8 Pretrials

Pretrials, as practiced in some states, are for the purpose of eliminating unnecessary court time and are an attempt to reach agreement on what are the issues in a case. The attorneys and the judge, in conference, discuss the issues and agree on which points evidence will be taken. At this time the judge or either party's attorney may ask for an expert witness.

16-9 Types of Court Trials

Courts are often divided into inferior, superior, and appeal. Inferior courts, so called, are police courts, small-claim courts, night courts, traffic courts, municipal courts, and other courts wherein transcripts of proceedings are not usually kept. Land litigation is generally assigned to superior courts, and a transcript of the entire trial is kept by a court reporter.

Trial court is a term used to designate the court wherein facts of a case are presented. There is no appeal from the trial court's decision about what the facts are, but there is an appeal from questions of law. As an example, Jones and Brown are in a dispute over their boundary as located along a creek. The trial court will decide in fact where the creek is located, and usually there can be no appeal from the facts. But whether the limit of ownership is on the bank, along the gradient line, or at the thread of the stream is a question of law that may be appealed. Any irregularities of trial procedure, inadmissibility of evidence, and similar questions of law are appealable. Appeals are based on the facts presented in the trial court, and, unless a new trial is ordered during the appeal, no new facts are presented or permitted.

Published court cases are confined to supreme court decisions, and, sometimes, courts immediately inferior to the supreme court. In law libraries superior court cases are not found; only final court of appeal records are in state reports.

The surveyor has little function beyond the trial court; the surveyor's testimony and usefulness are confined to presenting facts. Since only questions of law may be appealed, a surveyor does not appear before the appellate or higher court. Surveyors may be used in preparing affidavits for trial courts and for cases that are appealed. Although new evidence cannot be presented above the trial court level, a wise attorney will seek the help of his expert witness in preparing and stating facts should a case be appealed to a higher court. This insures that no salient facts are omitted.

16-10 Oath, Questions, and Answers

Prior to testifying, a witness is placed under oath to "tell the truth, the whole truth, and nothing but the truth." This does not mean that the witness is to volunteer information and ramble along on any subject that he thinks might have a bearing on the case. He is to answer questions as presented to him. If the answer given to a question results in a half-truth or could leave the court with a false impression and the witness feels that further information should be given, and yet the information is not called for by the question, he may ask permission of the court to clarify his answer. The reason testimony in general is limited to the answering of questions is that the court wants to know in advance whether the subject matter the witness is to talk about is admissible evidence. Also the opposing side then has opportunity to object if the question asked is improper.

16-11 Direct Examination and Cross-Examination

The examination of a witness by the party producing him is called *direct examination*; the examination of the same witness, on the same matter, by the adverse party is called *cross-examination*. The direct examination must be completed before the cross-examination begins, unless the court directs otherwise.[1]

In direct examination the surveyor's usual function is to answer questions directed at introducing certain evidence. The sources and nature of evidence

---
[1] Sec. 2045 CCP Calif.

and the relationships of evidence (usually by measurement) are the foundation of testimony.

If a question is not understood by the witness, he may and should ask for clarification. The witness is not bound to answer any question he does not understand.

After a question is asked and is understood, prompt answers should be made; long delays and hedging create the impression of indecision. It should be realized that the answer to the question is being recorded by the court recorder. Answers of "from this point to that point" mean nothing on the record; answers of "from point C on plaintiff's exhibit A to the centerline intersection of Main and Johnson Streets" are positive and clear. Nods of yes or no are not recorded. If either attorney or the judge asks the witness to place a symbol on an exhibit, he should do so and say, "I am now placing an 'A' at the point previously referred to." The record must state what was done.

> We call attention to the fact that in the record there are more than 50 instances where the witness indicated points on exhibits without any mark whatever thereon to enable us to ascertain therefrom the point to which the witness was directing attention, and in several instances the exhibit about which the witness was testifying was not even identified. This laxity on the part of the trial court and counsel made it impossible for us to give consideration to such evidence.
> —*Everet* v. *Lantz* 126 Colo. 504

The attorney's rule is "do not volunteer information; answer only questions asked."

In cross-examination the opposing attorney has two objectives in mind: (1) to discredit testimony in controversy with his claims and (2) to use the testimony to his advantage. If the witness is testifying to factual truth, his testimony cannot be discredited. All factual questions must be answered, and if some fact does help the other side, it is of no concern to the surveyor. If an answer cannot be avoided, a prompt normal answer leaves less impression on the mind of the jury, but it should not be so prompt that the attorney does not have time to object.

16-12 Leading Questions

A question suggesting to the witness the answer that the examining party desires is called a *leading* or *suggestive question*. On a direct examination, leading questions are not allowed, except in the sound discretion of the court, under special circumstances, making it appear that the interests of justice require it. The opposite party may cross-examine the witness about any facts stated in his direct examination or connected therewith and, in so doing, may ask leading questions. If the opposite party examines the witness about other matters, such examination is to be subject to the same rules as a direct examination and may be objected to and excluded.

16-13 Hearsay

Evidence not proceeding from the personal knowledge of the witness but from the mere repetition of what he has heard others say is called *hearsay evidence*.[2]

Evidence based on rumor, common talk, or the statement of someone else is hearsay and, with few exceptions, is inadmissible. In the surveying practice the surveyor often bases his opinions on hearsay facts, such as a monument found and commonly reported by many as being in a correct position. Obviously, a surveyor could not be present in person when all the original surveys were made; hence his opinion of where property lines belong may be based entirely on hearsay evidence of where the original surveyor set his monuments (see Sections 2-48 and 2-49).

Expert opinions are admissible even though they may be based on inadmissible hearsay evidence.

This was reinforced in two 1980 decisions by the Supreme Court of Georgia in the cases of *Cheek et al.* v. *Wainwright* and *King* v. *Browing*.

> As an expert witness, the surveyor is entitled to testify as to his opinion on the facts as proved by other witnesses. Code Ann. § 38-1710. An expert witness, such as the surveyor here, may testify as to his opinion on the ultimate issue in the case without invading the province of the jury so long as the subject is an appropriate one for opinion evidence. *Metropolitan Life Ins. Co.* v. *Saul*, 189 Ga. 1 (5 SE2d 214) (1939); Agnor's Ga. Evid., § 9–3. An expert may base his opinion on hearsay and may be allowed to testify as to the basis for his findings. *State Hwy. Dept.* v. *Peters*, 121 Ga. App. 167 (173 SE2d 253) (1970); *State Hwy. Dept.* v. *Parker*, 114 Ga. App. 270 (150 SE2d 875) (1966). When an expert's testimony is based on hearsay, the lack of personal knowledge on the part of the expert does not mandate the exclusion of the opinion but, rather, presents a jury question as to the weight which should be assigned the opinion. The evidence should go to the jury for whatever it's worth. *Durand v. Reeves*, 219 Ga. 182 (132 SE2d71) (1963); *Napier v. Little*, 137 Ga. 242 (73 SE3) (1911); *State Hwy. Dept.* v. *Peters*, 121 Ga. App. 167, supra.
>
> Similarly, documentary evidence illustrative of oral testimony and authenticated by oral testimony is admissible. *Fountain v. Bryan*, 229 Ga. 120 (189 SE2d 400) (1972); *Savannah Ice Delivery Co.* v. *Ayers*, 127 Ga. App. 560 (194 SE2d 330) (1972). Further, where the admissibility of such evidence is doubtful, it should be admitted and its weight left to the jury. *Savannah Ice Delivery Co.* v. *Ayers*, supra; *Durden v. Kerby*, 201 Ga. 780 (41 SE2d 131) (1947).
>
> —*King* v. *Browning*, 246 Ga. 47.

> 3. Appellants' third enumeration of error, that the testimony and plat of appellees' surveyor were based on hearsay and should be excluded, is also without merit. An expert, such as a surveyor, may base his opinion on hearsay. The presence of hearsay does not mandate the exclusion of the testimony; rather, the weight given the testimony is a question for the jury. *King v. Browning*, 246 Ga. 46 (1980). Accordingly, the third enumeration of error is without merit.
>
> —*Cheek et al.* v. *Wainwright et al.*, 246 Ga. 171.

16-14 Jury

All questions of fact, when the trial is by jury, are to be decided by the jury, and all evidence thereon is to be addressed to them (Sec. 2101 CCP Calif.).

---

[2]*Black's Law Dictionary*, West Publishing Co., St. Paul, 1979.

A jury trial may be requested by either side. If there is a jury trial, the jury decides questions of fact. If a government section corner is lost and there is a dispute about which of several monuments represents the true corner, the jury would decide which monument represents the true corner. If a fence has been in existence for a period of time and the question arises about how long the fence has been in existence, the jury would decide. But if there is a question of the legal interpretation of the meaning of a deed term, the judge would decide.

Few attorneys desire jury trials in boundary litigations; hence the judge usually tries both facts and law. Most of the testimony about the facts concerning boundaries are testified to by the surveyor. Any fact the surveyor observes and measures is a fact to which he may testify.

16-15 Lay Witness

Lay witnesses are those who may testify about facts within their knowledge and may not state their opinions.

This is the rule. But there are many, many exceptions to the rule. In fact, practically all of the discussion of the rule centers around instances that are exceptions.

Often there is no way of extracting the facts from a lay witness, other than by allowing him to express an opinion on things derived from his own perception. A statement that there was an "old" fence between the two properties is an opinion. "Hot" or "cold" may only be an opinion, especially when a person says how hot or cold a thing was. The lay witness testifies about subject matter readily understood by the court. Preventing the lay witness from stating opinions, especially on subjects on which he is not qualified, tends to prevent fraud and perjury and is one of the strongest safeguards of personal rights.

Specifically, the California Code (Sec. 1845 CCP) is this: a witness can testify to those facts only which he knows of his own knowledge—that is, which are derived from his own perceptions, except in those few express cases in which his opinions or inferences, or declarations of others are admissible.

And, of course, a witness is presumed to speak the truth. This presumption, however, may be repelled by the manner in which the witness testifies; by the character of the witness' testimony; by evidence affecting his character for truth, honesty, or integrity; by his motives; or by contradictory evidence. And the jury members are the exclusive judges of his credibility.

In analyzing lay testimony conclusions and opinions usually are inadmissible, except when they are specifically derived from personal observation of the facts in issue and when no better evidence is available to the jury.

Like all other evidence, for lay testimony to be admissible the judge must find that the testimony is based on personal observation and that the resulting opinion will be helpful to a clear understanding of the witness' testimony or the determination of the fact in issue. This last area includes areas in which normal people usually form opinions: "He looked sick." "He was as drunk as a hoot owl." "The fence was as old as Moses." Statements of this sort convey to the jury an impression by a manner of speech.

16-16 Expert Witness

In most instances in which a surveyor is called as a witness, he usually will appear as an expert witness. Before a witness is permitted to appear as an expert, the judge determines the witness' qualifications in relation to the ultimate issues. In arriving at his decision to either permit or exclude the expert witness, the judge must determine that the subject matter to be presented goes beyond the everyday knowledge of surveyors of ordinary experience and education. *Not all surveyors will qualify for expert witness.* A surveyor whose practice is predominately in urban subdivision layout would not qualify as an expert in rural section retracements. If a situation should occur in which a surveyor was determined to be unqualified, he could possibly be held negligent for practicing in a field in which he is unqualified.

Any opinion expressed by the expert surveyor will go beyond what is normally expected. The court must consider if the witness is specially qualified by examining special knowledge and skills and the variety and depth of experience. It is the responsibility of the proponent who wants to introduce the expert to persuade the judge that the expert possesses the necessary qualifications. If the opposing party objects, the burden then rests on the proponent to convince the judge about his witness's capabilities. Factors used to convince the judge may include, but are not limited to, the witness' training and education, experience, familiarity with standard references and authorities in his field, membership in professional associations and societies, and his general association within his profession.

> A land surveyor testified that he had run out the lines of lots surveyed by a former surveyor, and was familiar with his mode of marking corners, and the witness then testified to certain marks upon certain alleged corners as having been made by the former surveyor. It was held that his belief that the marks were those made by the former surveyor, was not evidence to be received by the jury as the opinion of an expert, but was merely the testimony of a witness to the effect within his knowledge, and was to be credited by the jury only so far as they believed him able, from his personal knowledge, to identify the marks in question.
> —*Barron* v. *Cobleigh*, 11 N.H. 557

Ordinarily, the surveyor, as an expert, confines his testimony to measurements, monuments found, and the interrelationship of monuments found. The case of *Curtis* v. *Donnelly* illustrates the evidence given.

> P. W. Warner, County Surveyor of Marion County testified he made and recorded a survey of Block 3, and record of his plat and survey was introduced in evidence. He testified that he dug into the earth and found the original corner stone in Block 2, which is 66 feet south of the southwest corner of Block 3; that he located the section line; that he measured the distance from the corner of Block 3, established as 66 feet from the corner of Block 2, to the section line; that he found a gain of 17 inches and of 12 inches; that he gave all the gain to the end lot; that he found certain corner stones and not others; that he did find the section corner stones; that fence located was 9½ feet off at one end and 19½ feet off at the other. In

> rebuttal, E. C. T., another surveyor testified that he measured the south side of Block 3 on the south line thereof beginning 14 inches east of the east side of the concrete walk on the east side of Green Street, which he treated as the southwest corner of said Block 3, and that from that point he measured east, and after allowing to each lot owner of said seven lots the number of feet as shown by the original survey of said block that belonged to each lot owner, and he found that the fence corresponded to the measurements. This is the full extent of his survey. He gave as his reason for beginning 14 inches east of the concrete walk in question, that he understood that the walk was built 14 inches west of the line of Green Street. The man who constructed the walk testified that it was built 10 or 11 feet west of the property line.
> —*Curtis* v. *Donnelly*, 273 Ill. 79.

On survey matters those not licensed may sometimes testify about boundary surveys.

> The court held that inasmuch as Furlong was not a licensed surveyor he could not testify. The statute is 4696, C.L. 1921 which makes it unlawful for any person "to practice or to offer to practice engineering or land surveying in this state, unless such person has been duly licensed under the provisions of this act." That statute has no application in this case. The testimony of Furlong was that he was engaged in land surveying in Minnesota, that he, at the time of his survey, was a clerk in the post office, and that he made the survey for his father who was one of the owners of the property. Such a survey is not practicing surveying. To practice a profession is to hold oneself out as following that profession as a calling, as one's usual business.
> —*Beverbrook Resort Co.* v. *Stevens* 76 Colo. 131.

16-17 Appointment of Expert Witnesses

Either side may call its own expert witness. In some jurisdictions the court may appoint an expert.

> Whenever it shall be made to appear to any court or judge thereof, either before or during the trial of any action or proceeding, civil or criminal, pending before such court, that expert evidence is, or will be required by the court or any party to such action or proceeding, such court or judge may, on motion of any party, or on motion of such court or judge, appoint one or more experts to investigate and testify at the trial of such action or proceeding relative to the matter or matters as to which such expert evidence is, or will be required, and such court or judge may fix the compensation of such expert or experts for such services, if any, as such experts may have rendered, in addition to his or their services as a witness or witnesses, at such amount or amounts as to the court or judge may seem reasonable.
> —Sec. 1871 CCP Calif.

The practice of the parties hiring their own expert witnesses has been severely criticized. There is an implication of bias toward the party calling the expert, especially when a sizable fee is involved. A precommitment or preconsultation with a party in itself presents the flavor of bias. A party may interview many experts before he can find one that is favorable to his cause and thus not present a true picture to the court. It is indeed unfortunate that many are often tarred by the same brush. Such ideas are an injustice to the conscientious expert whose only concern is the truth of the principle involved.

Today, there is a trend toward liberalizing the rules governing the qualification of expert witnesses.[3] The ultimate determination is "would the testimony be likely to assist the jury."

16-18 Duties of Expert Witnesses

In summary, the duties demanded of an expert witness are these:

1. Answer all questions clearly and intelligently.
2. Be absolutely unbiased and honest.
3. Have expert knowledge of the particular subject.
4. Be prepared to discuss the opinions of other authorities and state why you agree or disagree with them.
5. Limit testimony to things and opinions that you can defend before experts in the particular field.

The opinions of an expert are generally only arguments in behalf of a litigant. Hence an expert's testimony is often valuable only because of the reasons and facts given to support his conclusions. Rejection of conclusions unsupported by facts or reasons can be expected.

Testimony that cannot be understood by the jury or judge is practically valueless. A statement that "I ran a traverse around the deed and found an error of closure of two feet" is simple to the surveyor but obscure jargon to the average person. It is doubtful if the judge or jury would know that the traverse might be an office calculation and not a field measurement. "Error of closure" certainly needs clarification to most people. A judge became a judge because he was an attorney, not a surveyor. One of the assets of a good expert witness is his use of simple, clear English. Don't try to awe the jury with your ability to "elucidate" with "ostentatious" terminology. Explain in simple terms.

At no time should the surveyor place himself in the position where he is obligated to take a particular side in his testimony. When the expert is on the stand he should make every effort to rid himself of any bias or prejudice resulting from who is paying the fee or who has previously consulted with him.

But this does not mean that the surveyor is to avoid expressing positive, sincere convictions. The surveyor is on the stand because he is supposed to know what to do under the circumstances. He must present the facts and opinions as he sees them and present his arguments or reasons so clearly that anyone can understand and believe them. A person who is biased or prejudiced tends to deliberately slant reasons without sound foundation. If a person has a firm conviction based on sound reasons, he should prove it. If a person's

---

[3]*Evidence*, 11th ed., Gilbert Law Summaries, Gardena, Calif., 1979.

testimony is doubtful, he should have avoided appearing as an expert in the first place.

No one can expect to remember in detail all the facts he is to present without refreshing his memory. Surveyors take field notes and develop evidence in writing. Maps are examined and deeds are studied. Prior to taking the stand the surveyor should refamiliarize himself with all the data; otherwise, embarrassment may result from a searching cross-examination. He should be prepared especially to explain the error of the contrary opinion.

16-19 Opinion Evidence

As previously stated, the lay witness is limited to testimony concerning facts and may sometimes give opinions on things that are easy for everyone to understand. The expert may give opinions on subjects that are beyond the knowledge of average people. But such right to give opinions is not unlimited.

The expert witness' opinions may be based on such elements as (1) facts or data made known to him outside the court (field books of prior surveyors, plats, and descriptions), (2) facts or data from personal observations (distances between monuments and descriptions of found monuments), and (3) facts or data made known to him during the trial (testimony of other witnesses, assumed facts to hypothetical questions, and offered evidence). Usually, an expert surveyor may be required to disclose his facts or data prior to his offering an opinion concerning the facts.

At times courts may hold evidence on which expert opinions are based as inadmissible. As long as the facts are of the nature on which experts usually rely in that field in forming opinions, the facts themselves need not be admissible in evidence. At times it is impossible for the expert to frame a postive answer and he must express himself in terms of probability. There is no legal requirement that an expert witness be positive or even reasonably certain in formulating his opinion. This testimony only goes to the "weight" or evidence and not to the admissibility (*Gichner* v. *Antonio Tile Co.*, 410 F.2d 238).

No witness can give opinions on the ultimate fact that is being tried. Permitting an expert to tell the members of the jury what they must decide is usurping their exclusive rights. If the north quarter-corner location is in dispute and any of three stone mounds might be the right one, the surveyor cannot tell the judge or jury which of the mounds to select. He can walk all around the subject and even answer hypothetical questions that almost give direct answers on the solution. He could describe in detail the shape and size of the mounds and describe any special markings found. If leaves were found under one of the mounds, indicating recent construction, such facts could be emphasized. What was found in other similar locations and whether the original surveyor consistently set a certain type of monument could be discussed. Needless to say, the examining attorney would have to be well versed on the facts observed by the surveyor; otherwise, he would not know which questions to ask. The surveyor is more or less limited in his response to the questions asked.

In *Kirby Lumber Company* v. *Adams*, in response to ruling on the question of whether the surveyor could be asked, "Where is the true line?" the court ruled:

> This was not a matter about which they could give their opinion. It was a matter capable of being fully stated to the jury. The witnesses had surveyed the land and had fully described all the facts as found by them on the ground. It was the province of the jury to conclude from all the facts proved whether the line was where contended for by appellants or where insisted upon by appellees. The matter sought to be shown by the opinion of the witnesses was the very question in issue between the parties and was one for the jury alone to determine.
> —*Kirby Lumber Company* v. *Adams*, 93 SW, 2d 382.

No expert opinions may be given to the members of jury if they are capable of forming their own opinion. The purpose of giving expert opinions is to advise the members of jury on matters beyond their knowledge, not on matters within their knowledge.

Fortunately, the expert witness does not have to decide when he can or cannot give opinions. The judge will tell him. A question is presented to him, and if the other side objects, the judge will rule on whether the question can be answered.

16-20 Hypothetical Questions

A *hypothetical question* is a question put to an expert witness containing a recital of facts assumed to have been proved or proof of which is offered in the case and requiring the opinion of expert witness thereon. Hypothetical questions have caused considerable criticism in many courts, but most jurisdictions still permit them.

Because witnesses cannot express an opinion on the ultimate fact in issue, hypothetical questions are asked. "Hypothetical" implies assumed without proof. Each side thinks or hopes that it can prove certain facts to be true. But neither side can be certain of what the jury will declare as being true. So a hypothetical question such as "Assuming that this, this, and this are true, could you express an opinion as to whether a person so injured could continue doing surveying work?" is asked. The facts assumed to be true are the facts presented, and the facts must be in the record of the court when the questions are asked, or the examiner must have made an offer of proof. If the jury finds that the facts are not as assumed, the opinion of the expert is without effect. Several hypothetical questions may be asked by adding extra assumed facts or subtracting assumptions.

Hypothetical questions are more frequently used for doctors in personal injury cases. Rarely, the surveyor will find such questions presented to him.

16-21 Cause and Effect

An expert may testify about what might have caused an event, but it is better not to say what did cause it. Thus a surveyor might testify that the improper location of a house might be caused by measuring from an incorrect monument located 10 feet from the correct monument. He should not say that it was the cause; that is for the jury to decide. To say that a 2 × 2-inch pine stake located at a property corner would completely disappear in 30 years because of decomposition, termites, and so on, would be improper. To say that at all locations where pine stakes were originally set none was found that was over 30 years old and that those older than 10 years were in a bad state of decomposition would be proper.

16-22 Textbooks

Because the author of books cannot be cross-examined and the author did not write his books under oath, the books are, by common law, excluded as evidence; however, there are exceptions that vary from state to state.

In most states, the right to use books of science or art and published maps or charts is permitted under limited conditions by statutes. In California, Section 1936, CCP states that "historical works, books of science or art, and published maps or charts, when made by persons indifferent between the parties . . . are *prima facie* evidence of facts of general notoriety and interest."

> Evidence of this sort is confined in a great measure to ancient facts which do not presuppose better evidence in existence . . . The work of a living author, who is within the reach of the process of the court, can hardly be deemed of this nature. Such evidence is only admissible to prove facts of a general and public nature, and not those which concern individual or mere local communities. Such facts include the meaning of words which may be proved by ordinary dictionaries and authenticated books of general literary history, and facts in the exact sciences founded upon conclusions reached from certain and constant data by processes too intricate to be elucidated by witnesses when on examination. Thus, mortality tables for estimating the probable duration of the life of a party at a given age, chronological tables, tables of weights, measures and currency, annuity tables, interest tables and the like, are admissible to prove facts of general notoriety and interest in connection with such subject as may be involved in the trial of a cause.
> —*Gallagher* v. *Market St. Ry. Co.*, 67 Calif. 13.

In some states the expert is allowed to read what other experts have written on the question being considered, especially if it is of a science matter. Since the judge is an expert on law, it is seldom wise to try to tell the judge what the law is. Stating "It is my understanding that the surveyor must accept monuments in preference to measurements" is much better than stating "The law is the surveyor must accept monuments in preference to measurements." Stating that I was following code Section 23:726 in making my determination is usually acceptable.

Under Federal Rules of Evidence (F.R. 803-18) if an expert has relied on statements from scientific texts standard in the profession or other learned treatises, such statements may be read into evidence under a special exception to the hearsay rule, either by the party calling the expert or by the adverse party in cross-examination. It is accepted that an expert may be subjected to rebuttal testimony by other experts. Thus it is always possible to show on cross-examination that the texts the expert relied on in forming his opinion in reality do not support him; other portions of the book may have contrary thoughts expressed.

The Federal Rules permit cross-examination of any expert witness about opposing views expressed in *any* text, whether the expert relied on it or not (F.R. 803-18). These same Federal Rules provide that the opposing statements from the text may be read into evidence under a special exception to the hearsay rule and thus may be used as proof of contrary views. If a surveyor wants to rely on texts to reinforce his opinions, he should be knowledgeable about all texts within his specific field.

Since the admissibility of reading texts in court is variable, the surveyor should obtain advice from his attorney in his state about what is permitted.

16-23 Use of Photographs

It is well established that photographs of the site to show the conditions existing, if proved to be correct, are admissible as evidence. The use of aerial photographs has been rejected in several cases and admitted in others. A picture taken directly below the airplane has a certain amount of distortion to the untrained person because of the "explosion effect," and straight lines appear bent where changes in elevation occur. In a photograph of a forest directly under the camera, the center tree is vertical, but all vertical trees away from the center appear to slope outward as would occur in a bomb explosion. For these reasons and others, the courts have held aerial photographs as not being accurate, especially when other evidence gives a better picture.

16-24 Power to Compel Expert to Testify[4]

The right of an expert to refuse to testify about matters of opinion, without compensation, is traced back to English Common Law. In *Webb* v. *Page* (1843), 1 Car. and K., Eng. 23, the right of the expert to demand compensation prior to testifying was upheld.

For the usual witness fee, a lay or expert witness is required to testify to facts seen or observed by him. If a surveyor observes certain facts or sets certain monuments, he may be compelled to testify about these facts. But when a party selects an expert to render an opinion on a peculiar subject, the expert can refuse until satisfactory arrangements are made for compensation.

Many jurisdictions hold that an expert must testify to facts within his knowledge, even though special study, learning, or skill was required to determine them. In *Finda* v. *Bolton* (1883), 6 NJLJ 240, a civil engineer was denied the right to recover compensation for services as an expert witness. He was called as an expert by the party who employed him to make a survey. Since

---

[4]ALR 2-1576, annotated, The Lawyers Co-operative Publishing Co., Rochester, N.Y., 1919.

he attended court under a subpoena, he was bound to testify about what he knew, however he acquired the knowledge. In *Summers* v. *State* (5 Tex. App. 365) a medical expert was compelled to report on the findings of a postmortem examination without extra compensation.

But there is nothing in the law that compels an expert witness to make a free preliminary investigation to prepare himself for expressing an opinion. A surveyor who is asked to perform certain surveys as a preparation for litigation cannot be compelled to do so without compensation. If a deed is presented to a surveyor in court and the surveyor is asked where it is located on the ground, he may refuse to read the document and give an opinion. But if he has already read the document and has already formed an opinion, he probably would be bound to express his opinion.

In *United States* versus *Cooper* (21 D.C. 491), the court observed that it was an obligation of an expert to serve on payment of a reasonable fee.

In summary, surveyors in many jurisdictions are bound to testify about the results of surveys performed in the past. They are not bound, without compensation, to express opinions on things that require professional preparation prior to expressing the opinion.

It is interesting to note that an expert witness may not refuse to answer a question on the grounds that it will cause him civil liability. Usually, the only grounds for refusal to answer a question is that the answer may cause the witness to incriminate himself for a crime. Refusal to answer a question because it may cause embarrassment, disgrace, or monetary loss from civil liability is not an excuse. The privilege of refusal to answer a question applies only to crime (*Corpus Juris Secundum*, Vol. 98, p. 98).

16-25 Cross-Examination of the Expert Witness

Once direct testimony is concluded few surveyors are prepared for the cross-examination by the opposing party. It is the duty and responsibility of the advocate attorney to discredit the opposing expert in the eyes of the court and to strengthen his client's case. An expert can be cross-examined like any other witness, and the areas of examination may include the expert's qualifications, experience, honesty, prejudice, affiliation with his client, and common knowledge of his field, including publications and people.

The courts hold that cross-examination is the most effective manner of testing credibility and authenticity of testimony, and to insure due process to parties in litigation, cross-examination is a right. Further, in cross-examination leading questions and innuendos may be used. But cross-examination does have some restrictions; the cross-examiner may not utilize misleading and compound questions; he cannot be argumentative; he cannot assume facts not in evidence, and above all he cannot require answers to questions in areas in which the witness is not qualified.

On cross-examination and direct examination it is of utmost importance that the expert surveyor be completely honest in his testimony, for if it is determined that a portion of his testimony lacks credibility, this could jeopardize his entire testimony.

When the cross-examination has attempted to impeach or discredit an expert witness, the party for whom the witness is testifying is permitted to redirect in an attempt to "rehabilitate" him. Basically, this involves restoring his credibility before the court.

16–26 Expert Witness Fees

The surveyor's fee for expert testimony is a contractual arrangement between the surveyor and the party engaging him.

The courts cannot compel a person to perform work. If a measurement or observation is required, the right of a litigant to hire an expert is recognized. Before commencing the work, the surveyor should have a clear understanding of what his fee will be. If the work is performed, and there is no understanding about the amount of the fee, the surveyor may be compelled to testify about what he knows without the benefit of a fee. It is advisable to include within the fee all expenses that will be incurred in court appearances, since a person is compelled to appear in court whether he receives a fee or not. The only collectible part of the contract is the work for preparation to appear in court.

The compensation of expert witness cannot be dependent on the outcome of the litigation, since it furnishes a powerful motive for exaggeration, misrepresentation, or suppression.

> We are aware that witnesses who are to be called to give expert testimony which involves the special knowledge and skill of the witnesses, and often requires examination and study upon a particular branch of science, are, from the necessities of the case, justified in demanding and receiving compensation for their time and labor devoted to the investigation of the particular science about which they are to testify; but this practice has been allowed from the necessities of the case, and the inability of courts and juries to determine questions without the benefit of such expert knowledge. Such agreements, however, can never be valid where the amount to be paid is to depend upon the testimony that is to be given, and where the right of compensation depends upon the result of the litigation.
> —Schapiro, 144 App. Div. 1, 128 N.Y. Supp. 852.

Since 1968 a law in California regulates the procedure used to pay for expert witnesses. A person who is not a party to an action and who is required to testify before any court or tribunal or in the taking of a deposition in any civil action or proceeding solely about an expert opinion that he holds on the basis of his special knowledge, skill, experience, training, or education and who is qualified as an expert witness will receive reasonable compensation for his entire time required to travel to and from the place where the court or other tribunal, or, in the taking of a deposition, the place of taking such deposition, is located, and while he is required to remain at such place pursuant to subpoena. The court may fix the compensation or it can be by agreement with the attorney. The surveyor should ask for a subpoena under this law; it makes payment of a fee certain.

16–27 Amount of Expert Witness Fee is Subject to Cross-Examination

Cross-examination may extract from an expert the fact that he is to receive a fee and how much the fee is.

> An engineer who testified for the petitioner may be asked, on cross-examination, if he had not been promised, "considerable money" if the proceeding went through, as it is always competent to ask a witness, on cross-examination, by whom he is employed and whether he is paid, in order to show his interest.
> —*West Skokie Drainage District* v. *Dawson* 243 Ill. 175.

> In a park condemnation proceeding, it was error not to allow a question asked of a real estate agent, who had estimated values for the city, as to whether he was interested in the same of land that would be benefited by the park.
> —*Oakland* v. *Adams*, 37 Calif. App. 614.

16–28 Discovery by Interrogatories, Depositions, Admissions, or Production of Documents

Within recent years the process of discovery has been added to the preliminary features of court trials; each attorney is given the privilege of asking the opposing attorney to produce certain documents and to ask for admissions, depositions of witnesses, and answers to certain questions of parties while under oath. In cases involving land boundaries the process of discovery is implemented by four devices: (1) interrogatory is a list of questions directed to a party for his answers under oath. (2) Requests for admissions are in writing; they ask a person to acknowledge the truth of certain facts or the authenticity of certain materials. (3) Requests for the production of certain documents for the purpose of inspection and study. (4) Requests for depositions of witnesses who may appear at the trial. Deposition hearings are conducted with a court reporter and attorneys present.

Very frequently the surveyor is asked to give a deposition, and since it is recorded under oath, it is very important that consistent answers be given at the trial.

The attorney is allowed privileged information; he does not have to reveal that which is privileged. When a surveyor works directly for an attorney and develops information directly for the attorney, the information may fall into the privileged class; during the deposition the surveyor should not answer the questions without a slight pause, thus giving his attorney time to object. If the information is privileged, the attorney will object. Also if the question is one of opinion, an objection will be made. Such questions as "In your opinion is . . ." and "Do you believe that . . . ?" need not be answered, since they are a matter of opinion. Questions of facts must be answered. An answer to the question

"What did you find at the Northwest corner of Wilson's property?" may be "I found a 1-inch iron pipe at the corner of Wilson's fence." The answer does not admit that you think the fence is the corner of Wilson's property; it may be that a measurement controls. Be sure you understand the implications of the question before you answer.

When being served a subpoena for either a deposition or a trial it may be a *subpoena duces tecum*, which means bring along with you the following requested items. The request may include field books, maps, reports, or other information furnished the client or anything bearing on the court case.

The surveyor may be asked to assist the attorney in the preparation of questions for discovery and may be called on to answer questions as may any other witness or party to the litigation.

16–29 Survey Must Be Done by Surveyor

The surveyor who testifies in a case must be the one who made the survey. In the case of *Hermance* v. *Blackburn* (206 CA 653) the court notes:

> ... it was error to permit a witness to testify that a certain arch protruded over the lot line where said witness testified that the survey was not made by him personally, but by men in his employ.

16–30 Preparation for Testimony

Perhaps the Boy Scout motto "Be prepared" is the best advice. It is assumed that all sources of information have been investigated, monuments have been searched for, calculations have been made, pictures have been taken, and the trial date is approaching.

Between the time of gathering evidence and the time of trial several months or a year has passed. A thorough and complete review of all data, evidence, and facts is the first order of business for "refreshing memory." No one can answer questions promptly and certainly without the benefit of review. Some trivial fact, a fact sufficient for the opposing lawyer to use as a discredit, will escape momentary memory. Rewalking the survey lines, reexamination of monuments, notations of peculiar markings, observation of topography and plants, and conditions of fences and possession are all cross-examination questions. Particular attention should be paid to the apparent age of possession. These questions may be for the purpose of confirming personal presence on the site.

The most important exhibits of the surveyor are plats, pictures, and monument material. Plats are for different conditions to place emphasis on different features. A working sketch of the entire area, an enlargement of some detail, or overlays are all used to advantage.

Maps are pictorial representations of the results of work. They must be large enough for the judge and jury to see and must be clear enough to be understood. Important points are given letters or numbers to facilitate ready reference in testimony. Numbering all corners, such as Corner No. 1, and so on is an aid to ease testimony.

Nothing should be put on a map that cannot be authenticated; it is a map of fieldwork actually performed and observed. The opposing attorney delights in and is very adept at discrediting and embarrassing a witness by revealing defects in a map.

The map should be complete with respect to what was done and sufficient to prove the points. This will show all measurements, points found, the names of the surveyor and chainman, date of survey, legend, scale, north arrow, and all pertinent facts.

> Maps, plats and diagrams explanatory of locations may be introduced in evidence in connection with the testimony of a witness and as explanatory thereof. Such drawings are not produced as evidence within themselves, but for the purpose of enabling the jury to understand and apply the testimony in the case. The surveyor had testified fully as to the manner in which he made the survey. It was therefore competent for the appellant to offer to produce before the jury, in connection with the testimony of the surveyor and for the purpose of enabling him to point out such places to the jury, a map or plan of a subdivision and of the adjoining subdivisions, so marked to indicate the places where the stakes in question were found, and evidence tending to show that such map or plat was a correct representation of the subdivision and correctly indicated the points where such stakes were found, was proper for the consideration of the jury together with such map or plat.
>
> —*Justen* v. *Schaaf* 175 Ill. 49.

If pictures are taken, they should be numbered on the map and the position and direction of the camera indicated. In testifying, the statement "I stood at Point Q on plaintiff Exhibit B and faced the corner in the direction indicated by the arrow and took picture #5 to reveal the condition of the corner as I found it" is clear and concise. Equally as important, the same facts should be labeled on the picture. Fumbling around and trying to identify where you were does not improve the opinion the jury or judges have of the witness.

Detailed drawings are cross-referenced to the plot plan by a general note "See Detail P," and on the detail, reference is made to the plot plan as "Detail of corner #11 on the plot plan."

The object in testimony is to prove a point, not just assume it. In the following case the judge ordered a new trial; evidence was insufficient for him to come to a conclusion. The dispute was over the location of the west one-quarter corner of a section in Missouri in 1932. Two surveyors testified.

> He testified that he determined the southwest corner of section 3, from which he commenced his survey, as follows:
>
> "Q. I understand that you never established any section corner down there. A. It was not lost and was not necessary to establish it.
>
> "Q. You never found it? A. I found the exact point where the corner should be and was from reference points.
>
> "Q. Where did you start your survey? A. At the intersection of the two fences that are known to be on the section line, and also the intersection of the two roads

that my notes in the office show were located on the section line by Pleas Stubblefield in 1894.

"Q. You never found that corner stone? A. I located that corner at the intersection of the two fence lines that are known to be on the section line, and the two roads known to be on the section line.

"Q. You just assumed that they were? A. It was not necessary to survey, and there is no better way to get about it to establish a corner than the way I did. There is no other way to get the corner than the theoretical way I did.

"Q. But you never made any survey to determine that. You determined that by looking at the ground? A. That was all that was necessary."

He established the quarter section corner from which he ran the line between the northwest quarter and the southwest quarter 39 chains and 10 links north of the southwest corner of section 3. He said: "I gave each section of land the measurement shown on the plats of the United States government—which was by proportionate measurement—I gave the half mile on the south side the forty acres govenment measure which was a little different from my chaining, and the north half mile thirty-eight chains and thirty-eight chains and ninety links, which is figured by government, which showed that their chain would equal ninety-seven point seven five six links as against my chain which would equal one hundred links, and at the point figured by proportionate measurement I drove an iron pin and measured into a number of trees for reference and recorded the survey. * * * Q. In other words, each half mile bears its percentage of the loss, rather than one half-mile taking all the loss? A. Yes, sir, it does."

His chain man testified concerning the determination of the southwest corner from which the last survey started as follows:

"Q. Did you find any evidence of a corner being there? A. We didn't find no corner but we found the fences. We had two fences and two roads to get the point of beginning.

"Q. You just assumed there was a corner there because those two fences there and two corners. A. There was two roads that seemed to have been there quite a while.

"Q. There was no evidence of that corner outside of these fences and roads. A. We didn't find no rock.

"Q. Did you look for one? A. Yes, sir. The road had been cleared around in a curve. * * *

"Q. You never measured that to any other corner to establish this corner did you? A. No."

Plaintiff's other surveyor, a former county surveyor, had made the survey of 1902. He said he made a record of his survey but he had lost the book and testified from recollection. He said that the west side of section 3 measured about 175 links short of what the government field notes gave it. He said "we didn't find the section corner, however, on the west side of the section." Concerning the corner from which he started he testified as follows:

"Q. You don't know whether you started from the corner or not? A. I started from the recognized corner. It was fifteen feet west of what is now Kimball's fence, and it was in the center of the road west from there and south from there.

"Q. You didn't try to establish the fact that was a corner by any further surveying? A. I run that line. I started from the corner. We dug and found the corner.

"Q. Do you know whether you found the established corner or not? A. No, sir.

"Q. You didn't go out to establish a corner? A. We don't do that. When we find a corner that is not disputed we don't want any trouble.

"Q. You assume everybody was abiding by that. A. Yes, sir."

As to how he then determined the boundary line between the southwest quarter and the northwest quarter, he said: "Just divided what belonged to each quarter as the shortage. I divided the shortage equitably. There was a shortage in the government survey. I determined that because there was a shortage between the distance across the section that the government field notes said should be there and the actual measurement."

However, we think defendant is correct in his contention that plaintiff's evidence was insufficient to re-establish the lost quarter-section corner and to show in which quarter section the land in controversy was located. It may be that the point at which plaintiff's surveyors commenced to survey is the southwest corner of section 3, but plaintiff did not prove that it was. The testimony of these surveyors, which is hereinabove set out, shows that the original corner monument could not be found and that they assumed that the place from which they started was the section corner from the roads and fences. If they had other information upon which to base this assumption, it was not introduced in evidence. Their testimony is replete with conclusions of fact or of law rather than evidence in regard to facts. If the roads were on the section lines, the section corner might have been either on the north side of the road, the south side of the road, or in the middle of the road. Evidence of a survey which is not definitely shown to have commenced from a corner established by the government or, if lost, re-established in accordance with our statutes, is of no probative force. Clark v. McAtee, 227 Mo. 152, 127 S. W. 37; Id., 253 Mo. 196, 161 S. W. 698; Atkins v. Adams, 256 Mo. 2, 164 S. W. 603; Nelson v. Cowles (Mo. Sup.) 193 S. W. 579; Wright Lumber Co. v. Ripley County, 270 Mo. 121, 192 S. W. 996. There is not even one word of evidence in this record that the northwest corner monument of section 3 was found or that this corner was otherwise properly established. A surveyor's testimony "is never receivable except in connection with the data from which he surveys, and if he runs lines they are of no value unless the data are established from which they are run, and those must be distinctly proven, or there is nothing to enable any one to judge what is the proper result." Jones v. Lee, 77 Mich. 35, 43 N. W. 855, 857. See also, Roberts v. Lynch, 15 Mo. App. 456; Dolphin v. Klann, 246 Mo. 477, 151 S. W. 956. Plaintiff, on another trial, may be able to show a survey between definitely established corners whch would prove that the land sued for or some part of it is in the northwest quarter. Furthermore, on another trial, the court should have the benefit of the United States government survey.

—*Cordell v. Sanders*, Mo., 1932, 52 SW Rpt. 2d 834.

The appearance of a surveyor in court is actually a team effort with the attorney. Prior to litigation it is important that the surveyor identify with his attorney his strengths and his weaknesses, his manner of testimony, and his frank opinions. Neither the attorney nor the surveyor should be surprised by the other during the course of the trial. It is imperative that each thoroughly understand the manner of presentation, the evidence, and any problems that could possibly present themselves.

16-31 View of the Site by the Court

At the discretion of the court, either party may request a view of the premises or parts thereof to better relate the testimony to the court. The view may be taken at the beginning of, during, or at the end of the trial.

The attorneys may point out certain things to the judge or jury, but at no time may any of the witnesses talk, unless asked a question by the judge. The court reporter is usually along to record what is said. Ordinarily, the surveyor accompanies the court by being available to answer questions regarding particular items of evidence.

REFERENCES

The Consulting Engineer as an Expert Witness, Consulting Engineers Association of California, Burlingame, 1972.

Kaplan, John, and Waltz, Jon R., *Evidence*, 11th ed., Gilbert Law Summaries, Gardena, Calif., 1979.

Mundo, Arthur L., *The Expert Witness*, Parker and Baird Co., Los Angeles,

Stern, Steven T., *Introduction to Civil Litigation*, West Publishing Co., St. Paul, 1977.

CHAPTER 17

Eminent Domain

17-1 Scope

Surveyors in private practice are rarely directly confronted with eminent domain and its ramifications, whereas surveyors employed in governmental, public, or quasi-governmental organizations are often required to determine descriptions of land and their respective property rights for legal proceedings founded on the doctrine of eminent domain.

The brief discussion in this chapter describes the elements of eminent domain, who may exercise the power, what property is subject to appropriation, and what constitutes compensation. The procedures differ from state to state and cannot be discussed in detail.

17-2 Definitions

The following words and phrases are defined to clarify the meaning of eminent domain and its effect. Since each state has its own laws regulating the procedures for eminent domain, some variations in the meanings of words can be expected.

One should not become confused between the words *eminent domain* and *public domain*. *Eminent domain* is a process whereby a person's rights to land are extinguished by the public or a public agency; *public domain* is a particular type of land owned by the government. The original public domain included those lands that were acquired by the Federal Government (1) from the colonial states, (2) from Indians, and (3) from foreign powers by treaties, purchase, or conquest. Today, the public domain lands constitute those residual public domain lands that have never been disposed of by the federal sovereign. Other lands owned by the federal sovereign are public lands.

Eminent domain, in most states, is identified as "the right of the state, through its regular organization, to reassert, either temporarily or permanently, its dominion over any portion of the soil of the state on account of public exigency and for the public good" (Laws of Georgia, 36-101). Eminent domain is usually perceived by citizens as a constitutional doctrine that allows a government to take private property; historically, this is not so. Each state or the Federal Government may constitutionally limit how eminent domain will be used.

Condemnation is the act of declaring property needed by the sovereign (or other body politic) and for which eminent domain power will be exercised. Court action usually follows and is called a "condemnation suit."

Due process of law is required by the federal constitution before any person or group of persons can be deprived of property or vested rights. In brief, the person is entitled to his day in court to protect his interests, and he must be given ample warning of the taking.

Just Compensation is provided for in the Constitution. Any person losing title, interest, or being damaged by the process of eminent domain is entitled to remunerative compensation as decided on by disinterested and unbiased persons (usually a jury). Much has been written about "just compensation," and many real-estate appraisers make their livelihood by ascertaining the value of condemned property so that the court can arrive at just compensation.

When only part of a tract is taken, it is termed a *partial taking*. This is quite usual in the location of a long narrow strip for a new highway running across whole sections or parts of sections of land.

The land in a tract that is not taken is called a *remainder*.

Since a partial taking in eminent domain proceedings is in reality a subdivision of land, it is termed a *severance*.

Damages are the detriments to the former title holder and adjoiners due to the taking. A person does not need to have title to incur damages, since other rights and enjoyments may have been usurped.

Quick take is the term given to the act of early possession permitted in some jurisdictions. In effect, it permits the sovereign or the agent of the sovereign to begin possession, construction, surveys, or other privileges before the eminent domain action has been completed and title secured.

17-3 Exercise of Eminent Domain

One of the earliest forms of eminent domain is reported in the Bible when Ahab, King of Asmonia, wanted to acquire a vineyard near his place that belonged to Naboth. Naboth refused. Jezebel, Ahab's wife, pressed false charges of blasphemy against Naboth. After he was stoned to death on the king's orders, Naboth's land reverted to the king. But the wrath of God punished the king, the sovereign, for his wrongful act (1 Kings, Chapter 21).

Some courts have held that eminent domain is an incidental right of sovereigns, inherent in and belonging to every sovereign government (*Daniels* v. *State Road Dept.*, 170 So. 2d 846). In fact, eminent domain is pointed out as being absolute over all but restrained by the Constitution (*City of Miami Beach* v. *Cummings*, 266 So. 2d 122). The power of eminent domain existed and was created in each of the original 13 states on independence and in each subsequent state on enactment of its constitution through enabling acts. Like all powers, the United States may take property pursuant to its power of eminent domain, either by taking physical possession without court order or through the various acts of Congress (*Best* v. *Humboldt Placer Mining Co.*, 371 U.S. 334). The eminent domain right of the Federal Government cannot be enlarged on or decreased by any state. This power of the Federal Government, although younger than several of the states, extends to and is superior to all boundaries of the United States (*Shoemaker* v. *U.S.*, 147 U.S. 282). It extends to all lands within each individual state, and is superior to state authority, provided the property is required for the purposes of the national government and is being used for all the people in all the states. The very fact that the property to be condemned belongs to a state is no deterrent (*Oklahoma ex rel. Phillips* v. *Guy F. Atkinson Co.*, 313 U.S. 508).

In most states, the power of the United States yields to the requirements set forth by the states. When the United States is a private individual within a state, it operates under the laws established by each state in relation to the requirements.

The supremacy of the power of eminent domain of the Federal Government knows no boundaries. This federal power can limit power of the states by treaties or laws pursuant to the needs of the people in general. Since treaties are considered the supreme law of the land, the stipulations in a treaty are binding on all the courts and the people of the nation (*American Jurisprudence*, Vol. 2, p. 657).

17-4 Police Power and Eminent Domain

Where police power ends and eminent domain begins is often the subject of litigation. Fundamentally, the two are distinctly different; but, like so many things, the exact line of demarcation between the two does not lend itself to precise definition. *Police power* is the right exercised for the purpose of regulating use of property; it is not for the purpose of taking property. Property taken by eminent domain must be compensated for, but property regulated by the police power requires no compensation. The right to enact legislation to protect the safety, welfare, peace, and lives of people comes under police power; and, although this does interfere with private rights and does seem to take away private rights, no compensation is necessary.

In general, building or zoning regulations, set-back ordinances, restrictions on signs, and like ordinances are enforced under the police power without compensation to the property owners. Such regulations cannot be unreasonable; and, where designed for an esthetic reason, they have been declared outside the scope of the police power. Regulation of the use of highways is a police power, but the denial of abutter's access must be condemned. Parking zones in which cars may park for a limited time along the curb is regulated by the police power, but taking property for the purpose of creating a parking place is outside police powers.

17-5 Delegation of Eminent Domain

Legislators have given the right of eminent domain to certain private companies as being in the best public interest, and the courts have upheld this right. However, private property is still inviolate, and the right of eminent domain by a private corporation is only a tolerant invasion of that right. Because the authority of eminent domain is inherent only in the state or governmental body, any political subdivisions, governmental and quasi-governmental agencies, or business maintaining a public interest can only acquire this authority by legislative action.

The fact that the power of eminent domain may be delegated by the legislature is well established; however, the courts consider eminent domain as "one of the harshest proceedings known to the law" (*Brest* v. *Jacksonville Expressway Authority*, 194 So. 2d 658, 1968). As a protection for the individual the courts have repeatedly held that delegated powers must be restrictive and identifiable. Since legislatures cannot become involved in each instance of eminent domain, they have sought to make enabling legislation sufficiently vague and encompassing to permit wide latitude. On the other hand, the courts have generally held that statutes confirming the power of eminent domain are to be strictly construed against the one exercising the power and in favor of the aggrieved landowner (*Birmingham* v. *Brown*, 2 So. 2d 305). The legislatures seek broad statutes, whereas the courts constantly seek to narrow the statutes as a form of protection of an individual's property.

As a defense against a taking, the attorney looks into these questions. Has the entity doing the condemnation been authorized by the legislature with the power to exercise eminent domain? Is the power to condemn issued for a specific and identifiable purpose of the particular agency? Has the agency followed the formal procedure set forth for the exercise of the authority? In Georgia the authority to use eminent domain for pipelines transporting gas could not be used to take a right-of-way for oil products (gasoline) (*Botts* v. *Southeastern Pipe-Line Co.*, 10 SE 2d 375 and *Georgia Railroad Co.*, v. *Union Point*, 119 Ba. 309). In another Georgia case the right of eminent domain of a railroad as granted by the legislature could not be extended to a lessee without express authority from the legislature (*Harrold* v. *Central of Georgia Ry. Co.*, 144 Ga. 199).

17-6 Conflicting Authorities

Once public interests are identified and these are then authorized by the state legislature, resulting conflicts that arise between a delegated agency and a governmental entity are left to the courts. The Department of Transportation sought to condemn land acquired by the Florida East Coast Railroad (*State Road Dept.* v. *Florida East Coast Ry.*, 262 So. 2d 480). The court determined that the Department of Transportation's use was superior because the railroad had ceased to use the property for the public purpose for which it was granted.

The city of Miami sought to condemn railroad property for a park. At the trial evidence was introduced to show that the railroad had ceased to use the land and had leased it for commercial purposes beneficial to the railroad. Since the railroad failed to show that the land was used for the original purpose for which it was condemned, the court permitted another condemnation (*Florida East Coast Ry.* v. *City of Miami*, 229 So. 2d 152).

17-7 Purposes for Which Property May Be Taken

The legislature, in general, may not authorize the taking of land for private use. The legislature determines the question of what is public or private use, but the decision of the legislature is not final; the courts are the ultimate authority. The courts try to respect the legislature's findings, and the presumption is in favor of the state.

> The power of eminent domain is inherent in the state, yet it lies dormant until called into exercise by express legislation authority. When the power to take private property for public use has been conferred by the Legislature, it rests with the grantee to determine whether it shall be exercised, and when and to what extent it shall be exercised, provided, of course, that the power is not exceeded or abused. Courts cannot inquire into the motive which actuated the authorities, or enter into the propriety of constructing the particular improvements. The question: What is a public use? is always one of law. Attention will be paid to legislative judgement as expressed in enactments providing for appropriation of property, but it will not be conclusive. The question of whether a proposed use is a public one is for the courts to determine as a question of fact.
> —*State* v. *Hawk*, 105 Ore. 319.

> The distinction between public and private use lies in the character of the use, and determination thereof cannot be made upon consideration of legal principles alone but economic conditions and needs of the people must have attention. To constitute a public use, the use must be in common, and not for a particular individual. It is essential to a "public use," that public must, to some extent, be entitled to use for enjoyment of property not by favor, but as a matter of right.
> —*Middlebury College* v. *Central Power Corp.*, 101 Vt. 325.

> Eminent domain is a term used to designate the power of the sovereign to take or to authorize the taking of private property within the jurisdiction of public use without the owner's consent. It is essential to the exercise of the power of eminent domain that the property be taken for a public use and not a private use, and any legislation that attempts to grant the right to take private property for private use is void. To justify the taking of private property for public use, the public must be entitled to use or enjoy the property, not as a mere favor or by permission of the owner, but by right. Whether the use to which it is sought to appropriate the private property is a public use, and whether such use will justify the exercise of the compulsory taking of private property under the statute and constitution, are questions to be detemined by the courts.
> —*Litchfield and Madison Railroad Co.* v. *Alton and Southern R.R.*, 305 Ill. 388.

A railroad usually cannot condemn property for warehouses or like private uses, but it may condemn property for maintenance, construction, or operation of the railroad. In addition to the customary public uses the following have, at least in one state, been declared public use: public marketplace, wharves, levies and dikes, ferries, drains, canals, mills, terminal buildings, cemeteries, elevators, and log floating.

Under powers delegated to municipal boards the necessity of opening a public street or way is a political question, and in the absence of fraud, bad faith, or abuse of discretion the action of such board will not be disturbed by the courts. It is urged that the street use is not public because only three or four farms will be served by its opening. The use is not to be restricted to a few persons but all persons may come, if and when they choose and use the street when opened. Whether land will be devoted to a public use is determined by the character of its use rather than the extent of its use. That is, if the way is open for use by all, it is a public use whether advantage be taken of the street by a few or many persons.
—*Town of Perry* v. *Thomas*, 82 Utah 159.

The use to which the condemned property is to be put, in some jurisdictions, must be in the best interest of the public. Public use, alone, is not always adequate cause for taking. It is sometimes the duty of the courts to determine if a use is in the best interest of the public, if there is a suitable alternate, and if a lesser title, lesser interest, or a smaller area would be adequate for the public need and use.

The Fifth Amendment of the Constitution provides that property shall not be taken for public use without just compensation. The implication then is that property cannot be taken for a private use. The question is "What is a public use?" Early judicial decisions were unanimous in their scope, holding that the public must actually use the property. During the 1930s a change came about; today, the interpretations lean toward "public benefit." "Public use" is now considered to be elastic and is interpreted by the courts to fit changing social conditions and needs. Public use is no longer a question of law but one of fact to be resolved for each individual case (*Timmons* v. *South Carolina Tricentennial Commission*, 175 SE 2d 805).

17-8 Types of Titles and Interest Taken

It has been the common-law rule in the past that the title taken is nothing more than an easement. The easement obtained is usually regarded as perpetual, and this forms the basis for the necessity of compensation. On abandonment of the use granted, the full use of the land reverts to the fee owner. A fee title may be taken where there are proper provisions in the statutes.

If a fee simple title is taken, the land may be used for any public purpose and may be sold after the use is abandoned. Land acquired for easement purposes, such as highways, railroads, transmission lines, and pipelines, cannot later be used for building erection; but land taken in fee may be used for any suitable public purpose, or it may be sold.

Modern public improvements require so many different construction, maintenance, and service procedures that a fee title is, at times, essential. Interstate highways not only carry traffic but also they provide surface drainage, control access, and limit advertising and have many other functions that cannot be legally performed on an easement.

The extent of title acquired will depend on the statutory authority and use intended; it is not a greater title than that warranted.

"Involuntary easement," secured through the exercise of the power of eminent domain, is granted by the state. In this country the right to take lands from the owner, although existing as an inherent power of the sovereign, can be implemented only through the machinery provided by statute and can be exercised only by the state and certain private enterprises having a direct relationship to the public welfare, necessity, and convenience. Generally, an easement satisfies, but more and more a fee title is necessary—without reversion rights. In some states, by statute, the highway can only acquire an easement.

Riparian rights may be taken by eminent domain; and covenants, easements, and the right to light and air may be extinguished by condemnation. Usually, restrictive covenants do not "run with the land."

17-9 Property Subject to Condemnation

All private property, tangible or intangible, is subject to condemnation. Any interest in land, fee or easement, may be taken; any material such as stone, gravel, or timber may be taken, provided that it is authorized by law. In some instances the statutes specifically prohibit taking certain items such as orchards, but these laws can be repealed at the will of the legislators.

Property devoted to a public use cannot, in general, be condemned by a delegated authority, but the sovereign may condemn even publicly devoted lands and a public utility.

One railroad may have the right to condemn a passage across another railroad. Streets may cross railroads by condemnation. Telegraph companies have condemned rights-of-way along and across railroads.

Navigable waters cannot be taken by eminent domain, except for federal purposes, but riparian rights may be taken. Lands under the jurisdiction of the United States, devoted to a particular purpose, cannot be condemned under the eminent domain power of the state.

Real estate belonging to an infant may be taken and materials for highways can be condemned, provided the statute permits it.

17-10 What Constitutes Damages?

Loss of title or interests in property is not the only type of damages considered in condemnation; intangible as well as tangible rights can be destroyed. The adjoining property owner to a limited-access highway may have lost his right of ingress and egress.

> An abutting owner has two kinds of rights in a highway, a public right which he enjoys in common with all other citizens, and certain private rights which arise from his ownership of property contiguous to the highway, and which are not common to the public generally. . . . An abutting landowner on a public highway has a special right of easement and use in the public road for access purposes, and this is a

property right which cannot be damaged or taken away from without due compensation.

—*Lane* v. *San Diego Elec. R. Co.*, 208 Calif. 29.

The depreciation of the value of adjacent real estate because of the location of an undesirable facility has been termed a damage and may be a subject for compensation. Damages are judged in the same manner as they would be for a private owner acquiring title. Unless an expressed right, privilege, or interest has been taken or destroyed, no damage has been incurred.

An owner adjacent to or in the vicinity of property may suffer even though he retains his property. Examples of damages are access cutoff, nuisance, and deprivation of livelihood; all are subject to "just compensation."

The award of damages is generally a function of a jury and may be more or less than the market value. It is based on the price an owner not forced to sell would receive from a buyer who is not required to buy.

17-11 Due Process of Law

In this country, it is rare that property can suddenly be taken by eminent domain. Effort must be made to allow interested parties time to act on their rights. After property to be taken has been decided on and a resolution or some official declaration of intention has been executed, it is necessary to serve notice. A notice may be posted along the proposed route (in the case of a highway), published in newspapers, or mailed to record owners. During the hearing, when the court entertains a remonstration about the project itself, the property owners have a chance to rebut property damages. After adoption of the taking, dissatisfied owners whose lands were taken have a right to appeal. It is not unusual that the settlement of such cases takes several months.

17-12 Possession and Quick Take

In some jurisdictions, such as Illinois, the state has the right, by statute, to enter on and take possession of property early in the proceedings. Generally, this is done as soon as the notice of intention has been filed and after legal descriptions have been checked for sufficiency. In highway construction, this enables the surveyors and other engineers to begin the design phase while advantage can be taken of favorable weather. If the property owner is successful in winning his suit, and the state does not obtain title to the land, the owner will be due just compensation for damages incurred.

In some jurisdictions, construction can be completed under quick-take action, but the actual use by the public of the property is barred until private titles are extinguished. Only certain agencies (usually highways) of the state governments have the quick-take advantage; utilities, hospitals, schools, and like agencies must wait until the condemnation suit is finished and until title is secured.

17-13 Sufficiency of Descriptions

The legal description of property being condemned must be accurately prepared, and someone must definitely determine who the owners are and what is the extent of their damages. Often the title-examining job is given to title insurance companies or qualified examiners. Descriptions are checked against the project map to make certain that the public body has indicated in its application all land falling within the right-of-way shown thereon. In the event that a description is legally insufficient, or title to any portion is not in the parties named, or certain portions of the land falling within the right-of-way are not covered by the description on the application, the public body must take corrective measures in its application. One fatal flaw in an application may render the public body's case void, and all the preparations would have been for naught.

17-14 Location of Title Acquired

When real property interests have been acquired by eminent domain, the title or easement as defined in the writings must coincide with the location on the ground. With highways, it is usually the responsibility of a public employee to locate the right-of-way or title lines before, during, and after construction. In most states, right-of-way markers are erected to mark the boundary between the highway and the adjoiner (Fig. 17-1). If the highway has encroached on private lands or the right-of-way markers are set in a location other than that called for in the condemned title, damages have been suffered by the adjoiner, and the adjoiner is due just compensation under the Fifth Amendment.

If it is discovered that a client's deed falls within an area where a taking took place and the client has not been paid damages, the client should not be lead to believe he can occupy the area; the solution is to seek damages. In the process of condemning land several John Does and Mary Roes usually are named to cover such omissions. Normally, the area of taking is properly described; if an error exists it is usually in naming who owns the area condemned. Probably the best advice is to inform the client that he should see an attorney about the situation.

17-15 Surveying Responsibilities

The surveyor has contacts relative to eminent domain in basically two areas: (1) the preparation of descriptions for the identification of easements, rights-of-way, and other purposes and (2) the location of property and title rights by descriptions already in existence.

Except when provided for by statutory law or when surveyors have obtained permission from the owner, surveyors may not enter on land being condemned. Such entry may constitute trespass, and the owner will be due compensation for damages. When quick take is in effect or a court order permits the surveyors to enter the lands to make surveys, special care should be exercised. The cutting of trees or limbs, disturbing livestock, trampling crops or flowers all may be acts of damage to the owner whether the title is acquired by eminent domain or not.

Adjacent owners may be damaged by acts of surveyors even though they lose no title. The problems of property owners should be considered in the surveying, engineering, and construction of public projects.

17-16 NATURE AND FORM OF PROCEEDINGS / 435

Figure 17-1 Right-of-way marker.

The preparation of the legal description of the property being condemned must be based on an accurate survey so that it identifies owners involved and assesses damages. In some cases it has been held that land is condemned by the written description and if errors exist in the description, it can cause problems embarrassing to a governmental agency. The failure of a surveyor to identify an adjoiner or to identify a fence could legally cause the condemnation in question to be void for not following the law concerning the Due Process Clause. A failure to place the easement on the roadbed could possibly result in future litigation to provide corrective measures.

17-16 Nature and Form of Proceedings

Condemnation proceedings are regulated by statute and are exclusively a product of the legislature. A failure of an agency or governmental body to follow explicitly the requirements as set forth by law could result in the entire condemnation being held void by the courts.

The general steps for condemning property for highway purposes are as follows:

1. Resolution or other official declaration of intention.

2. Notice posted along the proposed route, or published in the local newspapers, or mailed to the owners of record.
3. An appointment of commissioners to locate and lay out the road or area. An engineer or surveyor is usually on this board.
4. Location of the road and filing of the report. The report contains preliminary plans and maps showing line ties and stating the amount of property damages.
5. Publication of the report.
6. Hearing. The court studies the appropriateness of the project, and the property owners have a chance to be heard on their damages.
7. Adoption or rejection by the governing body and the payment of damages.
8. A right of appeal by dissatisfied owners whose lands are affected.
9. The final survey to locate the highway or other project for construction within the title area.

17-17   Measure of Damages

No hard and fast rule exists as a measure of the extent of damages created by the taking of land. The owner is entitled to the market value of the land as of the date of notice of intent to condemn.

> The market value of land is the price it would probably bring after fair and reasonable negotiations where the owner is willing to sell but is not compelled to do so and the buyer is willing to purchase but is under no necessity of buying the property. In the absence of an agreement to the contrary, usually property taken under the power of eminent domain must be paid in money.
> —*Mandel* v. *City of Phoenix*, 41 Ariz. 357.

> In the instant case the business conducted on the land in question was not taken, and it may be moved to and conducted in another location. Compensation for the property actually taken for public use is provided for and any incidental loss or inconvenience to the business (for moving) must be borne by the owner in the interest of the general public. The Legislature may, and in some instances has, provided for loss to a business arising from the condemnation of land on which said business is conducted but such additional compensation rests in the discretion of the Legislature, and failure to provide for the same (costs of moving) in the statute authorizing the taking does not render the statute in conflict with the constitutional provision as to just compensation.
> —*State Airport Commission* v. *May*, 51 R.I. 110.

17-18   Compensation

Compensation means full remuneration for loss or injury sustained by the owner of the property taken for public use. Damages applicable is considered the actul cash value of the thing taken as of the date of the notice of intent to condemn.

Compensation must be provided for in any statute authorizing the taking of private property; the compensation must be certain and adequate; otherwise, the statute will be void. Generally, compensation must be for every kind of right in property that has a market value; this would include easements, lateral support, light, air, discharge of sewer, pollution of water, flooding, or other items. But in some circumstances, as in *State Airport Commission* v. *May* (Section 17–17), some items of damages are not compensated for.

In general, the change of the grade of a street is not compensative. The abutting fee owner cannot claim compensation for underground use in streets for such items as waterlines, sewer lines, drainage lines, gas lines, and like things. But the abutting fee owner to a railroad bed may collect compensation for additional use by telegraph poles and other double-use items.

17-19 Summary

Eminent domain in this country is the power of the sovereign or agent to take private property without the owner's consent under due process of law but with just compensation. The right of eminent domain is a superior right of all sovereigns over all citizens to take property, whether real or personal; and regardless of the origin of title, any title, interest, property, or enjoyment may be taken.

In their day-to-day practice surveyors will have little contact with eminent domain. However, they must realize that lands owned by or under the jurisdiction of governmental bodies, utilities, quasi-government corporations, and others may come under survey rules that are different from those of adjacent lands. A condemnation, properly conducted under the enabling act, clears title to that parcel and also affects all contiguous properties. Surveyors who utilize public domain corners, right-of-way monuments, and corners to other public properties must be knowledgeable about their full liabilities and responsibilities.

REFERENCES

American Jurisprudence, Vol. 18, Bancroft-Whitney Co., San Francisco, 1938, p. 621.
Corpus Juris Secundum, Vol. 29A, American Law Book Co., Brooklyn, 1965, p. 764.

Index

Abandonment, 106
Abbreviations in descriptions, 393
Abjuration, 148
Abstract, 320
Abstract companies, 321
Abstract of title, 320
   effect of, 325
   example of, 321
Abstractor, 321, 325
Abstractor plats, 361
Acceptance of dedication, 96
Access, 328
Accessories, corner, 199
Accidental errors, 282
   random, 282
   residual, 282
Accretion, 247
   apportionment, 250
   area of, 252
Accuracy, 281
Acquiescence, 92
Acts of nature, 80
Actual possession, 104
Adjoiners, call for, 375
Adjustment, 303
Admissibility of evidence, 20
Adverse actions, 98
   surveyor's duty for, 115
Adverse relationships, 80
Adverse rights, 98
   character of title of, 100
   effect of survey on, 101
   elements of, 103
   statutory character of, 99
Adverse title, character of, 100
Advertising, 142
Aerial photographs, 274
Agreement, 80, 84
   by adjoiners, 87
   by deed, 81
   followed by possession, 90
   by formal documents, 216
   by inference, 93
   parol, 80
   by recognition and acquiescence, 92
   unwritten, 80, 84
Alaska, 201, 202
Alluvion, 217
Alpha Daconis, 271
Altitudes, errors in, 293
Ambient conditions, 288
Ambiguity, 30
   extrinsic, 30
   patent, 32
   practical location with, 32
American Title Association, 356
Ancient survey plats, 33
Apportionment, 242, 255
Appurtenant easement, 394
Arbitrary plat, 362
Arbitration, 315
Area, 386
   conveyances by, 367
   uncertainty of, 295
Attending observations, 289
Attorneys' opinions, 325
Augmenting clause, 373
Authority of property surveyor, 223
Authority to perpetuate evidence, 307
Azimuth, 383

Baltimore, Lord, 164
Base line, 185
Basis of bearings, 352, 381
   changing, 383
Beach, 96
Bearings, basis of, 352, 381
   magnetic, 271
Bearing tree, 192, 197
Beginning, point of, 388

Bering, Vitus, 202
Best available evidence, 34
Biblical references to surveying, 146
Blaze, 44
Blunders, 282
Body of description, 375
Bosum rule, 249
Boundaries defined by writings, 217
Boundary, gradient, 176
Boundary survey, 3
Bounds, 365
Bounds description, 366, 398
Breach of contract, 127
Burden of proof, 14
Bureau of Land Management, 195
Burt's improved solar compass, 194

Cadastration, 2
Cadastre, 2
California, land grants in, 177
  platting laws in, 307, 311, 336
Calls, adjoiner, 177
  directional, 385
  passing, 57
  plat, 28
  reference, 366
  superior, 392
  survey, 377
Caption, 374
  check list for, 374
Care, standard of, 125
Carolina, 164
Cause and effect, 415
Centimeter, 266
Certificates, 359
  liability for, 128
Certification of survey, 344, 359
Certified tapes, 267
Cessions by original states, 181
Chain, Gunter, 270
  four-pole, 151
  of history of monuments, 52
  of title, 320
  two-pole, 194
Chamber survey, 75
Changing basis of bearings, 383
Check list, caption, 374
  description, 400
Chimney surveys, 75
Circular curved lines, 389
Circumferenter, 151
Circumstantial evidence, 12
City Engineer's monuments, 54, 66

City Engineer's subdivision examination, 347
Clarke's spheroid of 1866, 272
Clause, qualifying, 373
Client, 224
  agreements with, 225
  conference with, 225, 240
  contracts with, 225
  initial contact with, 224
  obligations to, 139
Closing corner, 169
Closure, 302
  error of, 195
  in sectionalized lands, 196
Colorado, establishment in, 213
  liability in, 122
Color of title, 103, 110
Commencement, 388
Common-law dedications, 94
Common report, 44, 60
Commons, 148
Comparison, 270
Compass, 272
  Rittenhouse, 191
Compensation, 432, 436
  just, 427
Compilation plats, 361
Compound confusion, 221
Computation, 239, 303
  survey, 277
  triangulation, 294
Concept of title, 152
Conclusive evidence, 12
Conclusiveness of evidence, 20
Conclusiveness of written words, 25
Conclusive presumptions, 18
Condemnation, 426
Conduct in estoppel, 83
Conference with client, 240
Confusion, 221
Connecticut, 159, 168
  establishment in, 212
Consistency, 278
Constant errors, 282
Continuous possession, 103, 106
Contractual liability, 129
Contract with client, 123, 225
Control monuments, 340
Conveyance:
  unwritten, 80, 82
  see also Descriptions
Coordinates, line defined by, 384
  use of, 342
Corner post, 191, 197

Corners, accessories to, 199
　closing, 191
　double, 189, 191
　material of, 197
　meander, 198
Corrective instruments, 327
Course, 73
Court appearances, 405
Court cases, 14, 203
Court questions and answers, 407
Court reports, 203
Court trials, 406
Covenants, 338
Cross-examination, 407
　on fees, 420
Cultural improvements, 355
Cuneiform tablet, 145
Curves, circular, 389
　compound, 389
　irregular, 391
　reverse, 390
　spiral, 391

Damages, 427
　eminent domain, 432
　measure of, 436
Datum, elevation, 352
Deceased person's sayings, 65
Dedicate, intent to, 94, 96
Dedication, 345
　acceptance of, 96
　common-law, 94
　donor of, 95
　effect of, 47
　intent of, 94, 95
　location of, 94, 95
　plat, 98
　purpose of, 97
　revoking offer to, 97
　statutory, 95
　unwritten, 94
Deed:
　circumstance of, 27
　date of, 35
　new start for, 320
　sequence of, 35
　warranty, 320
　see also Descriptions
Deficiency, 255
Deflection angles, 384
Degree of curve, 389
Delaware, 164
Delivery of plat, 240

Depot, 96
Deputies' instructions, 187
Descriptions, abbreviations in, 393
　ambiguity in, 371
　area, 367, 386
　body of, 373, 375
　bounds, 366, 398
　caption in, 374
　changing wording of, 372
　check list of, 400
　construction of, 27
　control of monuments in, 41
　coordinates in, 399
　dedication, 94
　exception, 398
　exceptions in, 394
　graphic, 364
　identification by, 24
　indispensable parts of, 24, 367
　intent of, 27, 34
　interpretation of, 34
　land, 364, 367
　linear, 367
　mathematical correctness of, 371
　metes and bounds, 365, 396
　monuments in, 378
　new start for, 320
　objectives of, 368
　parts of, 373
　proportional, 366
　recording, 370
　referral, 395
　sequence of, 35
　simultaneous, 35
　stationing in, 392
　strip, 367, 392
　subdivision, 364
　sufficiency of, 24, 368, 433
　survey as basis of, 371
　surveys for, 35
　technique of writing, 373
　use of coordinates in, 399
　whole, 395
　written, 364
Deviation, standard, 300
DeWitt, Simeon, 182
Dimensional data on plats, 353
Direct evidence, 12
Direct examination, 408
Directional calls, 285
Directions, 271
　basis of, 281
　errors in, 287

procedures for, 287
uncertainty of, 287
Discovery rule, 120
Disputable presumptions, 18
Dispute, property line, 86
Distance, 266, 270
  consistency of measurement, 280
  errors, 284
  uncertainty of, 287
Distance angle, 294
Division line, fractional parts, 259
Documents, recording, 311
Donation tract, 168
Donor, of dedication, 95
Doomsday Book, 148
Double corners, 189, 191
Double exception, 394
Due process of law, 427, 433
Duke of York, 161
Duties of surveyors as expert witnesses, 413
  to client, 139
  in court, 405
  to public, 137
  to surveyors, 139

Early survey systems, 154
Earth's magnetism, 272
Easements, 82, 94
  condemnation of, 431
  location of, 229
  unrecorded, 337
Education of surveyors, 134
Egyptians, 145
Elevation datum, 352
Ellicott, Andrew, 183
Ellicott's line, 183
Eminent domain, 126
  conferring, 428
  damages, 436
  delegation of, 428
  exercise of power of, 427
  proceedings, 435
  purpose of taking in, 430
  superiority and, 429
  use of condemned land in, 430
Empresario system, 173
Enclosure, 104
Encroachments, 318
  detection of, 315
Ending point, 387
Entryman, 201
Erie triangle, 161
Erosion, 247

Error of closure, 195, 199
  in altitudes, 293
  classification of, 282
  in direction, 287, 288
  of other surveyors, 140
  in position, 290
  relative, 282
  in traversing, 293
  in triangulation, 294
Errors, accidental, 282
  altitude, 293
  blunders and, 282
  classification of, 282
  constant, 282
  measurement, 284
  original monument, 335
  permissible, 235
  position, 290
  random, 282
  relative, 282
  residual, 282
  systematic, 282
  theoretical uncertainty, 299
  triangulation, 294
Establishment, 209
  by arbitration, 215
  in Colorado, 213
  in Connecticut, 212
  by county surveyor, 215
  in Illinois, 212
  by processioners, 210
Estoppel, 83
  by conduct, 83
Ethics, 137
Evidence, 1, 9
  admissibility, 20
  authority to perpetuate, 307
  best available, 22, 26
  chain of, 52
  circumstantial, 12
  common report as, 60
  compilation of, 239
  conclusions from, 11
  conclusive, 12, 23
  definition, 11
  direct, 12
  effect of, 13
  encroachments on photographs as, 316
  extrinsic, 12, 19, 34
  field, 231
  field notes as, 29
  finality of, 21

importance of, 9
indispensable, 9, 24
inferior, 18
of intent, 34
law of, 9, 13, 22
loss of, 22
measurement, 68
monuments proven by, 41, 44
office survey, 75
opinion, 411, 414
oral, 11
parol, 31, 57
partial, 13
perpetuation, 306
photographs to preserve, 312
possession, 59, 90
practical location, 32
preponderance of, 16
preservation of, 237, 305, 312
prima facie, 12, 82
primary, 12
proof by, 11
real, 11
relative value of, 76
reputation as, 60
responsibility to perpetuate, 307
riparian, 313
secondary, 12
sufficiency of monument, 41
of survey, 35
surveyor's duty regarding, 77
types of, 11, 12
value of, 13
vanishing, 305
witness, 47, 311
of witness objects, 44
writings as, 24
Examination, direct and cross, 408
Exceptions, double, 394, 398
in descriptions, 394
Excess and deficiency, 355
Exclusive possession, 103, 109
Expert witness, 411
appointment of, 412
cross-examination of, 416
duties, 413
fees of, 419
testifying of, 417
Extrinsic ambiguities, 30
Extrinsic evidence, 13, 19

Fairfax, Thomas, 150
Federal Ordinance of 1785, 179

Fees, 138
expert witness, 419
lay witness, 417
Feudal land system, 148
Field notes, 35, 183, 192, 238
as evidence, 29
public, 30
First Principal Meridian, 185
Florida, title survey in, 356
Foot, 256
Four-pole chain, 151
Fourth Principal Meridian, 186
Free survey, 220
French grants, 168

Gadsden Purchase, 181
General Instructions of 1843, 193
General Land Office, 195
Geographer of the United States, 182
Geographer's Line, 183
Georgia, 165
establishment in, 210
survey laws of 1854, 206
Gizeh, Great Pyramid of, 145
Good faith, 103, 110
Gore, 222
Gorham purchase, 162
Gradient Boundary, 176
Great Mahele, 202
Great Pyramid, 271
Grid north, 273
Groma, 179
Guadalupe-Hidalgo Treaty, 170
Guarantee, 128
expressed, 128
of possession, 319
of title, 318, 319, 339
Guide meridians, 196
Gunter, Edmund, 151
Gunter chain, 151

Hack, 49
Half sections, 185
Hawaii, 202
Hearsay, 56, 408
Hiatus, 221
History, importance of, 143
Holland purchase, 161
Homer's Iliad, 147
Homestead Right, 201
Horizontal distance, 266
Hostile possession, 79, 103
Hostile relationships, 103, 108

Hostilities, 99, 108
Hutchins, Thomas, 150, 183
Hypothetical questions, 415

Identification by photographs, 312
Identification of property descriptions, 24
Illinois, 154
    establishment in, 212
    land registry in, 330
Implied guarantee, 128
Improvements, cultural, 355
Indiana, 186
Indians, 177
    effect of, 177
    Kickapoo, 177
    Modoc, 144
Indispensable evidence, 12
Inference, definition of, 19
Ingress and egress, 432
Initial point, 195
Instructions of 1843, 193
    to deputies, 187
    to surveyors, 189
Intent of conveyances, 34, 377
Interpretations of writings, 27, 377
Interruption, 106
Irregular curves, 391

Jefferson, Peter, 150
Jefferson, Thomas, 150
Judgment, independent, 136
Judicial notice, 9
Junior rights, 35
Jury, 409
Just compensation, 427

Kentucky, establishment in, 212
Kickapoo Indians, 177
Kidder, Arthur D., 17
Krypton, 86, 266

Lake apportionment, 253
Land, 5
    certainty of location of, 335
Land Court, Massachusetts, 209, 319, 330
Land data systems, 7
Land description, 364
    objectives of, 368
Land grants, Louisiana, 169
    Mexican, 170
    Spanish, 169
Land location, 229
    certainty of, 338
Land ownership, 144

Land registry, 328
Land surveying, 2
Latent defect, 19
Law, due process of, 427, 433
    requirements for survey by, 41
    understanding of, 22
Law of evidence, 9, 13, 22
    definition of, 13
Lay witness, 410
Leading questions, 408
Legati, 148
Length, units of, 268
Liability, 119
    avoiding, 130
    contractual, 128
License numbers, 311
Limitation title, 93
Linear conveyances, 367
Lines, circular curved, 389
    defined:
        by coordinates, 384
        by dimensions, 380
        by geometric relationships, 386
        by monuments, 380
    definition of, 379
    straight, 379
Litigation, objects of, 405
    stirring up, 138
Livery, in deed, 148
    in law, 148
    of seisin, 148
Location, certainty of, 335
    guarantee of, 327
    land, 229
    practical, 89
    title, 434
Louisiana Territory, 169
Lozano, Mario, 175

Magnetic bearings, 70, 273
Magnetism, 272
Maintenance, 138
Mansfield, Jared, 186
Manual of Instructions, 192
    of 1855, 193
    of 1903, 195
    of 1947, 195
    use of, 200
Maps, *see* Plats
Maryland, 164
Mason and Dixon, 164
Massachusetts, 102
Massachusetts Bay Company, 155
Massachusetts Land Court, 209, 319, 330

Meander corners, 198
Meandering, 198
Measurement errors, 281
Measurement evidence, 68, 69, 234
Measurement index, 205
Measurements, consistency of, 278
  photogrammetric, 274
  search and, 231
  types of, 266
  uncertainty of, 234
Memorial, 199
Meridian, astronomic, 272
  determination of, 274
  guide, 196
  principal, 196
  reliability of observations of, 288
Mesopotamia, 145
Mete, 365
Meter, 266
Metes and bounds, description of, 365, 396
  survey of, 220
Mexican land grants, 173
  seniority of titles in, 170
  survey of, 175
Mexican roads, 172
Mile, 143
Military reservation, 168
Military tract, 168
Minerals, 170
  owners of, 170
Missionary tracts, 185
Missouri platting laws, 336
Model registration law, 3
Modoc Indian War, 144
Monumentations, requirements for, 340
Monument calls, 399
  passing, 57
Monuments, 41, 185, 303, 354
  acquiescences in, 92
  bounds, 365
  calls for, 11, 41, 378, 399
  chain of history of, 52
  city engineers, 66
  control, 231, 340
  density of, 341
  error of position of, 335
  evidence of, 41, 43
  evidence of being disturbed, 46
  identifying marks, 311
  intent controlled by, 43
  line defined by, 380
  location of, 43, 335
  measured position of, 69
  nonoriginal, 233
  obliterated, 41
  original, 41, 52
  physical characteristics of, 44
  platting of, 354
  possession as, 59
  proving, 44
  record, 355, 380
  reputation of, 60
  requirements for, 340
  search for, 231
  testimony to prove, 234
  uncalled for, 233
  undisturbed, 46
Mortgage policies, 326
Mortgagor, 95

Nature, acts of, 80
Navigable streams, 176, 248
  condemnation of, 432
Negligence, 127
New Castle, 163
New England, 155
New Hampshire, 157, 160
New Jersey, 162
New York, 160
  Dutch streets in, 2, 160
Nonperennial streams, 171
North, 271
  indicating, 351
  true, 272
North arrow, 351
North Carolina, 164
Northwestern Ohio, 169
Notches, 49
Notes, evidence as, 29
  field, 238
  sectionalized land, 192
Notice, judicial, 9

Oaths, 55
Oaths in court, 407
Obliterated monuments, 42
Odometer, 271
Office surveys, 75
Ohio Company Purchase, 168
Ohio lands, 167, 184
Opinion, attorneys', 325
Opinion evidence, 414
Oral evidence, 11
Ordenanza de Intendentes, 172
Ordinance of 1785, 183
Ordinance of 1796, 184
Original monument, 41, 43, 44
Original plats, 360

Original states, 134
Original survey, 35
    regulation of, 334
Overlap, 221
Ownership, 144, 229, 318, 319
    registration of, 328
Owner's policy, 326

Paper surveys, 75
Parallel, standard, 196
Parol agreement, 80
Parol evidence, 31
Partial evidence, 13
Passing monument calls, 57
Patent, 170, 320
    ambiguity, 19, 32
Pennsylvania, 163
Photographic measurements, 276
Photographs, 274
Plats, 364
    abstractor, 362
    ancient, 40
    basis of bearing on, 352
    compilations, 349
    contents, 360
    contrast of writings and, 29
    datum on, 352
    dedication by, 96
    delivery of, 240
    dimensional data on, 353
    drawing, 356
    effect of, 356
    features, 349
    laws regulating, 361
    north on, 351
    original, 348, 360
    paper, 75
    purpose, 349
    recording, 346
    reference calls for, 36
    scale of, 351
    symbols on, 351
    title association, 356
    title of, 351
    types, 348
    unrecorded, 14
Platting, 239
Platting law, 333
    California, 336
    features of, 335
    Missouri, 337
    monuments required in, 335
    objectives of, 334

Point, ending, 387
    initial, 195
Point of beginning, 387
Point of confusion, 222
Police power, 428
Porciones, 173
Position, 266
    errors in, 290
    uncertainty of, 335
Possession, actual, 104
    adverse, 98
    continuous, 106
    evidence of monuments as, 59
    exclusive, 109
    following agreement, 90
    guarantee of, 319
    hostile, 108
    importance of, 233
    monuments as, 60
    open and notorious, 105
    quick take of, 433
    statute of limitations and, 109
    survey of, 222
    title policies and, 327
Practical location, 32, 80, 89, 95
Precision, 281
Preponderance of evidence, 16
Prescription, 98
Prescriptive title, 100
Preservation of evidence, 237, 305, 312, 343
Presumptions, 18, 100
    irrebuttable, 18, 22
Presurveys, 180
    without recording, 347
Pretrials, 406
Prima facie evidence, 12, 215
Primary evidence, 12
Principal Meridians, 185
Prior surveys, 180
Privity of contract, 123
Processioners, 210
Profession, attributes of, 132, 134
    definition of, 132
    discredit to, 141
    education and, 134
    gaining eminence in, 135
    reputation of, 140
    service to the public in, 135
    trust in, 135
Professional reputation, 140
Proof, burden of, 14, 99
Proof by evidence, 11
Property, condemnation, 432

Property line dispute, 86
Property surveying, 3
Property surveyor, activities of, 4
  authority of, 223
  duties, 115
    as to adverse possession, 115
    to client, 139
    in court, 405
    as to easements, 229
    as to evidence, 22
    as expert witness, 413
    as to field search, 232, 233
    as to location, 327
    as to measurements, 235
    as to ownership, 229
    to public, 137
    as to research, 227
    to surveyors, 139, 140
Property surveys, documents used for, 228
  history of early, 144
  legal authority for regulating, 222
  in New World, 154
Proportional conveyance, 366
  rights, 242
  shore line, 282
  statutory, 366
Protraction, 199
Public, services to, 135
Public Domain, 179
  small tracts in, 202
  source of, 181
Public records, 321
Putnam, Rufus, 186

Qualifying clause, 374
Quarter sections, 185
Questions, hypothetical, 415
  leading, 408
Questions and answers, 407
Quick take, 427, 433
Quit claim, 320

Railroads, condemning, 429, 432
Random accidental errors, 282
Reading uncertainty, 287
Realengos, 173
Real evidence, 11
Reckoning, 272
Recognition and acquiescence, 80, 92
Recording, 23
Record monuments, 355
Records, private survey, 312

Rectangular system, 179
  beginning of, 182
Red River Case, 176
Redwood, 45
Reference calls for plats, 28
Reference calls for writings, 28
Registration of ownerships, 319
Registration of titles, 319, 328
Reliction, 248
  area of, 250
Reputation as evidence, 60
  professional, 5–21
Research and surveyor's duties, 277
Reservation, military, 6–14
Res gestae, 65
Residual errors, 282
Resurveys, 217, 219, 223
Reversion rights, streets, 242
Review of other's work, 141
Revoking dedication, 97
Rhode Island, 160
Ring count, 49
Rio Grande River, 173
Riparian evidence, 212
Riparian rights, 247
Rittenhouse, David, 182
Rittenhouse compass, 191
Roads, Mexican, 172
Roll-over wood, 50
Rope stretchers, 145

Sayings of deceased persons, 65
Scale, 351
Scrivener, 371
Secondary evidence, 11
Second order traverse, 297
Sections, half, 185
  numbering system, 184, 189
  quarter, 185
  running lines of, 189, 196
  survey of, 189
Seisin, 148
Seniority of titles, 23, 170
Senior rights, 23, 35, 375
Seven Ranges, 182, 184
  beginnings of, 182
  subdivision, 183
  surveyors of, 182
Severance, 427
Sexagesimal system, 272
Significant figures, 278
Sims, Captain Barlett, 177
Simultaneous conveyances 35

Spanish era, 169
Spanish lands, 169
Spanish water laws, 171
Specifications for survey, 296
Spiral curve, 391
Squatter, 83, 106
Squatter's sovereignty, 83
Standardization of tapes, 266, 267
Standard of care, 125
Standard parallels, 196
Standards, 297
Starting point, 379
State plane coordinates, 316
State's seaward limit, 254
Stations, 292
Statute of Frauds, 1, 81, 86, 149
Statute of Limitations, 91, 92, 99
Statutory dedication, 95
    proceedings for, 82
Straight lines, 266, 279
Streams, medium line, 248
    nonperennial, 171
    perennial, 171
    thread, 248
    torrential, 171
Street, reversion rights in, 245
    vacated, 242
Street condemnation, 432
Strip conveyances, 367
Subdivision apportionment, 255
Subdivision descriptions, 364
Subdivision of townships, 196
Subpoena, 417
Sufficiency of descriptions, 368
Sufficiency of monument evidence, 44
Superior call, 392
Survey based on the record, 221
Surveying instructions, 192
Surveyor-attorney relationship, 406
Surveyor general, 186
Surveyors, instructions to, 187
    qualifications of, 138
    records of, 54
    research responsibilities of, 227
    responsibilities of, 434
    rules of, 19
    unexpressed intent of, 40
    unworthy, 139
Surveyor's certificates, 359
Surveyor's chain, 270
Surveyor's conduct on stand, 405

Surveyor's duties, see Property surveyor, duties
Surveyor's fight, 177
Surveyor's liability, 119, 327
Survey plats, ancient, 40
    definition, 348
Surveys, 154
    adverse rights, 101
    basis of bearing of, 352
    boundary, 340
    called for, 35, 40, 377
    certification of, 344
    completion of, 240
    conclusions from, 240
    documents used for, 228
    early systems of, 154
    establishment, 209, 210
    free, 220
    intent of, 40
    law requiring, 35
    Mexican Grant, 173
    New England, 154
    New York, 160
    Ohio, 167
    original, 219
    plats of, 348, 356
    possession, 219
    prior, 180
    private records of, 312
    Seven Ranges, 184
    Spanish Grant, 169
    systematic procedures for, 218
    Texas, 175
    Tiffin's instructions for, 189
    title, 356
    township, 196
    unrecorded, 377
    Virginia, 154
    written conveyance, 217
Swedes, 163
Symbols, 351
Symmes Purchase, 168
Systematic errors, 282

Tacking, 107
Taking, partial, 427
Tapes, standardization, 267
Taping errors, 284
Taping specifications, 267
Taxes, 103, 110
Tennessee, 167
Tent survey, 75

Terrestrial photographs, 312
Testimony, 55, 234
    cause and effect in, 415
    conclusions from, 405
    incompetent, 25
    negative, 19
    preparation for, 421
Texas, empresario system, 173
    gradient boundary in, 176
    Mexican titles in, 169
    minerals, 170
    Red River Case, 176
    road beds in, 172
    surveys in, 173
    thirty-foot rule, 171
    water appropriation in, 171
Textbooks, testifying from, 416
Theoretical uncertainty, 282, 299
Thirty-foot rule, 171
Thuban, 271
Tiffin's instructions, 189
Timber Culture Act, 200
Title, abstract of, 320
    chain of, 320
    claim of, 106
    color of, 110
    concept of, 152
    condemnation of, 426
    corrective, 327
    definition of, 152
    guarantee of, 318, 319, 339
    identity of, 355, 369
    limitation, 93
    registration of, 319
    riparian, 247
    senior, 23, 170
    Torrens, 328
    unmarketability of, 327
    unwritten, 79, 116
    written, 81, 83
Title association plats, 356
Title insurance policy, 326
    wording of, 327
Torrens system:
    adverse rights in, 329
    characteristics of, 328
    disadvantages of, 329
Torrential streams, 171
Townships:
    error of closure in, 199
    subdivision of, 196
    Tiffin's instructions for, 189

Traverse, second-order, 297
Tree growth, 50
Trees, characteristics of, 49
    marking, 197
    witness, 491
Trespass damages, 130
Trial court, 406
Two-pole chain, 194

Ultimate fact, 414
Uncertainty:
    of areas, 295
    of measurements, 234
    of position, 234, 290
    theoretical, 299
Uncertainty expression, 299
Uncertainty of reading, 287
United States Military District, 169
Units of length, 266, 268
Unwritten agreement, 80
    elements of, 85
    property surveyor's duties with respect to, 115
Unwritten conveyances, 82
    and the government, 82
Unwritten dedication, 94
Unwritten title, 80

Vacancies of public lands, 82
Vanishing evidence, 305
Vara, 173
Vega, 271
Vermont, 159
Virginia, 154
Virginia Military District, 168

Warranty deed, 320
Washington, George, 150
Water boundaries, 247
Waters:
    appropriation of, 171
    changing shore line of, 46
    condemnation of, 432
    navigable, 248
    Spanish laws of, 171
Western Reserve, 168
Wharf, 96
William and Mary College, 150
Williams, Roger, 160
Wills, 264
Witness:
    appointment of expert, 413
    compensation of, 419

expert, 411
   duties of, 413
   lay, 410
Witness evidence, 55, 57, 311
   to prove monuments, 55, 57
Witness objects, 44
Wood, roll-over, 47
Words, conclusiveness of, 34
Writings, best available evidence as, 24, 25
   conveyances in, 22
   indispensable, 24
   interpretation of, 27
   plats contrast with, 29
Written descriptions, 364
Written evidence, 11, 41
Written title, 79, 82, 83

Yard, 266